DARWIN'S
BACKYARD

DARWIN'S BACKYARD

HOW SMALL EXPERIMENTS LED TO A BIG THEORY

JAMES T. COSTA

W. W. NORTON & COMPANY
Independent Publishers Since 1923
NEW YORK   LONDON

The experiments described in this book are not intended for use by children. Readers are cautioned to use common sense when undertaking any experiment. Use of appropriate tools and protective gear is essential, as is knowledge of one's own limits and abilities. References to specific products, services, service providers, and/or organizations are for illustration only and none should be read to suggest an endorsement or a guarantee of performance. Neither the publisher nor the author can guarantee the accuracy and completeness of this book for all purposes or make any representation with respect to the outcome of any project or instruction described. Web addresses appearing in this book reflect existing links as of the date of first publication. No endorsement of, or affiliation with, any third-party website should be inferred. Author and publisher are not responsible for third-party content (website, blog, information page, or otherwise). This book is sold without warranty of any kind, and none may be created or extended by sales representatives or written sales or promotional materials. Readers are solely responsible for ensuring that their activities comply with applicable law.

For Randal Keynes—
Darwin-inspired friend and mentor

But I love fools' experiments. I am always making them.

—CHARLES DARWIN

CONTENTS

PREFACE

We can only imagine that Emma Darwin had the patience of Job. At one point in the 1850s, sheets of damp paper stuccoed with frog eggs lined the hallway of her house, pigeons cooed boisterously in a dovecote in the yard, row upon row of glass jars with saltwater and floating seeds filled the cellar, and malodorous pigeon skeleton preparations permeated the air. And that was only the beginning: there was a terrarium of snails with suspended duck feet, heaps of dissected flowers, and

Down House, Darwin's home of 40 years in Kent, south of London. Photograph by the author.

the fenced-off plots in the lawn where the grass was carefully scraped away to study struggling seedlings. Of course, being married to Charles Darwin over a dozen years by then, she was undoubtedly used to it. Charles, she might have said to friends, was *experimentising* again.

Darwin's experimentising, which appeared to some as merely the odd pursuit of an eccentric Victorian naturalist, turned out to push the envelope on his—and our—understanding of the biological world and our place within it. Darwin was laying the empirical groundwork for key elements of his revolutionary ideas on evolution.

This book introduces a Charles Darwin that few people know. His evolutionary ideas were not pulled out of thin air. He was an observer and experimentalist, and his clever and quirky investigations were not the schemes of some solitary eccentric sequestered in a lab. No, Darwin's home was his laboratory, and his and Emma's large family of seven surviving children often worked with him as his able field assistants. Darwin even published with his kids, in a fashion: the very year *On the Origin of Species* was published, a notice on rare beetles appeared in the *Entomologist's Weekly Intelligencer* by Darwin, Darwin, and Darwin—the authors being sons Franky, Lenny, and Horace, ages 10, 8, and 7, written with a wink by their proud dad on their behalf. He also had a talent for roping the butler and governess into his field studies, along with his cousins and nieces. He signed up legions of friends and strangers alike to make observations, try experiments, send him specimens, and serve as sounding boards. Yet, while Darwin may be a household name and his work equally well known, most are unfamiliar with Darwin the scientist, let alone the person.

His landmark books are appreciated as astonishing compendia of information, yet even many Darwin enthusiasts have little sense of Darwin the inveterate observer and correspondent, ingenious synthesizer and experimentalist, or family man. The rich array of experimental projects carried out by Darwin and his family reveal a very different—very *human*—side of a person too often seen as a cardboard icon. Whether enthusiastically taking up one son's suggestion to test the viability of seeds in the crops of dead birds, staking out and chasing oddly buzzing bumblebees in the garden, or venting his frustration to friends over uncooperative fish spitting out the seeds he was trying to feed them, Darwin's experiments are often humorous and always instructive. They may have been "fools' experiments," as he liked to call them, but as a Darwin friend pointed out, "fools' experiments conducted by a genius often prove to be leaps through the dark into great discoveries."

Through his experiments and other investigations Darwin systematically gathered data testing his evolutionary ideas. Beginning with his geological works of the 1840s, his experimentising and other pursuits provided invaluable material that bolstered his arguments. His pace picked up considerably in the 1850s, when his experiments became behind-the-scenes efforts to look into nature as no one had before, through the lens of evolution and natural selection. Darwin referred to the *Origin of Species* as "one long argument," but we should step back and consider his entire post-*Origin* opus in precisely the same way: one *longer* argument. After the *Origin*, he published myriad papers and some 10 books: on orchids, domestication, human evolution, climbing plants, animal behavior, carnivorous plants, flower structure, and earthworms. The topics are far-ranging yet all of a piece in support of a grand vision, many involving a prodigious number of homespun experiments and other projects.

Beyond the fun of coming to know Darwin as worm whisperer, chaser of bees, and flytrap fancier, this book too has a serious point to make. Darwin's experiments instruct as well as entertain. Novel, amusing, at times hilarious, yes,—but they also shine a spotlight on science as a process. Darwin was a prototypical MacGyver figure: sleuth of the sandwalk—his gravelly thinking path at Down House—he shows how real insights into nature can be gained with simple tools at hand in yard, garden, or woodland. Modern visitors to Down House, ably managed by English Heritage, can see several of these experiments replicated in the meticulously restored gardens, greenhouses, and grounds. The deeper message here is that Darwin's experiments provide object lessons and blueprints for how science works. By and large these exper-

A view of Darwin's gravel thinking path, the sandwalk. Photograph by the author.

iments can be done here and now, in any schoolyard, backyard, classroom, or kitchen.

Anyone can become an experimentiser like Darwin and learn how to look a bit more closely at the natural world. In this regard, Darwin's experiments are an untapped resource that has been staring us in the face for a century and a half. At a time of much hand wringing over the teaching of evolution and critical thinking in science, one invaluable resource for helping communicate the essence of scientific inquiry has been all but overlooked. It is none other than the field's founder: Darwin himself. In *Darwin's Backyard*, I show how we can draw upon Darwin in exploring nature and better understanding evolution and how science works. Taking a thematic approach, this book thus has dual goals. First, I aim to take readers on a journey to see Darwin and his remarkable insights through the lens of his family life: his expansive curiosity at work and how family, friends, and a wider circle of naturalists were an integral part of this process. This very human Darwin with his homespun experiments is not the Darwin that most people are familiar with. Yet without appreciating this side of him, neither the man nor his achievements are to be fully understood. Second, I aim to show how Darwin's method has relevance today: how his backyard experiments can be *your* backyard experiments. To this end I offer up a menu of Darwin-inspired experiments, using that term inclusive of Darwin's observational projects as well as those more experimental in method.

Where did Darwin's penchant for experimentising come from? Though his no-nonsense father once despaired that his bug-collecting, horse-riding son would never amount to much, Charles Darwin clearly came by his philosophical turn of mind honestly—after all, his grandfather Erasmus Darwin was a famed physician and poet, with an inventive mind so fertile that the poet Samuel Taylor Coleridge coined the term "darwinizing" to describe his brand of wild speculation. Coleridge considered Erasmus to be "the most inventive of philosophical men," and his grandson was certainly cut from the same cloth. To understand the evolution of this experimentiser, we will start with his first forays into science as a kid, in league with his beloved older brother Erasmus, his grandfather's namesake, and their sometimes disastrous chemistry experiments. We'll meet, too, Darwin as a college student, making sense of the natural world from Edinburgh to Cambridge, and the exhilarating experience of the *Beagle* voyage. Here we see a working method born, inspired by Charles Lyell's landmark *Principles of Geology*. Within months of his return from his formative voyage around the world, Darwin was convinced of the reality of evolution and became

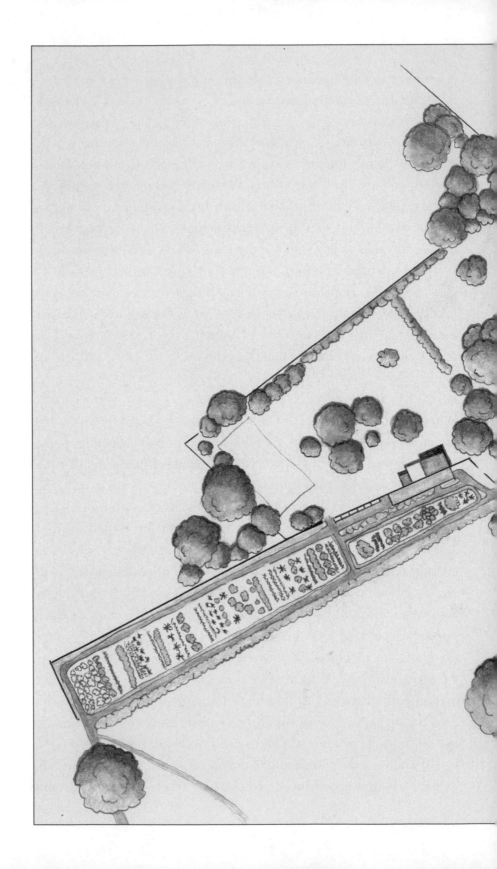

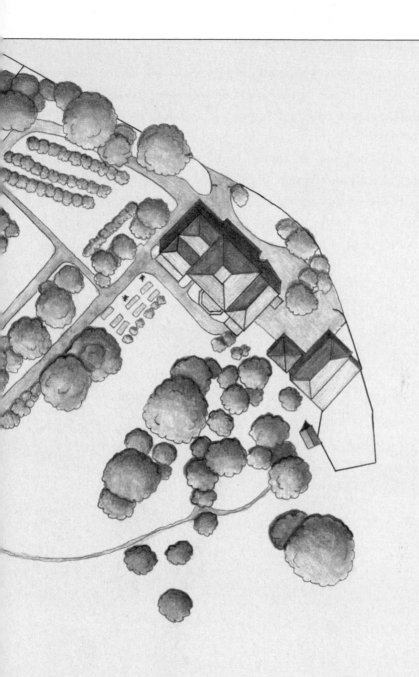

Down House and grounds, site of many of Darwin's investigations.
Drawing by Leslie C. Costa.

the Lyell of biology. Over time his curiosity grew and his experimental eye was cast further and further afield, ferreting out the secrets of barnacles and bees, primroses and pigeons, weeds and worms. And yet in a sense he didn't look further than the gardens, woodlands, and meadows of his beloved home, morphed into a kind of *Beagle*, a ship on the Downs with family and friends his dedicated crew. Marcel Proust once wrote that "the real voyage of discovery consists not in seeking new landscapes but in having new eyes." Darwin did travel to new landscapes, but for most of his life he simply learned to see what was before him, with new eyes. In getting to know Darwin the experimentiser, we, too, can learn to discover the unfamiliar in the familiar.

Cullowhee & Highlands, North Carolina

DARWIN'S
BACKYARD

1

Origins of an *Experimentiser*

When Charles Darwin was a kid of eleven or twelve his nickname was "Gas." No, not for flatulence. His schoolmates gave him the nickname for his penchant for the noisy and smelly chemistry experiments that he and his older brother Erasmus would conduct in their makeshift "laboratory" way at the back of the garden of their handsome red-brick house. The Mount, as the home was called, was built by their physician father Robert Darwin high on a hill overlooking the River Severn in Shrewsbury, a prosperous market town in the west country of Shropshire, England. Following the fashion of the time the Darwin boys boarded at Shrewsbury School despite its proximity to home. Erasmus—dubbed "Bones," but always 'Ras to his siblings—left the school to continue his studies at Cambridge in 1822, and his younger brother, then 13, dutifully kept up their investigations on the home front. The "Lab," as they called it, being a short walk from the school, it was easy for Charles Darwin to spirit a blowpipe and chemicals to his bedroom where he would "experimentise" at the gas lamp after bedtime—until, that is, the headmaster, Samuel Butler, got wind of it. Darwin never forgot the very public (and, he thought, unjust) rebuke of branding him a "poco curante"—a superficial dabbler.[1]

Chemistry was all the rage at this time, and had been since the late 1700s when the likes of Joseph Priestley in Britain and Antoine

Lavoisier in France made stunning discoveries on the nature of matter, Humphry Davy dazzled standing-room-only crowds with chemical demonstrations in London, and their own grandfather, Josiah Wedgwood, achieved renown across Europe for his chemical innovations in creating new forms of porcelain and glazes at his pottery, the Etruria Works.[2] Beyond the spectacles that chemistry afforded, the transformations inherent in dissolving, melting, and burning substances were terribly exciting for their philosophical implications about matter and the nature of all things, living and nonliving. The Darwin brothers may have found the classical curriculum of Shrewsbury School mind-numbing (Charles later said "nothing could have been worse for the development of my mind than Dr. Butler's school"[3]), but through their chemistry investigations they experienced some of the excitement and wonder of that scientific age. "The subject," Darwin later wrote in his autobiography, "interested me greatly, and we often used to go on working till rather late at night. This was the best part of my education at school, for it showed me practically the meaning of experimental science."[4]

"Experimental science" was to Charles and Erasmus *hands-on* science: manipulate, dissect, mix, probe, poke, prod, heat, and carefully observe the results. Then repeat, with suitable tweaking. Experimental science in their day was broadly conceived to include demonstrations and dissections, close observations and collections, unlike today's rather strict definition of "experiment" that involves well-framed hypotheses and careful attention to controls, replication, and sample size. Practitioners of science (*natural philosophers* until the English polymath William Whewell coined the term *scientist* in 1840) became increasingly self-conscious during the nineteenth century about the procedures and methods they deployed in figuring things out. Induction was at the heart of scientific inquiry in Darwin's day: amassing enough facts to connect the dots and draw inferences about general principles. Deduction, the use of general principles to make predictions about facts of nature, and frame concise and testable questions to refine the principles, came on the scene largely in the twentieth century. The so-called "hypothetico-deductive" method of modern science is a misnomer, however: pure deduction never replaced earlier induction; rather, induction and deduction are used together in a fruitful iterative and mutually reinforcing process. As we will see throughout this book, Darwin's approaches were varied, sometimes more inductive and at others more deductive by modern standards, and rarely meeting modern standards of rigorous experimental design. But for all that, Darwin managed to learn an awful lot about how the world works. To appreciate his broad-minded approach to "experiment" we must cast the net equally widely and consider the diversity of investigations he undertook over the years, in field (his yard, meadows, and woodlands) and lab (his study and greenhouse), and from methodical data-yielding experiments in the modern sense to tests, dissections, demonstrations, and comparative observations.

In subsequent chapters, then, we'll meet Darwin the gleaner, observer, and synthesizer in addition to Darwin the experimentiser. These were all of a piece to Darwin, because to him even observational studies were hands-on. And it all began in that smelly boyhood laboratory at the bottom of the garden.

Edinburgh: Flustered over *Flustra*

Robert Darwin saw that Shrewsbury School was becoming a waste of time (and money) for his son. In 1825 Erasmus finished up a medical course at Cambridge and was sent to the University of Edinburgh, the center of medical education in Britain, to continue his preparation for a medical career. Robert took the extraordinary step of removing 16-year-old Charles from the local school and packing him off to Edinburgh too. Clearly Charles had no aptitude for languages or the classics, so law was out of the question, and neither did he have the mathematical acumen to study physical sciences. Robert figured his sons would follow the family tradition and become doctors, so why not send Charles to medical school sooner rather than later, where he and his older brother might help each other out? It turned out to be disastrous in some respects: Charles and 'Ras got on as warmly as ever, but Charles discovered he had little stomach for the horrors of the operating theater—or the patience for certain mind-numbing lecturers. In other ways, though, it was an auspicious commencement to a deeper education: the spirit of the Scottish Enlightenment was still alive in Edinburgh, a more dynamic and forward-looking place than most in Britain at that time, its medical faculty up on the latest ideas from the continent.

A central idea was "philosophical anatomy," which took as its expansive subject the material basis of animal structure and function, growth and decay, health and disease. Partially an outgrowth of "vital materialism," it held that the properties of life were inherent in matter itself. It was married to the exciting implications of Luigi Galvani's demonstrations back in the 1780s and '90s of the commonality between electricity and the animating spark of life. Galvani, who was professor of surgery at the University of Bologna in Italy, had discovered "animal electricity" (called electrophysiology today) in his famous experiments with frog's legs showing that electricity could induce muscular motion. (His 1791 treatise on the subject, *De viribus electricitatis*, was an important source of inspiration for Mary Shelley's *Frankenstein*.) The mysteries of animal motion, long attributed to some divine animating spirit, was yielding to the empirically testable material phenomenon of electricity.

Jean Baptiste de Lamarck was a proponent of these ideas. A sometime botanist, Lamarck was assigned to analyze and classify the ill-defined menagerie of invertebrates (a term he coined) at the Muséum d'Histoire Naturelle in postrevolutionary Paris. These animals inspired his doctrine of evolution, or "transformism." "Nature," he wrote in 1801, "in producing in succession every species of animal, and beginning with the least perfect or simplest to end her work with the most perfect, has gradually complicated their structure." Lamarck saw the ubiquitous, abundant, "least perfect" of animals—invertebrates—as lowest in the scale of organization, and therefore closest to the origin of living organisms. In his thinking, microscopic organisms ceaselessly arise through spontaneous generation, and through their vital energy or life force, complicate their structure by extremely slow degrees over generations in response to their adaptive needs. Insofar as invertebrates represented an early stage in the gradual complication of structure, in Lamarck's view, by virtue of their superabundance and diversity they also represented the best group with which to document transitions. If, as Wordsworth rhapsodized in his poem on revolutionary France, "Bliss was it in that dawn to be alive, But to be young was very heaven!"—to be an invertebrate zoologist then, too, "was very heaven," moving the likes of Charles Darwin's grandfather Erasmus Darwin—brilliant Enlightenment physician, inventor, political radical, and poet—to verse of his own:

> *Organic life beneath the shoreless waves*
> *Was born and nurs'd in ocean's pearly caves;*
> *First forms minute, unseen by spheric glass,*
> *Move on the mud, or pierce the watery mass;*
> *These, as successive generations bloom,*
> *New powers acquire and larger limbs assume;*
> *Whence countless groups of vegetation spring,*
> *And breathing realms of fin and feet and wing.*[5]

A better poetic expression of Lamarckian transformism cannot be found. By the time Charles was born in 1809 his grandfather's poetry was decidedly out of fashion in England, and his politics and sympathy for Lamarck reviled. But not in Edinburgh: even a decade and a half later, when Charles joined his brother in the Scottish capital in 1825, radical politics and radical science were very much part of the scene (though certainly not necessarily approved of).[6]

By 1827 Erasmus completed his course and moved on to anatomy school in London, leaving his increasingly unhappy brother behind. Charles began working under Robert Edmond Grant, sometime doctor and now an invertebrate zoologist of growing reputation at the university. Grant had lived in Paris for a time, attending lectures by Lamarck and studying under other notable *savants*: embryology with Étienne Geoffroy Saint-Hilaire, and comparative anatomy with Georges Cuvier.

Geoffroy became a friend and defender of Lamarck while Cuvier heaped ridicule upon them both; their animosity eventually erupted into very public verbal duels before the French Academy of Sciences in 1830—an episode deeply interesting to Darwin, which I'll relate in the next chapter.

Grant took Darwin under his wing. He introduced him to the Plinian Society, established a few years before for students keen on natural history, and took Darwin as his guest to meetings of the Wernerian Natural History Society, ordinarily off limits to undergraduates. Perhaps most importantly, Grant took Darwin with him to explore the tide pools of the Firth of Forth for marine invertebrates. A fan of Lamarck and Geoffroy, Grant studied minute gelatinous sea creatures, seeking in them evidence for "transformism, the search for a unified single set of laws uniting the organic and inorganic domains."[7] Grant was studying tiny colonial polyps, coralline algae, and similar marine organisms, and was sure these creatures represented intermediate forms linking plants and animals. Coralline algae had actually been (mis)classified as plants by Lamarck, who rejected the "animal-plant" linkage idea. Grant, however, held a view more akin to Geoffroy, seeing the unity of all life, plant and animal, as being in one great system. He called them "zoophytes," or "animal-plants" (mainly consisting of what are called Bryozoa and allied groups today, a form of colonial invertebrate). Grant showed that they formed a free-swimming larval form (decidedly animal-like), yet they looked like plants with blade-like "leaves" undulating in the currents, and could reproduce by budding. "The Animal and Plant Kingdoms are so intimately blended at their origins," Grant declared, "that Naturalists are at present divided in opinion as to the kingdom to which many well-known substances belong." In fact, "several organized bodies . . . which have neither roots, nor capillary vessels, nor a digestive stomach, nor other distinct organs or plants or animals, connect the Vegetable and Animal Kingdoms by imperceptible gradations."[8]

Darwin marveled at the beauty and intricate complexity of these squishy denizens of the sea, with their pulsations, tiny waving tentacles, and colorful patterns. At Grant's urging he made observations of his own with a borrowed microscope, learning more than he bargained for in the process. There was the excitement of discovery, to be sure, as when he revealed the animal nature of the enigmatic seaweed-like *Flustra* by showing that it, too, produced free-swimming "ova" (the larval form) by the thousands, proudly reporting his results in a paper to the local Plinian Society on March 27, 1827. This was Darwin's first scientific contribution. But his discovery also held another lesson, one that left a bad taste in his mouth: scientific rivalry. Grant felt territorial about his marine invertebrates, and co-opted Darwin's discovery, evidently feeling that the research program was his alone. Without Darwin's knowledge, Grant read a similar paper to the Wernerian Society a few days ahead of Darwin's. Adding insult to injury, he reported the *Flustra* discovery without acknowledging his student's role. Darwin was hurt, and also defiant; he decided to carry on with his investigations, documenting free-swimming larval forms in several other species. But his enthusiasm for Grant cooled. When Grant tried to engage Darwin some weeks later, the professor was met with stony silence; as Darwin later described it: "One day, when we were walking together, [Grant] burst forth in high admiration of Lamarck and his views on evolution. I listened in silent astonishment, and as far as I can judge, without any effect on my mind."[9] It's hard to believe Grant's declaration made no impression on Darwin, but he wasn't about to engage in intellectual intimacy with his professor again.

Barely a year later Darwin left Edinburgh; it was clear that medicine was not for him, and neither were intellectual thieves. His father was probably at wit's end. What to do with a med school drop-out son? Why, send him to Cambridge University, of course. The Anglican church was the obvious alternative: with his cheerful demeanor and penchant for dogs, natural history, and countryside rambles (besides dabbling in his makeshift laboratory), surely Charles would be eminently suited to the respectable and not-too-taxing life of a country vicar. England had (and still has) a long and venerable tradition of vicar-naturalists, personified by the likes of John Ray and Gilbert White. He arrived in January 1828, and although Grant and Edinburgh were left far behind, those "zoophytes" stuck with him.

Cambridge: Beetling, Botanizing, and Geology 101

Charles Darwin spent three years at Cambridge, in comfortable rooms in the front quad of Christ's College said to have once been occupied by the Rev. William Paley (1743–1805). His plan was to take Holy Orders, and to that end Paley's writings truly delighted him— especially Paley's celebrated *Natural Theology; or, Evidences of the Existence and Attributes of the Deity, Collected from the Appearances of Nature*, a work that Darwin later said "was the only part of the academical course which, as I then felt, and as I still believe, was of the least use to me in the education of my mind."[10] *Natural Theology* opened with the famous metaphor of the watch and watchmaker: intricacy of structure and adaptation in nature compel us to infer a divine designer, just as the structure and functionality of a watch compel us to infer an intelligent watchmaker. For a time, he fell in with a group of well-heeled young men high in spirits but low in ambition. He later lamented that his time at Cambridge was "worse than wasted," but he was being a bit hard on himself. After all, he discovered the joys of beetle collecting through his cousin William Darwin Fox, also attending Cambridge at the time, a pursuit that might seem like a pointless hobby but surely gave him firsthand experience with diversity and variation. Many years later his fellow evolutionist Alfred Russel Wallace would look back on his and Darwin's successes and suggest that it had something to do with the fact that both were ardent beetle collectors as young men. The extraordinary diversity of these insects led the twentieth-century evolutionary geneticist J. B. S. Haldane to quip that the Creator clearly had "an inordinate fondness for beetles." They impressed Darwin in ways that only made sense later. In the meantime it was fun: "no pursuit at Cambridge was followed with nearly so much eagerness or gave me so much pleasure as collecting beetles," he wrote in his autobiography.[11] His friend Albert

Albert Way's cartoon of Darwin going a-hunting for beetles. Reproduced by kind permission of the Syndics of Cambridge University Library.

Way poked gentle fun at Darwin's obsession in his cartoon of Darwin astride a giant beetle.

Actually, beetling was more than fun for Darwin: it was practically a blood sport. He threw himself into beetle collecting with competitive abandon, almost coming to blows with a collector whom he had been paying to collect beetles for him, until he discovered the rascal was giving a rival first pick of the rarities he found.

His beetle mania could be dangerous too—as in this episode he related in a letter to his friend Leonard Jenyns:

> I must tell you what happened to me on the banks of the Cam in my early entomological days; under a piece of bark I found two carabi (I forget which) & caught one in each hand, when lo & behold I saw a sacred Panagaeus crux major; I could not bear to give up either of my Carabi, & to lose Panagaeus was out of the question, so that in despair I gently seized one of the carabi

between my teeth, when to my unspeakable disgust & pain the
little inconsiderate beast squirted his acid down my throat & I
lost both Carabi & Panagaeus![12]

But in beetling, too, we get a glimpse of one aspect of Darwin's
modus operandi that he was to later develop into something of an art
form, contributing mightily to his later success: the art of procuring
help. This likely started with his cousin Fox, an equally keen beetle
collector though perhaps not as consumed as Darwin, who came up
with the idea of hiring attendants to carry their equipment on long
beetling rambles, and collect for them too. What a great idea—well-
heeled golfers and hunters have their caddies and gun-bearers, so why
not well-heeled beetlers with porters to lug nets, boxes, and traps? The
distinguished Darwin biographer Janet Browne insightfully noted that
"there were always other people hidden behind Darwin's immediate
achievements." Darwin's researches went off in various directions but
he never lost the ardor for beetles he developed in his Cambridge years:
"Whenever I hear of the capture of rare beetles," he later reminisced,
"I feel like an old war-horse at the sound of a trumpet."[13]

But if beetling taught Darwin a certain stick-to-itiveness when it
came to collecting as well as the finer points of anatomical structure
and variation, these were surely reinforced by his other Cambridge
passion: botanizing. In his second year there, Darwin met Rev. John
Stevens Henslow, who had been appointed Professor of Botany a few
years before. Just 13 years older than Darwin and in many ways quite
the opposite of Professor Grant, Henslow was warm and encouraging
toward his pupils. Darwin became a regular at Henslow's weekly countryside botanical rambles, where he was known as "the man who walks
with Henslow," and at his table as a frequent dinner guest, welcomed
by Mrs. Henslow and the children. There was no hint of Grantian
materialism or transmutationism in Henslow's teaching, nor jealous
sniping over intellectual turf. Henslow was up on the latest thinking
from the French "philosophical anatomists," but put this firmly in the
context of natural theology: structure and function, classification and

relationships, could be understood in terms of a divine plan. His fellow Cambridge dons approved—scientific Cambridge at that time had a fully institutionalized and doctrinaire "intelligent design" mindset that informed practically all that was taught or studied at the university. Yet there was no conflict between their religion and science: scientific insights were seen as part of the mystery of the Creator's plan. There was no desperate flailing in the quagmire of biblical literalism here— that had long since been jettisoned, and the Cambridge divines were no less devout for it. Darwin's eventual discoveries would put him on a collision course with some of the most cherished tenets of the natural theology tradition, but that was later. For now, he fully appreciated and accepted natural theology, eagerly memorizing whole passages from Paley—required reading at the time.

Darwin's time under Henslow's tutelage was transformative in several respects, perhaps first in rekindling and stoking his passion for science beyond collecting and identification. A year earlier fellow student John Maurice Herbert, Darwin's admiring friend and companion on many a beetling jaunt, bowled Darwin over with the gift of a fine Coddington microscope, given anonymously. Henslow revived Darwin's earlier excitement over microscopic analysis and discovery. But if Henslow's approach to botanical instruction taught Darwin to "get his eye in" and appreciate intricate structure and individual variation, it also taught him to appreciate the big picture: a philosophical botany, literally global in scope, striving to understand underlying patterns in the geographical distribution of species. Grand insights were articulated by such continental thinkers as the Swiss botanist Augustin de Candolle in his *Essai élémentaire de Géographie botanique* (1820), and the rockstar of explorer-naturalists, Alexander von Humboldt, in his *Essai sur la géographie des plantes* (1807). Such works would later be seen as classics establishing new sciences; we'll explore Humboldt's and de Candolle's ideas in Chapter 5, as well as Darwin's eventual research program and remarkable experiments connected with geographical distribution. That work and more grew from seeds planted by Henslow in 1830 and 1831.

It was at Henslow's weekly soirees that Darwin was introduced to the luminaries of Cambridge science, notable among them the astronomer John Herschel and philosopher William Whewell, both of whom had a profound influence on Darwin and his growing passion for natural philosophy. At a time when the practice of science was only just becoming self-conscious, these thinkers were helping lay the framework for a methodological philosophy of science: How do we know we know something about nature? What is the best way to figure things out? Herschel received acclaim for his 1831 *Preliminary Discourse on the Study of Natural Philosophy*, a treatise that introduced a roadmap for inductively gaining insight into nature's *vera causae*—"true causes"—and argued for the primacy of natural law governing nature, laws set up by the Creator as the means by which the clockwork universe was regulated. Darwin was flush with excitement over the clarity of Herschel's luminous treatise: "If you have not read Herschel," he wrote Fox, "read it directly."[14] Within a decade of Herschel's treatise, Whewell made his own profound contributions to the philosophy of science, publishing *History of the Inductive Sciences* (1837) and *Philosophy of the Inductive Sciences* (1840). The latter not only introduced the concept of *consilience*, an intuitive and powerful means of gaining insight into *vera causae* that complemented the Herschelian approach, but also introduced the term "scientist" as an apt replacement for the time-honored but imprecise term "natural philosopher." "As we cannot use physician for a cultivator of physics," wrote Whewell, "I have called him a physicist. We need very much a name to describe a cultivator of science in general. I should incline to call him a Scientist. Thus we might say, that as an Artist is a Musician, Painter, or Poet, a Scientist is a Mathematician, Physicist, or Naturalist."[15]

The very pursuit of science was being literally defined, analyzed, and systematized by these thinkers before Darwin's eyes. As an undergraduate he almost surely did not grasp this at the time, but he knew he longed to follow in their footsteps. Literally so, in the case of Humboldt, whose *Personal Narrative of Travels to the Equinoctial Regions of America* documenting his 5-year voyage of exploration in remotest

Central and South America with Frenchman Aimé Bonpland electrified Darwin and other young naturalists of his generation. In April 1831 we see the first indication of a plan in the offing when Darwin wrote excitedly to Fox: "At present, I talk, think, & dream of a scheme I have almost hatched of going to the Canary Islands—I have long had a wish of seeing Tropical scenery & vegetation: & according to Humboldt Teneriffe is a very pretty specimen."[16] Initially the jaunt was to include Henslow and Marmaduke Ramsay, a tutor at St. John's College, but it was more wishful thinking than realistic on Henslow's part, given his professorial and parish duties (not to mention his growing family). Darwin worked himself into a frenzy over the trip, rhapsodizing to his sister Caroline:

All the while I am writing now my head is running about the Tropics: in the morning I go and gaze at Palm trees in the hothouse and come home and read Humboldt: my enthusiasm is so great that I cannot hardly sit still on my chair. . . . I never will be easy till I see the peak of Teneriffe and the great Dragon tree; sandy, dazzling, plains, and gloomy silent forest are alternately uppermost in my mind.[17]

The Canaries trip was not to be, but had, as one happy outcome, his study of geology. To truly emulate Humboldt, Henslow counseled, he must become adept at identifying rocks and minerals, and understanding geological formations and processes. Darwin attended some lectures, and although Henslow personally taught him a bit, the most significant contribution he made toward Darwin's geological education was introducing him to another Cambridge don, the Rev. Adam Sedgwick.

Sedgwick was Professor of Geology and then-president of the Geological Society of London. That spring of 1831, Sedgwick introduced Darwin to the concept of reading the landscape like a book, finding the stories in the landforms by learning to read them. Darwin sat in on Sedgwick's lectures, and like the big-picture philosophical issues inherent in botanical geography or the structure of zoophytes, geology awed him. Much discussed at the time was the age of the earth and its vast oceans and continents. Few took seriously the idea of an earth mere thousands of years old, but was it tens or hundreds of thousands, or even millions of years old? After one lecture where Sedgwick held forth on the deep antiquity of the earth Darwin exclaimed to a friend, "What a capital hand is Sedgewick [sic] for drawing large cheques upon the bank of time!"[18]

Like Henslow, Sedgwick was fond of field excursions with his students, at times leading parties of 60 or more on horseback through the countryside, stopping to lecture at various outcrops and other formations along the way. During his summer holidays he did his own fieldwork, and as luck would have it that year he would be ranging into north Wales, not terribly far from Darwin's home. Sedgwick was very amenable to having Darwin accompany him. Darwin headed home to Shrewsbury in June 1831, and Sedgwick was due to pick him up in August for a 3-week excursion. Darwin was still mooning about Tenerife, reporting to Fox: "The Canary scheme goes on very prosperously. I am working like a tiger for it, at present Spanish & Geology."[19] In the meantime, Darwin did his best to come up to speed so as to impress Sedgwick. On Henslow's instructions he bought a good-quality clinometer (a device to measure the angle of slopes, strata, and other formations) and practiced using it for hours on an array of furniture tipped every which way in his bedroom. He even tried his hand at a bit of fieldwork, writing to Fox that he was trying to make a geological map of Shropshire ("don't find it so easy as I expected"), and trekking 16 miles to study a prominent limestone formation called Llanymynech Hill. His notes from that trip survive, and they reveal a Darwin who consistently had difficulties keeping his compass bearings straight. The Darwin scholar Michael Roberts has suggested that Darwin may have been dyslexic judging by how often he reversed compass readings in his notes.[20]

Sedgwick arrived at The Mount on August 2nd, and geologized locally for a couple of days. While at Shrewsbury one episode is recorded that gives some insight into the development of Darwin's thinking as a scientist-in-training, a revelation about science as a process. Darwin related to Sedgwick that a quarryman had shown him the shell of a tropical volute gastropod he had found in a local gravel pit, thinking the professor would be amazed. He was shocked when Sedgwick dismissed the notion that the shell naturally occurred in the gravel pit, concluding it was surely discarded there by someone. "I was then utterly astonished at Sedgwick not being delighted at so wonderful a fact as a tropical shell being found near the surface in the middle of England," he wrote in his autobiography. Significantly, he continued: "nothing before had ever made me thoroughly realise, though I had read various scientific books, that science consists in grouping facts so that general laws or conclusions may be drawn from them."[21] This episode surely resonated with the philosophy of John Herschel, and put the zealous researches of Humboldt and others into a new light. Darwin's many subsequent investigations—including his flair for experimentation—were always aimed at "grinding general laws out of large collections of facts," as he wrote in his autobiography.[22]

Darwin and Sedgwick headed off for Wales and Darwin's extended "Geology 101" lesson on August 5th. One of the main objects of the excursion was locating and mapping outcroppings of Old Red Sandstone, a sedimentary formation that Sedgwick had identified as Devonian (he had in fact coined that name for the geological period), dated by modern technology at about 419 to 359 million years old. The two were unsuccessful in locating the Old Red, but Darwin learned a great deal of practical field geology. On the last leg of the excursion they separated to cover more ground. Darwin made his way south to Barmouth to see friends, stopping briefly at the breathtakingly beautiful Cwm Idwal hanging valley in the Glyderau Mountains of Snowdonia, with its crystal-clear lake backed by the imposing wall of Devil's Kitchen and dotted with massive erratic boulders. He made some observations—once again reversing several compass readings—and dutifully reported them to Sedgwick by letter. Sedgwick wrote back correcting some of what he took to be Darwin's misinterpretations of certain geological formations, but all in all he performed admirably as a geological trainee (though when he returned to Cwm Idwal ten years later he realized how much he missed, as we will see). Then, after seeing his friends he headed home, where a letter from Henslow awaited him. The trip to the Canaries was cancelled. Instead, it offered him a much more extraordinary adventure: to serve as a gentleman-naturalist on a voyage around the world.

Darwin Abroad

In some ways it made little sense that Darwin should be invited to take such a post. He was barely out of college, and not a seasoned naturalist. Indeed, he wasn't the first to be approached: Henslow had been the one initially invited and was tempted. But he had a family and growing responsibilities to the university and his parish. The invitation next went to Leonard Jenyns, Henslow's brother-in-law and a good naturalist, but he was a newly minted priest just appointed to a parish. Jenyns thought about it for a day and reluctantly turned it down. They both then thought of Darwin. Once the objections of Darwin's father and the initial misgivings of the captain, Robert Fitzroy, were overcome, Darwin took his place aboard the HMS *Beagle* as gentleman companion to the captain and informal naturalist on a Royal Navy surveying mission. They departed Plymouth on December 27, 1831, not to return for almost 5 years.

I will not give an overview of all Darwin did on the *Beagle* voyage here—that's beyond our scope.[23] For this book's purposes, we'll explore how the *Beagle* voyage led to Darwin's later experimentation. Darwin, like any scientist undertaking a research program, whether experimental or observational, proceeded to form questions in the context of the theory he had learned. As a college student Darwin was only just aware of the big outstanding questions of natural science—the nature and age of the earth and its formations, the patterns of the geological record, the diversity and distribution of life, and (not least) that "mystery of mysteries," as John Herschel put it, the origin of species and varieties. At the time and through most of the *Beagle*'s voyage Darwin was pondering those questions and their possible solutions within the framework of natural theology. What released the inner experimenter in Darwin after the *Beagle* voyage was seeing the world through a new lens, permitting him to ask fresh questions that only arose in contemplating his observations in a new light.

The process of seeing the world anew that began with Grant and his zoophytes and was amplified in the inspired company of Henslow, Sedgwick, and others late in his Cambridge career was now helped mightily along on the *Beagle* voyage, in particular owing to the watershed *Principles of Geology* by Scottish geologist Charles Lyell (1797–1875), the first volume of which was published in 1830. Encouraging his soon-to-be-companion's geological education, Captain Fitzroy made a present of Lyell's book—a book which, Darwin recalled later, Henslow had recommended that he acquire and study, "but on no account to accept the views therein advocated." That seems like an odd recommendation, perhaps, to read a book but not believe it, but this may reflect the fact that Lyell's new model of earth processes and history were being hotly debated in the learned societies at the time. Lyell, who trained as a lawyer and knew how to argue, championed the vision of fellow Scot James Hutton, who saw the history of the world as one of ceaselessly acting natural forces that slowly but surely shape the earth's surface in an endless cyclical process. No miraculous interpositions, no divine catastrophes, just the incessant mundane forces exerted by water, volcanism, ice, and other natural processes combined with the inexorable forces of the uplift and subsidence of land. So it ever was, in Hutton and Lyell's view, and so it ever shall be. Hence the subtitle of Lyell's book: "An attempt to explain the former changes of the earth's surface by reference to causes now in operation."

Darwin probably dipped into Lyell's *Principles* in fits and starts. The *Beagle* didn't make it out of Plymouth Harbor before he was overcome with seasickness. His shipmates likely told him that he'd get his sea legs soon enough, but sadly he was some shade of green throughout the voyage—which in retrospect may have been a good thing from the viewpoint of the history of science, as it encouraged Darwin to go ashore at every opportunity. (In fact, of the nearly 5 years of the voyage, Darwin was actually on board the ship just a year and a half altogether.) Thus he was surely doubly excited at the prospect of landing at Tenerife—relief from vomiting, and at long last an opportunity to see sights so rapturously expressed by Humboldt. And there it was, looming over the port of Santa Cruz. Alas, the *Beagle* crew were prevented from landing owing to a cholera outbreak back in England. The local authorities insisted on a 12-day quarantine for the English ship, but, impatient, Fitzroy didn't want to lose time and so hoisted anchor immediately. The captain noted in his memoir of the voyage: "This was a great disappointment to Mr. Darwin, who had cherished a hope of visiting the Peak. To see it—to anchor and be on the point of landing, yet be obliged to turn away without the slightest prospect of beholding Teneriffe again—was indeed to him a real calamity."[24] But the concern at Santa Cruz was well-founded: through the nineteenth century epidemic after cholera epidemic raged in different parts of the world, killing tens of millions of people.

Still, Darwin didn't have to wait much longer to experience his first tropical isles: the ship called at the Cape Verde Islands in mid-January, and it was there, at St. Jago, that he noticed something remarkable: a horizontal white band of shells embedded 45 feet above sea level within a cliff face. He now saw this with Lyellian eyes: the sea shells must have gradually risen, as an abrupt, violent movement surely would have broken up the nearly horizontal layer. This was consistent with Lyell's theory of slow change over great periods of time. Darwin wrote to Henslow: "The geology [of St. Jago] was preeminently interesting & I believe quite new: there are some facts on a large scale of upraised coast (which is an excellent epoch for all the Volcanic rocks to [be] dated from) that would interest Mr. Lyell."[25]

It was a moment he never forgot: the very first locality he examined showed him "the wonderful superiority of Lyell's manner of treating geology, compared with that of any other author, whose works I had with me or ever afterwards read."[26] For the rest of the voyage he pointedly examined local geology with Lyell's *Principles* as his guide. He answered Lyell's call, understanding, as Lyell admitted, that the insights and generalizations of the new science may be limited as yet, but "they who follow may be expected to reap the most valuable fruits of our labour. Meanwhile the charm of first discovery is our own . . . as we explore this magnificent field of inquiry."[27]

With characteristic enthusiasm Darwin threw himself into geology, making the rocks ring with his eagerly wielded hammer and getting more and more excited about reading landscapes and formations new to him. What wasn't to love about this new science?—it dealt with grand, awe-inspiring questions about origins and antiquity, scientific sleuthing that required diligent fieldwork and close observation, representing Hershelian induction at its best. He filled notebook after notebook with observations, musings, and interpretations, and his admiration for Humboldt only grew. On seeing a tropical forest for the first time he rhapsodized to Henslow: "I never experienced such intense delight—I formerly admired Humboldt, I now almost adore him; he alone gives any notion, of the feelings which are raised in the mind on first entering the Tropics."[28]

While Darwin increasingly saw himself as a geologist, that's not to say he was less diligent in other departments of natural history. Those mysterious marine invertebrates that he first saw in Edinburgh captivated him yet and were always a dip net away. When Darwin wasn't prostrated with seasickness on those long stretches of open ocean between anchorages he often deployed a plankton net of his own making to see what marine life he could find—just the second recorded instance of the use of one, which means he likely learned to make one from Robert Grant. He pulled in an astonishing menagerie of minute marine organisms—"Many of these creatures so low in the scale of nature are most exquisite in their forms & rich colours," he wrote in his diary. And then a perhaps unexpected reflection: "It creates a feeling of wonder that so much beauty should be apparently created for such little purpose."[29]

Those tiny marine organisms had outsized significance in Darwin's thinking. His zoological notes contain page after page of drawings and observations made with the aid of his microscope, foremost among them the "zoophytes," now recognized to be a grab bag of groups now separated into Cnidaria (several orders, including solitary and colonial corals), Bryozoa (the "moss-animals"), sponges (Porifera), and corallines (Rhodophyta, red algae that secrete a hard calcareous casing). In one significant notebook entry from June 1834 he recorded detailed observations of a curious species of *Flustra* (a bryozoan), related to the one he studied in Edinburgh, that was coating the sinuous fronds of kelp growing in dense undersea "forests" near Port Famine, in the Strait of Magellan. "I examined the Polypus of this very simple Flustra," he noted, "so that I might [correct] at some future day, my imperfect notions concerning the organization of the whole family of Dr Grant's Paper."[30] The paper referred to is a translation Grant published of an 1826 work by the German naturalist August Friedrich Scheweigger, arguing that corallines are plants, not animals, which was the common view in the mid-1830s. Darwin took up the study of these organisms to see if he could find any evidence of animality—microscopic sensitive polyps, perhaps, or the production of "ova" like those he found in *Flustra*. In a letter to his sister Catherine, written aboard ship "a hundred miles south of Valparaiso," he commented that "Amongst animals, on principle I have lately determined to work chiefly amongst the Zoophites or Coralls: it is an enormous branch of the organized world; very little known or arranged & abounding with the most curious, yet simple, forms of structures."[31] Writing to Henslow from Valparaiso, he reported that he examined some corallines that had "quite startled" him. He was sure that the classification given by Lamarck, Cuvier, and others was incorrect. "Having seen something of the manner of propagation, in that most ambiguous family, the Corallinas: I feel pretty well convinced if they are not Plants, they are not Zoophites."[32] What *were* they then?

Over the course of the voyage his thinking on this, well, evolved. It was all very confusing; some true corals like the *Caryophyllia* he collected in the Galápagos seemed to reproduce like encrusting *Flustra*, a zoophyte, while others like the scleractinian reef coral *Madrepora* (now

Acropora) that he found in the Cocos Islands were more like coralline algae. So for that matter was the puzzling *Millepora*, a hydrocoral ("I cannot help suspecting that their nature is allied to Corallina rather than to Polypiferous corals"), while a so-called nullipore coralline alga that Darwin found in Tasmania seemed awfully plant-like, reproducing via bud-like segmentation. In his zoology notes he commented that this novel method of growth in corallines "calls to mind the propagation of trees." Reproduction by budding or creeping runners was a decidedly plant-like thing to do. Now coralline algae seemed to do it, and, more remarkably, so did some zoophytes and actual corals: "There is an analogy between the Corall-forming Polypi & turf-forming plants," he mused—"turf-forming plants" that propagate by runners or stolons. A few lines later he reiterated: "I think there is much analogy between zoophites & plants, the polypi being buds, the gemmules the inflo-rescence which forms a bud & young plant."[33] These entries say more than they seem; the distinction between the plant and animal worlds was beginning to blur in Darwin's mind. These simplest of animals and proto-plants (if that's what coralline algae were) shared some fun-damental similarities in structure and mode of reproduction, a capti-vating idea that just a few years later would dovetail beautifully with his nascent transmutationism. We'll see in later chapters how Darwin eventually sought evidence for not only common ancestry of plants and animals but all life, and also the idea of an ancestral unisexual state that gave rise to separate sexes. Those were not ideas he held while on the *Beagle* voyage, but clearly the seeds—or buds?—were planted.

No, Darwin was no transmutationist during the voyage—not even, as legend would have it, upon stepping ashore in the Galápagos Islands. In fact, judging from his notes and comments, he was struggling to make sense of the structure, adaptations, distribution, and relationships of organisms within the familiar framework of natural theology. Yet we continually see Darwin pondering the significance of what he's seeing during the voyage. Thus in Australia we find him musing over the sig-nificance of finding ant lions, those diminutive but ferocious insect lar-

HMS *Beagle* Entering Rio de Janeiro Harbor on April 3, 1832. © 2007 by Jay Matternes. Photo by Richard Milner. Courtesy of Jay Matternes and Richard Milner.

vae that create pitfall traps to prey upon unwary crawling insects. Here they were on a sunny bank in New South Wales, so similar to the ones at home yet here they were amidst the otherwise exotic animal life of Oz. Did Australian zoology represent a wholly separate creation? Oddities like kangaroos and platypuses suggested maybe so, and the possibility of single versus multiple "centres of creation," in the jargon of the day, was a topic of serious debate. The close similarity of Australian and European ant lions suggested to him a link between these widely separate faunas. Australia did not represent a separate creation, however odd its mammals: about the ant lions, "what would the *Dis*believer say to this? Would any two workmen ever hit on so beautiful, so simple & yet so artificial a contrivance? It cannot be thought so—The one hand has surely worked throughout the universe."[34] But toward the end of the voyage we also see him pondering the relationships of Galápagos mockingbirds, seemingly different from island to island yet as a group rather different too from the common mainland species: Are they mere varieties of the mainland species? Sister species? Why so similar yet distinct, and what is it about islands anyway that makes them so productive of novel forms? "If there is the slightest foundation for these remarks the Zoology of Archipelagoes will be well worth examining: for such facts would undermine the stability of species."[35]

As a voyage of exploration there was, of course, plenty of hands-on collecting and observing on the *Beagle* voyage. In Darwin's *Journal of Researches* (1839) the words "observation" or "observe" appear 73 times, "examine" another 48. Experiment? Just 4. But there is evidence of a questioning Darwin who wonders if such-and-such is true and devises an experiment on the fly to find out. Thus, for example, we find him confirming that terrestrial flatworms cut in half lengthwise can regenerate into two new individuals ("In the course of twenty-five days from the operation, the more perfect half could not have been distinguished from any other specimen. . . . Although so well-known an experiment, it was interesting to watch the gradual production of every essential organ, out of the simple extremity of another animal").[36] He also tested whether carrion-feeding condors find their food by sight or smell: "I tried . . . the following experiment: the condors were tied, each by a rope, in a long row at the bottom of a wall; and having folded up a piece of meat in white paper, I walked backwards and forwards, carrying it in my hand at the distance of about three yards from them, but no notice whatever was taken." He pushed the package closer and closer until it practically touched the condor's beak, whereupon the bird immediately tore into it. "Under the same circumstances, it would have been quite impossible to have deceived a dog," he wrote in his notebook, but reviewing similar experiments by Audubon and others he had to admit that "the evidence in favour of and against the acute smelling powers of carrion-vultures is singularly balanced."[37] Another experiment was aimed at the nature of corallines: undertaken at Bahia, Brazil, in August 1836 on the last leg of the journey home, he placed a number of them in sunlight to determine if they emitted gas like photosynthesizing plants. "On several occasions having kept vigorous tufts of articulated Nulliporae in sea-water in sun-light, it appeared as if a good deal of gas was exhaled; it would be curious to ascertain what this is."[38] This was reminiscent of experiments done in the 1770s by Joseph Priestley and Jan Ingen-Housz showing that oxygen bubbles were produced by plants under similar conditions. In other cases his experiments bordered on the goofy, as when in the Galápagos Islands he tossed a marine iguana "several times as far as I could, into a deep pool left by the retiring tide," to see if it would persist in returning to the spot where he picked it up. (The imperturbable iguana invariably did so, we learn, and a giant tortoise was similarly unimpressed when Darwin unsteadily rode it, in what must have been a spectacle recalling Albert Way's cartoon of a beetle-riding Darwin.)[39]

Darwin's youthful "experiments" on the *Beagle* voyage may have been larks done quickly in some cases, but they also reveal a questioning mind eager to investigate, experience, and learn. The iguana-tossing Darwin is also the Darwin filling notebook after notebook with his finds, thoughts, and observations, alert to anything interesting or unusual. Constantly tossing out his plankton net (easier than capturing iguanas), for example, and marveling over the bounty of his catch, imagine his surprise to one day haul in a bevy of beetles far out at sea. No big deal, one might think, but what were those beetles doing in the open ocean? Later he would come to see this and similar innocuous observations as bearing profoundly on an understanding of geographical distribution (which we take up in Chapter 5). Or again, near Rio de Janeiro Darwin collected a fungus similar to one he knew at home. Musing on how the European species is attractive to beetles, as if on cue a beetle flew in and landed on the fungus in his hand. As with those Australian ant lions, "We here see in two distant countries a similar relation between plants and insects of the same families, though the species of both are different."[40] What did it mean?

Through it all there is Darwin's sense of wonder—as when he beheld the "aërial voyages of spiders," when myriad minute spiderlings swarming over the ship's rigging spun gossamer silk strands to be borne away on the breeze. Or the astonishing evening of "snowing butterflies" off the coast of northern Patagonia, when "vast numbers" of the insects "in bands or flocks of countless myriads, extended as far as the eye could range." Or the "splendid scene of natural fireworks" witnessed in the south Atlantic, with St. Elmo's fire on the mast and yard-arms and the sea "so highly luminous, that the tracks of the penguins were marked by a fiery wake."[41] The words "beautiful" and "beauty" appear no less than 105 times in Darwin's *Journal of Researches*, and "delight" and "delightful" some 37. And although "sublime" appears just six times, each is expressive of rapturous Humboldtian moments. How inexpressibly gratifying, then, that when Darwin sent with humility (and perhaps some modicum of trepidation) a copy of his newly published *Journal of Researches* to Humboldt himself with the author's compliments, he received warm praise and thanks from the great man—along with far-ranging questions and comments on subjects from glacial phenomena to volcanism and climate to sea currents. An elated Darwin replied with enthusiasm, including data on the temperature of the sea in different locations and closing with an expression of his admiration: "That the author of those passages in the Personal Narrative, which I have read over and over again, & have copied out, that they might ever be present in my mind, should have so honoured me, is a gratification of a kind, which can but seldom happen to anyone."[42]

A House Burnt by Fire

Darwin returned from the *Beagle* voyage in October 1836 to find himself something of a celebrity, thanks to Henslow and others giving regular updates to the learned societies on his discoveries and impressive efforts in collecting and observing. Darwin had come a long way by the fall of 1836. His physical journey around the world may have concluded, but his intellectual journey was about to get even more interesting. A month after his return he was elected to the Geological Society of London, and a month after that he rented a small house on Fitzwilliam Street, in Cambridge, to sort out his prodigious collections and equally prodigious plans. He sought out experts to analyze and describe his specimens, and it didn't take long before papers on his remarkable discoveries were heard. Richard Owen, a few years older than Darwin and a fast-rising star of comparative anatomy, soon began analysis of Darwin's fossil mammals from South America, declaring before packed

meetings of the Geological Society that Darwin had found remarkable
new species, gigantic forms of groups even now found on that continent. These fossils supported the Law of Succession, the fundamental link between extinct groups and related ones currently living in an
area—findings deemed so important that Lyell, then president of the
Society, dedicated his annual presidential address to discussing them
in February 1837. (At the same meeting Lyell tapped Darwin for the
Society's Council, another indication of the high regard he was held in.)
The ornithologist John Gould was almost as speedy as Owen: in January
1837 he read not one but two papers at the Zoological Society regarding
Darwin's curious South American birds, including one on the enigmatic
finches of the Galápagos Islands. In February he followed with another
paper on the Galápagos mockingbirds. Meeting with Gould in March—
the month he moved from Cambridge to take up residence not far from
Lyell in London—Darwin was astonished to learn that in fact nearly all
of the Galápagos bird species he collected were unique to those islands,
yet bore a close relationship to South American species.

    That was just the beginning: 1837 was a watershed year for Darwin.
It opened with his first scientific paper on January 4th ("Observations
of proofs of recent elevation on the coast of Chili"), and by the year's
end he had read two more papers that were well received, written
the third volume of the official multiauthored narrative of the *Beagle*
voyage (produced in a frenzy of writing between January and September 1837; the resulting *Journal and Remarks* was soon reissued as the
stand-alone *Journal of Researches*), and received a £1,000 grant from
the Admiralty to serve as editor for a multivolume treatment of the
zoology of the voyage. Little did he know what he was getting himself
into with that last project: the volumes duly came out between 1838
and 1843, during which time he experienced all the headaches editors
do today: missed deadlines, less-than-careful authors, endless corrections, and illustrators working on their own time. First he lined up the
authors: his fish would be treated by Leonard Jenyns, Richard Owen
had his fossil mammals, Thomas Bell the reptile and amphibian collection, John Gould agreed to cover the birds, and George Waterhouse

took on the living mammals. The insects were too large and diverse a group for any one specialist, so they got farmed out to various entomologists for separate publication, and although kindly Henslow meant well in agreeing to take on the plants, in the end he was much too busy.

Darwin wanted to work up the marine invertebrates himself—another indication of the special status these organisms held in his thinking—but he soon gave up that idea and they never were treated in a dedicated volume. In the meantime he was consumed with writing introductions to the various volumes, adding notes and observations, and supervising the engravings and printing. But as important as those volumes were, they became a backdrop to other, more exciting developments. For the watershed moment in that watershed year came in March, when the insights of Owen (regarding his South American fossils and geological succession) and Gould (regarding his South American and Galápagos birds and their essential geographical relationships) suddenly dovetailed and crystallized for Darwin the heretical idea that species change over time would at once elegantly explain these seemingly unrelated observations, and more.

Darwin's subsequent work then proceeded along two avenues: geologist and closet transmutationist. Darwin the geologist read paper after paper at the Geological Society, in good Lyellian fashion reporting on areas of elevation and subsidence in the Pacific and Indian Oceans deduced from coral formations, volcanic phenomena associated with earthquakes and a model of mountain chains and volcanoes as effects of continental uplift, and even on the "geological force" exerted by humble earthworms—a subject he was to return to in his very last book some 40 years later (see Chapter 10). He expanded some of his papers into full volumes: coral reefs in 1842, volcanic islands in 1844, and a sweeping geology of South America in 1846. These books were testaments to his industry in those heady years; magnanimous Lyell was delighted that Darwin's new theory of coral atolls, which suggested an evolution of coral formations from fringing reef to atoll as volcanic islands eroded to sea level, upended his own undersea crater-rim theory.[43] Lyell's *Principles of Geology* had its controversial aspects, but its overarching theme—explaining "the former changes of the earth's surface by reference to causes now in operation"—was taking the geological world by storm, and Darwin was now Lyell's chief apostle. In 1837 the *Principles* was already in its fifth edition, just 7 years after the first volume appeared, and Darwin was still as excited by Lyell's vision as he was when he read that first newly published volume given to him by Fitzroy at the start of the *Beagle* voyage. By this time the two had become close friends, but Darwin also saw Lyell as a mentor. A comment he made a few years later to Leonard Horner, Lyell's father-in-law, sums up Lyell's effect on him nicely: "I always feel as if my books came half out of Lyell's brains . . . for I have always thought that the great merit of the Principles, was that it altered the whole tone of one's mind & therefore that when seeing a thing never seen by Lyell, one yet saw it partially through his eyes."[44]

Darwin supported Lyell's view of a dynamic earth characterized by slow and steady uplift and subsidence, a seesaw oscillation of different parts of the crust at different times. It was assumed that continents could not move laterally, but by virtue of the forces below, the comparatively thin and flexible crust could be elevated or depressed as pressure increased or decreased. Volcanism and earthquakes can seem locally catastrophic, but in the grand scheme of things these resulted in incremental changes to the landscape. Over the millennia small changes add up to great ones, Lyell maintained, and so landforms inexorably evolve. Darwin concurred, and when he read papers about Lyellian uplift, volcanism, and earthquakes he spoke with the voice of experience—no armchair naturalist, he had seen Lyell's theories confirmed for himself, as when he found clear evidence of not only *past* uplift on the west coast of South America (in the form of high-and-dry marine deposits—strata bearing shells and coral meters above the present level of the sea), but also *present* uplift. He was at Valdivia, Chile, when an earthquake struck miles away near Concepción in February 1835: "I happened to be on shore, and was lying down in the wood to rest myself. It came on suddenly, and lasted two minutes; but the time appeared much longer. The rocking of the ground was most sensible. . . . There was no difficulty in standing upright, but the motion made me almost giddy." "A bad earthquake at once destroys the oldest associations," Darwin mused: "the world, the very emblem of all that is solid, has moved beneath our feet like a crust over a fluid—one second of time has conveyed to the mind a strange idea of insecurity, which hours of reflection would never have created."[45]

Nearly 2 weeks later when the *Beagle* reached Concepción the crew witnessed the terrible devastation. Darwin gathered as much information as he could about the earthquake and ensuing tsunami: "I feel it is quite impossible to convey the mingled feelings with which one beholds such a spectacle," he wrote. "Several of the officers visited it before me, but their strongest language failed to communicate a just idea of the desolation. It is a bitter and humiliating thing to see works, which have cost men so much time and labour, overthrown in one minute."[46] He noticed, too, the geological aftermath:

> The most remarkable effect (or perhaps speaking more correctly, cause) of this earthquake was the permanent elevation of the land. Captain FitzRoy having twice visited the island of Santa Maria, for the purpose of examining every circumstance with extreme accuracy, has brought a mass of evidence in proof of such elevation, far more conclusive than that on which geologists on most other occasions place implicit faith.[47]

It was clear to Darwin that "the principles laid down by Mr. Lyell" are correct, and we may "fearlessly maintain that the problem of the raised shells . . . is explained." Periodic earthquakes and their associated uplift would slowly but surely keep raising the newly exposed marine detritus higher and higher, ultimately to become another stratum high in the hills of the Andes but always bearing the unmistakable evidence of its origin beneath the sea. Lyell's views as laid out in the *Principles* were as ascendant as these geological formations, and Darwin's astute *Beagle* observations made him the first among a new breed of Lyellian geologists.

But, if seeing the world through Lyell's eyes had its triumphs for Darwin, it's important to point out that it also had its failures—failures perhaps understandable in the context of the science of the time, and in any case instructive in teaching us something about science as a process, about the difficulty sometimes in seeing things in a light altogether different from what we expect. The old joke about looking for missing keys under a streetlight not because they were lost there but because the light is better contains a truth about science as a human endeavor: we often look for things where we are comfortable looking. Or, put another way, the questions we ask are often shaped by the framework we are used to working in. In important respects Darwin's ultimate triumph lay in seeing the organic world with new eyes—he became biology's Lyell, and certainly pursued an idea of nature—including, shockingly, of humans—outside of his society's comfort zone. But at times he was also guilty of being wed to preconceived ideas, leaping to expected and comfortable conclusions.

Lyell (and Darwin) thought that crustal uplift and subsidence occurred in a balanced way over the earth's surface, so that elevation in one region was more or less compensated by subsidence in another. Evidence of past uplift went hand in hand with evidence of subsidence, as testified by the thick limestone and sandstone formations often found far inland (think marine fossils in Kansas), indicative of a previous encroachment by the ocean. Flush with the success of his interpretation of South American uplift he decided to try his hand at solving a geological mystery closer to home: the origin of the so-called "parallel roads" of Glen Roy, Scotland. The expansive and green Valley (*Glen*) Roy, near Lochaber in the west Scottish Highlands, is ringed by three closely spaced parallel white terraces running along the hillsides. Straight as can be, from a distance they look very much like man-made roads, hence the name. The "parallel roads" had puzzled leading geologists of the day, but Darwin saw them as analogous to those strata of marine fossils he had seen high in the Andes. The young Turk made an imaginative leap. In late June 1838, just a few months after his paper on volcanic phenomena, he traveled to Glen Roy and spent 8 days mak-

A drawing by Darwin's college friend Albert Way illustrating the "parallel roads" on the hills of Glen Roy, Scotland. From Darwin (1839b), plate II.

ing careful observations. He worked up his analysis in a long paper he read the following February. The "parallel roads," he declared, were clearly marine in origin. He was sure that the terraces represented ancient marine beaches, formed during a period of subsidence of the land when the sea flooded the wide valley. As the land was raised in fits and starts, he imagined, different "roads" would have been cut at slightly different elevations, the lapping waves leaving their mark on the hillsides as indelibly as a bathtub ring marks the former level of the bathwater. "The whole country has been slowly elevated," he wrote, "the movements having been interrupted by as many periods of rest as there are shelves."

Darwin's theory was elaborate, involving arguments about the fluidity of the earth's crust, the structure of its interior, and the action of powerful elevational forces. Along the way he considered and dispensed with rival theories positing that the parallel roads were lake, not marine, shorelines. He all but scoffed at this suggestion, arguing that there was no sign whatsoever of the remains of a barrier large enough to dam such an expansive area. A miles-long wall of rock nearly a half-mile high would have been necessary; vestiges should remain as with any blown-out man-made dam. What's more, his ancillary theory of iceberg transport of boulders reinforced the marine incursion theory. Again following Lyell's lead Darwin sought to explain so-called erratic boulders—large boulders situated far from the parent material yielding that rock type—as having been dumped by melting rock-strewn icebergs that had floated south from arctic seas. The very year Darwin read his Glen Roy paper he also read a paper to the Geological Society entitled "Note on a rock seen on an iceberg in 61° south latitude"—an observation made on the *Beagle* voyage and now cited as further support for the idea of an ancient marine incursion. Otherwise, how else could erratics have gotten so far inland?

At the same time there were nagging difficulties with his favored theory—notably an absence of any marine fossils in the area whatsoever, and the absence of "roads" throughout the greater region even though topography showed that any ancient marine incursion could not have been confined to Glen Roy alone. But he argued that lack of such evidence was not the same as negative evidence. Riding high with his election to the Royal Society in January 1839 (following in the footsteps of his father and grandfather before him), Darwin decided that his inaugural paper before the premier London scientific society would be on Glen Roy, reading the first part of a hefty two-part paper barely 2 weeks after his election.

He was very pleased with himself, but the feeling didn't last long as it became increasingly clear that Darwin had misinterpreted Glen Roy. Just a year later, in 1840, the famous Swiss naturalist Louis Agassiz took the scientific societies of London and Glasgow by storm with papers propounding his glacial theory, a radically new way to see the landscape. Agassiz, building on the work of fellow Swiss Jean de Charpentier, first developed the theory positing a period of worldwide glaciation in 1837 based on the glaciers and associated landforms in the Swiss Alps. It was met with much skepticism at first, but steadily gained traction as geologists learned to read the evidence for the action of massive, slowly flowing rivers of ice. Seeing simi-

lar formations associated with present-day glaciers in the Alps helped
them connect the dots for areas that are now ice-free. In 1840 Agassiz toured Britain in the company of Oxford geologist William Buckland (Lyell's professor) and other notable geologists, showing them
how many of the landforms and other phenomena they thought they
knew could now be understood as products of glacial action. How
astonishing that must have been, to come to terms with the idea of
towering glaciers covering vast areas of Britain and Europe. Darwin's
marine theory held on for a few years, but the glacial theory, spur-
ring new geological fieldwork, brought together under one explanatory umbrella a host of observations: rounded cobbles and other signs
of alluvial deposition, scored bedrock, kettle lakes, erratic boulders,
moraines, and more. It became clear that there was indeed a barrier
damming the great Glen, and that barrier was ice.

Darwin always considered his Glen Roy paper his "greatest blunder," one he was thoroughly ashamed of. It makes for a case study in
scientific self-deception; his successes interpreting the South American landscape had manifestly gone to his head, leading to a selective
reading of evidence for and against his marine incursion idea. He soon
became a convert to the glacial theory. Returning to Snowdonia's Cwm
Idwal in 1842, he realized how he (and everyone else before Agassiz)
had missed the real story of that valley too: "The ruins of a house burnt
by fire do not tell their tale more plainly, than do the mountains of
Scotland and Wales, with their scored flanks, polished surfaces, and
perched boulders, of the icy streams with which their valleys were lately
filled,"[48] he wrote years later. Our interest in Darwin's blunder in Glen
Roy lies not so much in the fact that he was wrong, but in how it made
him more cautious in subsequent scientific studies—at least publicly.
Not only must evidence be carefully and (as best we can) objectively
weighed, but Darwin came to adopt something of a "shock and awe"
approach to marshaling overwhelming evidence to bear on his arguments. His later so-called delay in publishing his species theory was
not so much due to any simple fear of going public, but rather can be
understood more as a determination to accumulate and interweave so

much evidence that his case would be as unsinkable as the indomitable *Beagle*. And then, too, Darwin performed more and more experiments in pursuit of the evidence he needed.

Experimentiser on the Downs

While Darwin the geologist had his ups and downs, Darwin the closet transmutationist kicked into high evidence-gathering gear. As he commented to a correspondent in 1861, "all observation should be for or against some theory if it is to be of use."[49] Indeed, that became the working principle he applied rigorously to his "species work," as he called it. And the theory he had in mind, well developed by the time of his ill-fated Glen Roy paper, was the idea of descent with modification driven by a process he dubbed "natural selection." He revealed his research agenda in a breathless notebook entry in 1838: once you grant that species may change, "the whole fabric totters & falls." And then: "Look abroad, study gradation study unity of type study geographical distribution study relation of fossil with recent . . . the fabric falls!"[50] In good Whewellian consilience mode Darwin realized that evidence for this heretical idea was to be found in disparate areas, and he was determined to pursue each line, ferreting out as much information as he could for or against his theory—a theory that shone brightly to him, guiding his investigations like a navigational star guides ships.

This brought out the experimenter in Darwin. Experimentation entails *questions*, and once Darwin became a transmutationist with a theory of change all sorts of new questions arose. He was suddenly looking at organisms and their structure, distribution, classification, behavior, etc. from a different angle, and his notebook reminder to "look abroad" was a call to action. We have seen how by this time Darwin had conducted a number of quick experiments, but now he became a dedicated "experimentiser" by necessity, owing to lack of data. Here too we see the beginnings of Darwin the canvasser and interlocutor pressing friends, family, and acquaintances for assistance and information—the original crowd-sourcer, to use a modern term. But there was only so much he could gain from others. Looking at the world in a wholly new way calls for asking questions about the world that no one had thought to ask before—questions that would need to be answered by careful observation in some cases, and experimentation in others.

From very early in his transmutation notebooks we see repeated references to the need for experiments: "Many interesting experiments might be tried by comparing [zoophytes] to plants," he noted in the A notebook; and "It would be curious experiment" to soak seeds in salt water to see if variation could be induced, and to "experimentise on land shells in salt water" he jotted in the B notebook.[51] These and other notebooks are peppered with ideas for experiments, and he eventually kept one for that purpose: the "Questions & Experiments" (Q&E) notebook was started in mid-1839. Most entries in this notebook were made by 1844, the year Darwin wrote a long (over 200 page) account of his evolutionary thinking to that point (now referred to as his species *Essay*), but he clearly used it in the composition of the 1842 outline of his theory (now referred to as the species *Sketch*). Historians have pointed out that the Q&E notebook served as a storehouse of ideas and questions about breeding, variation, crossing, and pollination, key subjects for Darwin.

Ironically, of the many animals that Darwin mentions in his Q&E notebook—indeed, in *all* of his transmutation notebooks—the one that was about to consume nearly all of his attention for the next 8 years merited precisely one entry: barnacles. These ubiquitous marine organisms soon became his go-to group as he grappled with the nature of variation. But Darwin's barnacle odyssey actually started years earlier in Patagonia, with a mystery specimen that he couldn't quite identify at first. That is the subject of the next chapter.

Experimentising: Going to Seed

Being an experimentiser means being an observer and an asker of questions. *Testable* questions—questions framed in a way that an experiment or careful observations might give a reasonably reliable answer one way or another. Those are the incremental steps of scientific progress. But, it all really starts with observing. Darwin was an inveterate observer—he noticed color, behavior, structure, and function, and often marveled at adaptation. Indeed, his life's work became one long study to understand how adaptations arise. As he expressed it in *On the Origin of Species*:

> How have all those exquisite adaptations of one part of the organisation to another part, and to the conditions of life, and of one distinct organic being to another being, been perfected? We see these beautiful co-adaptations most plainly in the woodpecker and missletoe; and only a little less plainly in the humblest parasite which clings to the hairs of a quadruped or feathers of a bird; in the structure of the beetle which dives through the water; in the plumed seed which is wafted by the gentlest breeze; in short, we see beautiful adaptations everywhere and in every part of the organic world.

Darwin's answer to the question of how exquisite adaptations arise was natural selection. But before evolutionary process and pattern can be studied, the adaptations themselves must be understood. That's where observation comes in. Your Darwin-inspired adventures as an experimentiser begin where his did, with observing. We'll take a close look at "Plumed seed . . . wafted by the gentlest breeze" and other wind-dispersed seeds—object lessons in adaptive structure and function.

A. Materials
- Notebook and pencil
- Hand lens or dissecting microscope, if available
- Tape measure or meter stick
- Fan
- Stopwatch
- Masking tape

- Marker
- Graph paper
- Scale, if available
- Obtain assorted wind-dispersed seeds, for example:
  * Plumed seeds, such as milkweed (*Asclepias*), Indian hemp (*Apocynum*), dandelion (*Taraxacum*), goldenrod (*Solidago*), sycamore or plane tree (*Plantanus*)
  * Winged seeds, such as maple (*Acer*), ash (*Fraxinus*), tuliptree (*Liriodendron*), basswood or linden (*Tilia*), pine (*Pinus*), or hemlock (*Tsuga*)

Note on obtaining seeds: Weedy species like dandelion are widespread and produce seeds year-round. The other plants listed will produce seeds seasonally; in the northern hemisphere their seeds will be mature by late summer or early fall. You can find these plants at local nurseries and garden centers. Seeds can also be purchased from online suppliers such as Prairie Moon Nursery (www.prairiemoon.com).

B. Procedure

1. Observe your seeds and their wind-dispersal mechanisms carefully with the hand lens or dissecting microscope.
2. Sketch the seeds, noting the similarities and differences in the wind-dispersal structure found in the different plumed seeds you collected. Repeat for the winged seeds.
3. Why do you suppose the seed dispersal structure is not identical for all of the plumed seeds or all of the winged seeds, even though in each group they are adapted to disperse their seeds in basically the same way?

    (In a word, the answer is *ancestry*.)
4. Carefully compare the winged structures of pine, hemlock, and ash seeds with a hand lens or dissecting microscope. Which two are the most similar to each other? If you answered "pine and hemlock" you are correct. Consider *why* this is true. You might say that these two are conifers while a tuliptree is not, and that

explains the observed similarity and dissimilarity. That is correct too, at one level. But how would Darwin explain this?

(He would express it in terms of ancestry. The pine and hemlock wings are *homologous* structures, and they are *analogous* with the wings of the ash seeds—the analogous structures are *convergent*: similar structures arising in different evolutionary lineages for the same adaptive reasons.)

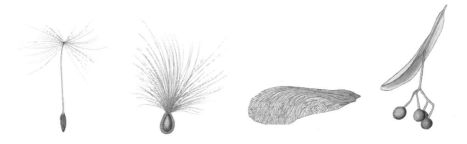

How do the different wind-dispersal mechanisms compare, in terms of their effectiveness in carrying seeds away from the parent plant? In this experiment we'll explore the relationship between size, structure, and dispersal efficiency.

1. Set up the fan 1 yd (or 1 m) above the floor and blowing horizontally; experiment with the fan speed setting to determine an effective speed for the available area.
2. Extend the tape measure in the direction the fan is blowing and use masking tape to affix it to the floor.
3. Drop a seed specimen from a standard height just in front of and above the fan. Mark with tape where the seed hits the floor, and record the distance with the tape measure. Repeat five times with the same seed type.
4. Repeat Step 3 with each type of seed in your collection. Calculate the average distance traveled for each. Which seed type went the farthest and which the shortest distance?
5. Weigh each seed used in the experiment, if possible.

6. Vary the experiment by dropping each seed type from a standard height of 2 yd (~2 m) just in front of the fan.
7. Record the time that the seed remains in the air with the stopwatch. Repeat five times with the same seed.
8. Repeat Step 7 with each type of seed in your collection. Calculate the average time aloft for each seed type.
9. Graph the average distance traveled and average time aloft for each seed type.

Is there a relationship between the distance traveled and time aloft? Does size or weight play a role in efficacy of wind dispersal? Plumed seeds tend to be light and ride passively on air currents. Most winged seeds are heavier, but some can generate lift, slowing their descent and increasing their dispersal distance.

See also:

"Early Days" and "Doing Darwin's Experiments" at the Darwin Correspondence Project: www.darwinproject.ac.uk/learning/universities/getting-know-darwins-science/early-days and www.darwinproject.ac.uk/learning/11-14/doing-darwins-experiments.

2

Barnacles to Barbs

January 1835 found the 26-year-old Darwin and his *Beagle* comrades along the west coast of South America, amid the splendid desolation of the endless rugged and densely forested islands, looming snow-capped mountains, and deep bays and channels of the submerged Chilean Coast Range. Despite being high summer the weather was far from balmy: it was a place where "it rained as if rain was a novelty," he commented ruefully in his diary; where the rain "never seems to grow tired of pouring down." A week in the Chonos Archipelago, January 8th through the 14th, "passed rather heavily," he wrote; "the climate is so very bad & the country so very uniform in its character."[1] Between downpours he managed to get out on shore and collect, though, and it was there that, poking around a windswept gravelly beach, a number of thick Chilean abalone shells of the genus *Concholepas* caught his eye. They were riddled with tiny holes, like shot; holes he recognized were caused by the drilling of a minute organism. Intrigued, he collected some.

Back onboard the *Beagle*, snug in his cabin-cum-lab, Darwin duly recorded his find as specimen no. 2495 and, under the microscope, carefully dissected adults and several early stages of what turned out to be an odd parasitic barnacle. Yellowish in color and upside-down in their holes, they were unlike any barnacle he had ever seen. "Who would recognize a young Balanus [barnacle] in this illformed little monster," he asked in his specimen notebook, concluding that "it is manifest this curious little animal forms new genus."[2] Darwin didn't know it then, but he had found the smallest barnacle species in the world and was to bestow, years later, the name *Cryptophialus minutus* upon it. He also didn't realize that there was far more to this barnacle than met the eye; a decade later the by-then confirmed but closeted evolutionist would take another look at these specimens and make an astounding discovery, one that he believed had profound implications for his secret evolutionary ideas. As if somehow auguring his incendiary views to come, a few days after collecting his burrowing barnacle, on the night of January 19th, the towering Volcán Osorno erupted. Plainly visible to the astonished crew of the *Beagle* despite a distance of over 70 miles, Darwin recorded the spectacle in his diary, "a very magnificent sight."[3]

The destructive power of the eruption and the earthquake that followed was awful, but at the same time Darwin could not help but notice the astonishing geological effects of the event: the beach was uplifted some 10 feet, stranding hapless nearshore marine life high and dry and gasping, a palpable demonstration of Lyellian processes at work. Two weeks later, along the same coast, he trekked inland to see oyster beds deep in a forest at an elevation of 350 feet: evidence of uplift past that resonated deeply with the newly raised shoreline. At this point he was not many years away from applying gradual Lyellian geological and landscape change to equally gradual change in species, and would come to appreciate a profound correspondence between his mystery barnacle and the mysteries of geological change.

But quite a lot was to happen before then. For one thing, nearly 2 years yet remained of his *Beagle* voyage; 2 more years of strikingly beautiful landscapes and endless, featureless ocean, fascinating peoples, intriguing collecting, and near-constant seasickness. "I hate every wave of the ocean, with a fervor, which you, who have only seen the green waters of the shore, can never understand" he wrote from Tasmania to his cousin William Darwin Fox a year later.[4] Yet there is no question that the voyage was a turning point. Darwin would declare in his diary that "nothing can be more improving to a young naturalist, than a journey in distant countries . . . the habit of comparison leads to generalization."[5] *Generalization*, the opportunity to see the big picture, connect disparate dots, inductively puzzle out the laws of nature. The collections and observations made at a given time and place on this or that species, fossil, or geological formation had great value, and collectively they yielded insights greater than the sum of the parts. So it was that organisms as seemingly dissimilar as cirripedes and domestic pigeons would eventually be united under a common explanatory framework: methodical analysis of structure and development within each of these groups, with all of their attendant oddities, would yield mutually reinforcing evidence for evolution by natural selection. Much later, discoveries of *Hox* genes would confirm the deep structural bond of invertebrates and vertebrates—of organisms as dissimilar as barnacles and barbs, the fancy pigeons of Barbary descent—in spectacular fashion that surely would have thrilled Darwin. The stage was set for this in South America, on a remote and windswept Patagonian beach.

In that same letter to Fox, Darwin commented how he looked forward "with a comical mixture of dread & satisfaction" to the scientific work awaiting him at home. "I suppose," he mused, "my chief place of residence will at first be Cambridge & then London"—scientific centers of choice for organizing and working up his extensive collections and notes, their museums and scientific societies lately displaying a "rapidly growing zeal for Natural History" that would be immensely useful, he was sure. That would come to pass, but as a transitional stage between his wide-ranging travels and settling down permanently at Down House—not unlike the barnacles that would come to captivate him. As Charles Kingsley, the clergyman-naturalist-historian and prolific author, wrote of them in his 1855 book on seashore life:

> This creature, rooted to one spot through life and death, was in
> its infancy a free swimming animal, hovering from place to place
> upon delicate ciliae, till, having sown its wild oats, it settled down

in life, built itself a good stone house, and became a landowner, or rather a *glebae adscriptus*, for ever and a day. Mysterious destiny![6]

Mysterious indeed, these barnacles—so like diminutive *glebae adscriptus*, the medieval laboring "adscripts of the soil" permanently attached to the land. But a deeper mystery was how the identity of these humble organisms eluded naturalists until just a few years before Darwin's walk on that beach in southern Chile: before the 1830s they had been classified as mollusks.

Barnacles are arthropods, classified today in the infraclass Cirripedia of the subphylum Crustacea, meaning their closest relatives are crabs, shrimp, lobsters, and their ilk. They take their name from their elongated, slender and gracefully curved legs, or cirri ("cirripede" is derived from the Latin *cirrus*, "curled" or "tufted," like cirrus clouds, and *pede*, "foot"), tipped with fine hairs for sweeping food particles from the water column. The shell-like armor and two-part valves of common encrusting barnacles, together with the erroneous belief that they did not undergo metamorphosis, may well have misled naturalists since time immemorial into thinking that they were odd relatives of limpets or clams. It was not until 1830 that John Vaughan Thompson (1779–1847), a British army surgeon accomplished in zoology and botany, revealed their remarkable structure and metamorphosis in a memoir published in his collection *Zoological Researches, and Illustrations; or Natural History of Nondescript Or Imperfectly Known Animals*. Seen through a good microscope, these organisms were far from nondescript, but they certainly were imperfectly known. Thompson worked out the surprising life history of barnacles, and how their anatomical structure pointed to a relationship with arthropod crustaceans. Piquing the interest of other naturalists, studies confirming and extending Thompson's findings appeared in due time, but so too did studies arguing against Thompson's findings. The matter was not fully settled until the early 1840s, but most naturalists acknowledged the organisms' true identity well before then.

Thompson showed that barnacles start out as a free-swimming one-

eyed larva he called the *nauplius*, compact with odd projecting structures for swimming, resembling a menacing alien spacecraft of the kind
you might see in a *Star Trek* movie. After six months or so these mature
into an adult affixed to one spot. In between is a short-lived transitional
stage of the barnacle, mobile but nonfeeding, which Thompson called
the *cyprid*, which plays the role of scout, feeling out potential real
estate for permanent settling using specialized antennae. Once a suitable spot is found, the cyprid attaches itself headfirst by producing a
glue-like substance from the antennae. The head-standing critter soon
metamorphoses into an adult, and an even stronger substance cements
the deal, literally, as the adult is permanently bound to the substrate
by its head. It also begins to secrete protective calcareous plates—a
fortified cell.

Now the barnacle takes on its familiar form: little cones or volcano-
shaped structures stuccoing rocks, boats, shells, even whales and
turtles. At least, that's the form of the common sessile or "acorn" barnacles; a whole other group of them are "pedunculate," or goosenecked,
perched atop a flexible stalk attached to the substrate. The adults can
continue to grow for a time, molting within the safety of their shelter
as all good arthropods do, and expanding the protective plates as they
get larger.

Thompson's well-thumbed
memoir on barnacles was among
the many books in the *Beagle*'s
reference library (which was
at Darwin's fingertips, lodged
in the same cozy 10 × 11-foot
poop cabin where he worked by
day and slung his hammock by
night).[7] With Thompson's book
at hand and his keen marine
zoology interests, inspired by
his Edinburgh mentor (and evolutionist) Robert Grant, Darwin

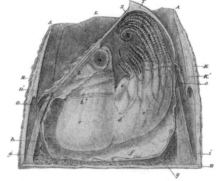

The sessile (acorn) barnacle *Balanus
tintinnabulum* encased in its protective
shell. Note the modified legs, or cirri,
which serve as the food-gathering structures of these suspension-feeders. From
Darwin (1854), vol. 2, plate 25.

was familiar with barnacles and their habits. Yet none that he knew of drilled holes. His cataloging and dissecting done, however, he packed the shot-through *Concholepas* shells away and moved on to the next intriguing collections as the *Beagle* made its way northward along the coast.

Cuidado

Barnacles were on Darwin's mind again in March 1837, the month that saw him turn into a transmutationist. He had just moved to London, renting a place on Great Marlborough Street in Bloomsbury for easy access to the British Museum, and conveniently just up the street from his brother 'Ras. The parallel Darwin perceived between geography and geology—the geographical distribution of related species and its uncanny correspondence with the relationship between living and extinct species of an area—got his adrenalin pumping, as he realized this correspondence spoke of species changing. The epiphany fired his imagination, sending him seeking more information, new observations, to test his hunch. Questions abounded, but among Darwin's earliest notebook entries as a transmutationist we find him especially concerned with reproduction, metamorphosis, and the unity of life, all somehow bearing, he was sure, on the nature of species, varieties, variation, and change—and barnacles were front and center.

In March 1837 he pulled out his old marine zoology notebook from Edinburgh, flipped it over, and commenced a series of notes from the back. The first of these, under the heading "Zoology," referenced the recent debate over barnacle metamorphosis.[8] John Obadiah Westwood, the Oxford arthropod specialist, challenged Thompson's interpretation of barnacle structure and development, and therefore their classification. His critique was based on the prevailing view of relationships between animal classes and phyla, which held that crustaceans are more closely related to vertebrates than to insects. Westwood reasoned that since many insects and no vertebrates exhibited metamorphosis—no vertebrate experienced anything at all like the radical transformation of caterpillar to butterfly, say—by analogy crustaceans should not be so insect-like as to metamorphose either. Westwood was not alone in this belief—odd as it may seem today, notable naturalists of the day could not see the evidence for metamorphosis that Thompson pointed out; such is the power of preconceived ideas, perhaps. Westwood concluded that the weight of evidence pointed to barnacles as some sort of anomalous mollusk, somewhere between the "articulates" (arthropods) and vertebrates. Thompson and Westwood read dueling papers at the Royal Society, published back-to-back in the Society's *Philosophical Transactions* in 1835. It was these papers that Darwin referenced in his notebook.

Darwin's money was on Thompson's interpretation, maybe owing to the stock he put in the argument based on the clear segmentation and other arthropod-like features of the barnacle's early stages. This may be the earliest indication of Darwin's interest in the value of development to inform relationships. In this case early stages were used to make decisions about classification, but later he would see how early stages, especially embryos, could give clues to evolutionary diversification. That idea, too, was in its infancy in the 1830s, and Darwin was abreast of the latest discussions. Richard Owen (1804–1892) would have impressed the utility of early stages upon the younger naturalist in the many conversations they had in his post-*Beagle* years. Darwin might have even heard Owen, a rising star and first permanent Hunterian Professor at the Royal College of Surgeons, hold forth on the subject in his acclaimed Hunterian Lectures in the spring of 1837. On May 9th of that year Owen declared to a rapt 400-plus audience that studying the structure of fully developed animals is inadequate for diagnosing true relationships. Early developmental stages were key: just as entomologists now gain insight into insect relationships by studying their immature stages, "What should we know of the zoological relations of the Barnacle, if we were acquainted only with its organization in the last fixed stage of existence?"[9] In this very same lecture, however, Owen took a dim view of such discreditable ideas as transmutation—a public condemnation that would have reinforced Darwin's resolve to keep his evolutionary speculating to himself had he been present.

Reproduction became as central to Darwin's transmutational musings as geography and geology. In July 1837, during a visit to the family
home in Shrewsbury, he started a fresh notebook with copious notes on
his transmutationist grandfather's work *Zoonomia*. In it he speculated
on the nature of species relationships over time: a branching pattern
captures the essential idea of change within lineages as well as links
between lineages, and he showed that he understood the dynamic of
lineages diverging over time, as in one comment pointing out that gaps
between groups like arthropods and vertebrates will grow over time,
as extant species become more divergent from their common ancestor.
"Heaven know[s] whether this agree[s] with Nature," he wrote, literally
ending on an underscored cautionary note: "Cuidado"—*caution*.[10]

And proceed with caution he did, at least publicly—that was only
prudent, as he was becoming something of a rising star. But in the pri-
vacy of his notebooks his speculations and questions ran rampant as
he immersed himself in literature of all kinds, from hard-core geology,
botany, and zoology to practical agricultural improvement and breeding, with philosophy, religion, and literature thrown in. He boned up
on the prevailing arguments against transmutation, clarifying the key
issues but also arming himself. Darwin knew only too well Lyell's
extensive anti-transmutation arguments in the watershed *Principles*.
The fifth edition came out in 1837, the very year of Darwin's conversion
to transmutation; reviewing Lyell's arguments he wrote in the margin
of his copy: "If this were true, adios theory."[11]

Another Volcano Erupts

From time immemorial those inclined to see transmutational change
cited links throughout the organic world as evidence: species or their
taxonomic groupings can be arranged in something of a sequence, it
was thought, generally in order of increasing "complexity" (however
that was defined). Schemes varied, but their common denominator was
continuity, the idea that all forms could be linked through a chain of
intermediates—versions of the ancient Greek Scale of Nature idea that

became codified in some strains of Christian thought. With increased knowledge of geological strata and the fossil record in the eighteenth and especially nineteenth centuries came the realization that to a large extent this chain seemed to map onto a temporal sequence. The Chain of Being and its development over geological ages were thus easily interpreted in terms of an unfolding divine plan marked by successive bouts of creation (and extinction), a teleological view that inevitably had as its aim the arrival of humanity, the pinnacle of creation made in the very image of the Creator. But such chains of relationship are just as easily interpreted in transmutational terms, as Lamarck (and later Darwin) came to believe.

One of the most serious arguments *against* transmutation, in contrast, was based on supposed *dis*continuities, morphological gaps between major animal groups. If there were such gaps, the rug was pulled out from under any possibility of transmutation. Thus was born the concept of *embranchements*, distinct, separate, decidedly unlinkable taxonomic groups reflecting distinct body plans, an idea introduced by the illustrious and formidable Georges Cuvier in Paris. Cuvier (1769–1832) recognized four basic body plans—*Radiata* (sea stars, jellies, and relatives), *Mollusca* (bivalves, slugs, cephalopods . . . and barnacles), *Articulata* (arthropods and relatives), and *Vertebrata* (all animals with backbones: fish, reptiles, amphibians, birds, and mammals)—each fully independent of the others. Sure, recognizing the diversity of forms within these categories, certain taxa in one *embranchement* could "approach" others in another *embranchement*, but there was no possibility of a real relationship or link between *embranchements*, let alone a passage—barnacles might have some traits that apparently "approach" insects among the articulates, for example, but they were essentially mollusks no matter how much "articulateness" they may exhibit, and one was certainly not derived from the other.

Cuvier's system was devised in explicit refutation of his (unfairly) reviled colleague Lamarck, with his notorious transmutational ideas. No linkage between the fundamental taxonomic groups, no possibility of some kind of "passage," no transmutation. *Why* transmutation was seen as such an affront is a bigger issue than we have space for here; suffice it to say that this became a signal issue for a larger battle over materialism and natural law versus divine providence, and thus over atheism versus received religion. In important respects this was a false dichotomy, not least because then, as now, some simply saw transmutation by natural law as *itself* a divinely ordained process. But the readiness with which the anticlerical revolutionaries of France and their hopeful counterparts in Britain in the late eighteenth and early nineteenth centuries brandished the banner of change, transmutation in all things—from societies to species—and the ensuing horrors this unleashed in the Terror would only harden the position of conservatives defending the social order through the central authority of church and state. Thus it was that transmutation was itself transmogrified into a byword for atheism and even sedition by Darwin's day.

That was certainly the way that Richard Owen saw it. He embraced Cuvier's system of *embranchements* with gusto, and his later term *archetype* (ideal generalized form of a given species or group) was inspired by Cuvier. Fervently opposed to Lamarck and transmutationism, Owen was determined to stamp out such heresy in England. Lyell's polemic against Lamarck and transmutation in the *Principles* was written in the same conviction, though not the same mean-spiritedness. Owen's antipathy and Lyell's disapproval would have been abundantly clear to Darwin, and concerns over the social implications of his theory were behind his note of *cuidado*.

Transformation was thus very much on Darwin's mind in the late 1830s—not least his own transformation as his thinking developed. In the fall of 1837, not long after his brief notes on the Thompson-Westwood debate over crustacean metamorphosis, he delved into another and more famous debate, one that took place back in 1830 and had implications that struck at the very heart of his transmutation theory. That was the year that Georges Cuvier squared off against his younger colleague and sometime protégé, Étienne Geoffroy Saint-Hilaire at the Académie Royale des Sciences in Paris. The issue was no less than the unity of life. For years Cuvier and Geoffroy argued, mostly in private, over the fundamental question of relationship between Cuvier's four *embranchements*. Geoffroy (1772–1844) was as brilliant an anatomist as Cuvier, and while not an out-and-out transmutationist, he was a friend and supporter of Lamarck, who had died the year before the debate. Geoffroy subscribed to an idea that might be described as a zoological uber-plan: he saw a single generalized body plan uniting all animals, all four *embranchements*, allowing that some creative force had fashioned each from the same basic starting point. Perhaps some had even been derived from others, rather than being created *de novo*. Geoffroy's evidence was anatomical structures that he traced through development— essentially what we call homologies today. Through a series of acclaimed studies he had built up a body of work establishing not only unity of the animal body plan, but also principles of connection between seemingly disparate forms, the significance of rudimentary structures, and his "law of balancement of organs." It was Geoffroy who discovered that the mammalian skull could be understood in terms of fusion of various bones, separate in so-called "lower" vertebrate forms and, significantly, early embryological stages of "higher" forms. "Nature constantly uses the same materials," Geoffroy maintained, "and is ingenious only in varying their forms." Early on Cuvier applauded his younger colleague's discoveries, but when Geoffroy sought to derive from them a visionary new philosophy of morphology proclaiming the unity of all animal forms, Cuvier balked.

The trigger of the great debate was a paper Geoffroy encouraged, read by a pair of younger naturalists presenting anatomical evidence that cephalopods form a linking group between mollusks and vertebrates. Cuvier did not take it well. The senior naturalist's vehement rejection of the paper provoked Geoffroy to enter the fray. The argument soon became public as the savants presented dueling papers at the Académie Royale through the spring and summer of 1830. This all unfolded against the backdrop of another drama developing in Paris at the same time: the political upheaval that culminated in *le trois glorieuses*, the Three Glorious Days in late July 1830 when the people rose up and, manning the barricades, forced the abdication of the repressive King Charles X. It was a matter of perspective which was the more significant battle: one Frédéric Soret, of Geneva, had occasion to visit the aging polymath Johann Wolfgang von Goethe in Weimar, Germany, just as news of the Paris uprising reached the city, setting "everyone in a commotion," as Soret recalled. He recounted his meeting with the German savant:

> "Now," [Goethe] exclaimed as I entered, "what do you think of this great event? The volcano has come to an eruption; everything is in flames, and we no longer have a transaction behind closed doors!" "A frightful story," I replied. "But what else could be expected under such notorious circumstances and with such a ministry, than that matters would end with the expulsion of the royal family?" "We do not appear to understand each other, my good friend," replied Goethe. "I am not speaking of those people at all, but of something entirely different. I am speaking of the contest, of the highest importance for science, between Cuvier and Geoffroy Saint-Hilaire, which has come to an open rupture in the Académie."[12]

Soret was at a loss for words. There is no question whom Goethe was rooting for: a visionary in the same mold as Geoffroy, Goethe himself had made fundamental contributions to comparative anatomy, and even advanced a theory of unity of form in botany that paralleled Geoffroy's work in zoology.

The Cuvier-Geoffroy debate is directly relevant to Darwin's development as an experimentiser, though it took time for him to feel the aftershocks of the seismic events of 1830 at the Académie. There is no evidence that he knew about the debate at the time, his letters and diary filled instead with excited talk of beetle collecting, fishing, hunting, and horses. *Le trois glorieuses* found him preparing for a trip to north Wales. Geoffroy had published an account of the debate in his *Principles of Zoology* (1830), including his and Cuvier's papers read at the Académie. Darwin laboriously made his way through Geoffroy's book in late 1837, filling four notebook pages with comments: "[Geoffroy] states there is but one animal: one set of organs—the other animals <u>created</u> with endless differences," reads one entry—note the word "created" underscored. He saw the transmutational significance, continuing: "does not say propagated, but must have concluded so."[13] Several passages are marked in Darwin's copy of the book, the margins littered with exclamation marks, comments, and notes to extract quotations. It is telling which passages got him excited: explaining about the "natural development" of animal bodies (a phrase underlined by Darwin), Geoffroy stated that "man, considered in his embryonic state, in the womb of his mother, passes successively through all the degrees of evolution of the lower animal species: his organization, in its successive phases, approaches the organization of the worm, the fish, the bird."[14]

Geoffroy, familiar with the embryological work of Karl Ernst von Baer, was convinced that modifications at various stages of embryological development were key to his idea of unity of all animal forms—an idea with clear evolutionary implications. Cuvier, dismissive of Geoffroy's "unity of composition" talk as "contrary to the simplest testimony of the senses," maintained that the resemblances that Geoffroy pointed to merely reflect how the "agreement of parts" simply follow from their "coordination for the role that the animal has to play in nature." Form follows function, in Cuvier's view, not the other way around. Darwin excitedly scored this passage, too. But where Cuvier asserted that "conditions of existence"—adaptation to environment, say—explained resemblances, Darwin disagreed: resemblance may also follow from common ancestry. As for the "unity of composition" that Cuvier so emphatically dismissed, it's clear from Darwin's marginal comment that it was a given: "The unity of course due to inheritance." Common ancestry was common sense to Darwin.

Darwin's interest here resonated with his interest in Thompson's barnacle work. He noted how Geoffroy acknowledged a wide gap between insects and mollusks, but also the variation within each group, certain organ systems being more or less developed in each and forming something of a series. "This explains the large hiatus that has been noticed between the families," the Frenchman asserted, "especially with regard to the beings at the center of each series, and also the very numerous relations that they exhibit at their extremes." Referencing this in his notebook that fall of 1837, Darwin noted that barnacles were just such a group at the extreme of the series, and wondered whether Geoffroy thought that such a series was linear or could be branched.

The Whole Art of Making Varieties

But in the late 1830s barnacles were filed away in Darwin's mind—or maybe mentally shelved is more like it, similar to his tagged *Beagle* specimens shut up in sturdy glass-doored museum cabinets. As we'll see, he returned to them nearly a decade later, but now he increasingly looked at the puzzle of reproduction, inheritance, and variation through the lens of domestic breeds—a knotty issue that would shed light on the species question. It was knotty because common sense and the prevailing view held that little could be learned from domesticated breeds. If anything they were evidence that species could *not* change, as Lyell so eloquently expounded in the *Principles*: domestic varieties were changeable only to a degree, it was claimed, and as soon as they run wild they revert back to their generic form. Patently unnatural, domestic animals couldn't survive in a state of nature; only a Gary Larson could imagine wild packs of fancy poodles. Darwin's first hunch was that the environment somehow interacted with individuals to engender variation and species change, and if kept separate from other individuals undergoing their own changes, wholesale transmutation would eventually be achieved.

Thinking along these lines, his initial interest in domestication had more to do with how people affect variation by controlling the environment of farmed plants and animals. But by summer of 1838 he began to have second thoughts: maybe there was more to domestic breeds than met the eye? Immersing himself in the breeders' craft, he pressed friends and acquaintances for information and struck up conversations with anyone and everyone having practical experience with one kind of

domestic group or another: beekeepers and horticulturists, pigeon fanciers and husbandrymen, poultry aficionados and dog breeders. There
was plenty of expertise close at hand, too, between the coachmen and
gardeners who worked for his father and Uncle Josiah. And, of course,
he pored over the writings of the likes of Bakewell, Yarrell, Youatt,
Wilkinson, Sebright, and other pioneering breeders behind Britain's
first agricultural revolution. The problem was that there was often con-
flicting information on phenomena like hybridization and heritability.
Soon their pamphlets yielded an unexpected insight.

Darwin was primed to appreciate the basic idea of selection applied
by people to improve animal breeds months before his reading of Malthus catalyzed his insight into selection in nature. It's hard to say *when*
he became conscious of the concept, but unmistakably it first appears
in entry 17 of transmutation notebook C, dating to February 1838:
"The changes in species must be very slow owing to physical changes
[being] slow and offspring not picked as man [does] when making varieties," he wrote in his usual shorthand.[15] This is a revealing comparison:
believing that physical changes in the environment induce changes in
species, and aware that "picking" offspring, *selecting* those that will or
will not be bred, is the way that people create new varieties, he is suggesting here that the rate of species change in nature is slower in part
because there's no one to do the picking and accelerate the process.
His thinking along these lines is evident in a number of other entries;
he was beginning to appreciate more and more the analogy between
domestic breeds and species and varieties in nature. It all had to do
with the power of picking.

"Picking" becomes methodical selection in his thinking through the
influence of two agricultural improvement pamphlets: Sir John Saunders Sebright's *The Art of Improving the Breeds of Domestic Animals*
(1809), and John Wilkinson's *Remarks on the Improvement of Cattle*
(1820). These works, which Darwin heavily annotated in March 1838,
drove home the power of selective breeding. In fact, the word "selection" in this very context is used repeatedly by Sebright and Wilkinson,
in passages that Darwin underscored. Consider, for example, Sebright's

assertion that the art of breeding consists "in the selection of males and females, intended to breed together, in reference to each other's merits and defects," or that traits like "the fineness of fleece, like every other property in animals of all kinds, may be improved by selection in breeding." Yes, climate, soil, and other environmental factors have some effect on wool quality, he acknowledged, "but not so much as is generally supposed."[16] Sebright's use of the word "selection" meant selective breeding, with some individuals chosen for breeding and others for the pot. He shook his head at the widespread misapprehension that improvements are achieved and new varieties made by hybridizing existing breeds: "The alteration which may be made in any breed of animals by selection, can hardly be conceived by those who have not paid some attention to this subject," he declared. "They attribute every improvement to a cross," when the actual cause was simple: "it is merely the effect of judicious selection."[17] The power of selection is what it's all about, a point underscored by Wilkinson:

> The worst must unquestionably be rejected, while the rest, and especially the best of these, are carefully to be preserved for future stock. . . . By such procedure [i.e., selection], animals have at length been produced, so different from the generality of the stock from whence they were originally taken, that none but such as are well acquainted with these matters, could have an idea, that there existed between them the least affinity. The distinction indeed between some, and their own particular variety, has scarcely been less than the distinction between that variety and the whole species.[18]

Darwin gave this passage an excited double score in the margin, as he did another passage where Wilkinson emphasized that such changes "are in general gradual," proceeding "but slowly through several generations." This was all electrifying. In his C notebook, he declared, "Sir J Sebright pamphlet most important," and that the "Whole art of making varieties may be inferred from facts stated."[19] He saw clearly a parallel between domestic and wild animals, and the power of selection to diversify. His ungrammatical notebooks make for difficult reading, but their almost stream-of-consciousness shorthand also conveys a sense of the energy and excitement of discovery Darwin must have been feeling.

In mid July, he opened a fresh notebook. About 25 percent of the entries pertain to domestication and record his ups and downs in relating what he learned from Sebright, Wilkinson, and others to species and varieties in nature. "The varieties of the domesticated animals must be most complicated," he lamented at one point. In selective breeding only a few individuals are picked out to reproduce, but how does that happen in nature? And he still sensed that separation played a role in the formation of new breeds by preventing intermixing: "The very many breeds of animals in Britain shows, with the aid of seclusion in breeding how easy races or varieties are made," he wrote a few pages later (emphasis Darwin's).[20] Always alert to observations or data bearing on species and varieties, he was amazed by the sheer number of rose varieties when he visited the famous botanic garden and arboretum of George Loddiges in Hackney, just north of London: "Loddiges garden 1279 varieties of roses!!! Proof of capability of variation," he recorded on 23 September 1838.[21] Variation enough to permit the cultivation of so many rose varieties suggested an inexhaustible supply—so much for Lyell's insistence that variation was limited! And unlimited variation meant an unlimited capacity for change.

That visit to Loddiges' garden came just a week before Darwin's re-reading of Malthus provided the spark that led him to natural selection—suddenly grasping nature's version of the picking and choosing process going on in farmyards and fields for centuries. The key was superfecundity. This was a vision of reproductive output on a staggering scale. Picture each bit of confetti in a ticker-tape parade as so many lottery tickets. There are scant few winners; the vast majority in the confetti blizzard are destined to be swept away. What determines winning or losing in the farmyard is the breeder calling the shots; in nature it is the demands of the immediate environment determining which of those hopeful propagules makes it. Variations—ubiquitous, abundant, small, and (very importantly) random—provide the raw material, with Malthusian population pressure the crucible for selection to act, slowly, inexorably changing species according to the demands of their environment.

In another notebook Darwin sketched out how he might explain the process to others, an outline for an argument if you will, building on domestication as an analogy. First, he noted to himself, point out that new varieties are made in two ways: one is by environmental change, while the other is by picking offspring and preventing free intercrossing. Next ask: "Has nature any process analogous?—if so she can produce great ends." Rhetorically asking "But how?" he instructs himself to "make the difficulty apparent by cross-questioning." Finally, he says, "Here give my theory—excellently true theory."[22] An excellently true theory it was, but there were so many details yet to work out. What causes all that variation? To what extent is it heritable, and how? What role does environment play? How common is hybridization, and what are its effects? Why do parental traits sometimes blend in offspring and sometimes seem unequally expressed? Some of these issues weren't resolved until the twentieth century, with the revelations of Mendel and the later insights into DNA structure and gene expression. Darwin could only grasp at straws.

Just as Darwin was entering his own new domestic life—he and Emma were married in the chapel at her home, Maer Hall, in January 1839, and soon happily residing on Upper Gower Street, London—he pondered domestic breeds ever more deeply. He pestered friends, family, and colleagues with questions at every opportunity, and gathered all manner of information bearing on domestic varieties and breeding, even starting a separate notebook on questions and experiments. A slender notebook of about 40 pages, it's divided into sections that tell us something about his working method between about 1839 and 1844: "Experiments in crossing &c. Plants," "Questions Regarding Plants," Questions Regarding Breeding of Animals," "Experiments in crossing animals," are some of the general headings for lists of questions directed to botanists and horticulturists like his Cambridge mentor

Henslow and the gardeners at Maer and Shrewsbury, medical men like his father and Henry Holland (Darwin's doctor in London), zoologists and museum men like John Gould and Edward Gray, breeders like William Yarrell, and others.[23]

As he wrote in his autobiography, with natural selection he had "at last got a theory by which to work." But although he was bursting with questions and did manage experiments and other investigations (see especially Chapter 6 for Darwin's work on crossing and pollination at this juncture), he was able to direct frustratingly little of his efforts to his species theory in the next several years. He was overwhelmed as a new husband, parent, and ambitious young scientist determined to make a name for himself. After William Erasmus ("Doddy," later Willy) was born in 1840, Emma was in a nearly continuous state of pregnancy for the next decade or more. On the scientific front he was in demand as an active participant in the learned societies, and there was the seemingly unending task of writing up his *Beagle* material. His output was prodigious, with three volumes on geology (coral reefs in 1842, volcanoes in 1844, and the geology of South America in 1846), plus his travel memoir (first out in 1839), introductions to the *Beagle* voyage's zoological volumes (fish, reptiles, birds, and more), and last but not least, various geological and zoological papers.

With a mysterious illness beginning to manifest itself on top of (maybe because of?) the work load, soon after their second child, Anne Elizabeth, was born in 1841 he resigned from the secretaryship at the Geological Society and stepped down from the council of the Royal Geographical Society. This illness was to worsen and plague him for the rest of his life, with periods of violent gastrointestinal distress and severe headaches that came and went. The precise nature of the malady is still debated today—was it the lingering effects of a parasite picked up on the *Beagle* voyage? A psychosomatic reaction to the stress and strain of harboring ideas he knew would be fiercely repudiated by society and threaten to estrange his devout wife? Overwork, or an allergy or metabolic condition like lactose or gluten intolerance? Perhaps some inherited disorder? We may never know.[24] What is certain is that he responded to what would be considered quack treatments today: the "water cure" (dousing in chill water and lying motionless for hours wrapped in wet towels) together with strict regulation of his exertions, mental and physical. The enforced idleness was maddening to Darwin, but it seemed to have a positive effect, at least for a time.

Prior to the worst manifestations of his illness the family got away to Maer for a month's respite each summer, and it was there, in 1842, that he finally managed to gather his thoughts and write out a brief sketch of his theory to date. He opened this brief account with the analogy from domestic breeds, entitling the first section "On variation under domestication, and on the principle of selection." It was a format he stuck with: from the expanded 230-odd page *Essay* version of this sketch written out 2 years later to his "big species book" manuscript *Natural Selection* to *On the Origin of Species* itself, he consistently set up his case for evolution by natural selection with domestication as a strong analogy with variation, selection, and slow, steady change in nature.[25] It was a philosophical and rhetorical approach that was likely inspired by the renowned John Herschel, whom we met in Chapter 1. In his *Preliminary Discourse on the Study of Natural Philosophy* (1830), the much-admired astronomer, mathematician, and philosopher pointed to the power of analogy as one way to gain insight into *vera causae*—true causes—in nature. This was a book that Darwin studied closely, one of just two cited in his autobiography many years later: the other was Humboldt, and together they fired up his zeal for science. He recalled, "No one or a dozen other books influenced me nearly so much as these two."[26]

By the time Darwin penned his lengthy *Essay* in 1844 he had a largely complete theory—there were some key details missing yet, as we will see in the next chapter, but in structure and content the *Essay* very much has the look of a condensed version of the *Origin* that appeared 15 years later. Darwin thought it was sufficiently complete at that stage that he hired the local schoolteacher to write out a fair copy (his own handwriting anything but fair), sealing it in an envelope with a poignant letter asking Emma to publish the essay without delay in the event of his untimely death. But if it was *that* complete, and he had that much faith in its essential correctness, why not just go ahead and publish it? That is a question that scholars have debated and discussed for over a century. Some have claimed that Darwin delayed publishing out of fear of the repercussions, or out of respect for the deep religious convictions of his wife. There may be something to the repercussions idea—it may not be coincidental that the scandalous *Vestiges of the Natural History of Creation* came out that same year, thrusting the idea of transmutation into the limelight in 1844 and provoking venomous condemnation by the leading lights among naturalists—many of them Darwin's scientific friends, mentors, and professors. But it's also true that he simply wasn't ready to publish: he had more work to do, probably more than he realized then—not least a return to barnacles, which Hooker had something to do with.

Darwin had struck up a correspondence with botanist Joseph Dalton Hooker (1817–1911) soon after Hooker's return in the fall of 1843 from a 4-year southern voyage under James Clark Ross, on which he served as both assistant surgeon and naturalist. (That was the acclaimed voyage that confirmed the existence of a vast new continent—Antarctica.) Knowing that Hooker was going to work on a flora of the southern lands, Darwin asked Henslow to forward his Galápagos plant collection to him in case they proved useful. Hooker thanked him warmly, and they started corresponding on points of common interest as scientists and travelers. Their friendship grew to the point where, in January of 1844, the year he penned the *Essay,* Darwin ventured to take Hooker into his confidence. "I have been now ever since my return engaged in a very presumptuous work & which I know no one individual who would not say a very foolish one," Darwin ventured. He explained how he had been struck by the Galápagos species he observed, and the fossil mammals of South America, and how he resolved to collect any and all facts that might bear on the question of the nature of species. "I have read heaps of agricultural & horticultural books, & have never ceased collecting facts," he continued. "At last gleams of light have come, & I am almost convinced (quite contrary to opinion I started with) that species are not (it is like confessing a murder) immutable." Darwin hastily added a reassuring line distancing himself from the unpalatable ideas of Lamarck: "Heaven forfend me from Lamarck nonsense of a 'tendency to progression' 'adaptations from the slow willing of animals' &c."—yet, he had to acknowledge, the conclusion he had come to was not so very different from the Frenchman's, though the mechanism of change is very different. "I think I have found out (here's presumption!) the simple way by which species become exquisitely adapted to various ends." How would Hooker take this? Darwin was a bit worried: "You will now groan," he wrote, "& think to yourself 'on what a man have I been wasting my time in writing to.'"[27] Far from contemptuous or dismissive, Hooker was intrigued and invited his new friend to tell him more.

The Barnacle Returneth

While Hooker was not dismissive of Darwin's species theory, he wasn't an instant convert either. His skepticism proved invaluable to Darwin: the botanist was a font of information and a fierce (but friendly) critic, constantly challenging Darwin to sharpen his arguments and defend his evidence. He helped Darwin recognize where the weaknesses of his theory lay. They exchanged long letters in which Darwin peppered Hooker with questions seeking data, observations, and Hooker's judgment on matters relating to geographical distribution, variation, classification, structure, and species relationships. Darwin pored over botanical manuals and turned to Hooker for assessment of this or that observation, ever on the lookout for facts that would bear on his theory. It was in this vein that Darwin wrote Hooker expressing interest in an article on species and their variation by Frédéric Gérard, in the newly published *Dictionnaire universel d'histoire naturelle*. Darwin picked up on intriguing language used by Gérard, such as how species can *develop* certain traits (sounding like thinly veiled transmutationism), and how the Frenchman questioned whether "species" per se even existed on the basis of their often unclear boundaries. Darwin's innocent query provoked a contemptuous reply from Hooker: Gérard was "evidently no Botanist," he sniffed, but one of those "narrow-minded studiers of overwrought local floras." He had little patience for armchair naturalists who make a big deal out of variable characters and questionable delineations of species that were, after all, issues discussed in cautious terms by careful and methodical botanists, yet "taken up by such men as Gerard, who have no idea what thousands of good species there are in the world." Hooker declared that he was "not inclined to take much for granted from any one [who] treats the subject in his way & who does not know what it is to be a specific Naturalist himself."[28]

Reading between the lines, Darwin realized that Hooker was indirectly referring to him and his grandiose theorizing, going on about species and varieties but really having little idea what these are like in nature. "How painfully . . . true is your remark that no one has hardly

a right to examine the question of species who has not minutely described many," he acknowledged. But, he continued, defensively: "I was, however, pleased to hear from Owen (who is vehemently opposed to any mutability in species) that he thought it was a very fair subject & that there was a mass of facts to be brought to bear on the question, not hitherto collected." Darwin would take comfort in the fact that he had "dabbled in several branches of natural history and geology," he said, and although he would get "more kicks than half-pennies," he would pursue his species theory. At least he would improve upon Lamarck,

Some of the acorn barnacles studied by Darwin. Courtesy of Richard Milner.

he ventured: "Lamarck is the only exception, that I can think of, of an accurate describer of species at least in the Invertebrate kingdom, who has disbelieved in permanent species, but he in his absurd though clever work has done the subject harm, as has Mr Vestiges, and, as (some future loose naturalist attempting the same speculations will perhaps say) has Mr D."[29]

Hooker was quick to say that he wasn't insinuating that *Darwin* was one of those armchair naturalists. Rather, he wrote, smoothing things over, "what I meant I still maintain, that to be able to handle the subject at all, one must have handled hundreds of species with a view to distinguishing them & that over a great part—or brought from a great many parts—of the globe." But Darwin felt the truth of Hooker's criticism. "All which you so kindly say about my species work," he replied, "does not alter one iota my long self-acknowledged presumption in accumulating facts & speculating on the subject of variation, without having worked out my due share of species." This exchange of letters took place in September 1845.[30] Within a year Darwin was deep into a systematic study of barnacles, an undertaking that he first thought might take up to a year but in fact ballooned to 8 years. By the end of it he had "minutely described" many species, earning "the right to examine the question of species."

Darwin's barnacle work snowballed in part because other naturalists cheered him on: Hooker, certainly, but also Richard Owen at the Royal College of Surgeons, and the famed zoologist Louis Agassiz at Harvard, who called the world of barnacles a "great desideratum" desperately in need of study. The real coup came when the zoological keeper at the British Museum, Edward Gray, agreed to lend Darwin the Museum's entire cirripede collection, much of it uncataloged. Imagine that: the whole national collection at his fingertips! But the other reason the project exploded was the unexpected discoveries he made—discoveries with profound implications for his species theory. As Janet Browne aptly put in her monumental Darwin biography, the barnacle studies would prove to be another of Darwin's "explorations into the reproductive unknown."[31]

Darwin's barnacle odyssey began with the curious specimen he found on the windswept beach in the Chonos Archipelago. He showed it to Hooker, who had come to Down House in the fall of 1846 to meet Darwin's old *Beagle* shipmate Bartholomew Sulivan. Just back from South America, Sulivan had completed a detailed survey of the Falkland Islands, among other accomplishments, as commander of his own ship. Back in 1835 when Darwin first found his burrowing barnacle, he commented in his notes that "it is manifest this curious little animal forms new genus." Now he asked Hooker to help him name it, and they came up with *Arthrobalanus*—or "Mr. Arthrobalanus" in their frequent discussions of the odd crustacean. In fact this creature seemed doubly odd: besides its burrowing habit, Darwin thought he could discern *two* penises. It took a while, but he eventually realized that the Mr. was in fact a Ms. The "penises" were the cirri, filament-like feeding appendages, of female specimens. He wasn't to discover the male of this species for several years yet—it turned out to be the smallest barnacle male known. In the meantime his skills at dissection got steadily better, helped by Hooker's recommendations for improving the optics of his microscope. In one letter he thanked Hooker thrice over: for helping with the microscope, for suggesting he use porcupine quills instead of glass tubes in dissection, and for the gift of "capital" chutney. Between the three, "I have many daily memorials of you," he joked, and proudly reported that he devised a set of wood blocks to support his wrists while dissecting, "a splendid invention." He was excited with his new cirripede hobbyhorse. "As you say," he wrote to Hooker, "there is an extraordinary pleasure in pure observation . . . After having been so many years employed in writing my old geological observations it is delightful to use one's eyes & fingers again."[32]

This was November of 1846, and the Darwin household was growing in more ways than one: he and Emma enlarged the house over the past year to include a schoolroom and two bedrooms, as their family had grown to four children (1-year-old Georgy, 3-year-old Etty, 5-year-old Annie, and 7-year-old Willy)—and counting, with baby Bessy joining the family in July of 1847. By this time Darwin was deep into his barnacle work, and although he took some time to attend scientific meetings in London and Oxford now and then what he loved best was the "absolute rurality" of Down. His friends loved it too, and he regularly invited them for scientific respites from London. It was an idyll compared to the noise, smoke, and grit of the city, though at times it must have seemed almost as crowded: besides growing family and domestic staff there was a near-constant procession of visiting scientific friends (often with their wives and children) and relatives such as Charles's brother 'Ras (despite professing to hate country life) and other siblings and their families. Their Aunt Sarah Wedgwood also made frequent appearances, having moved to the village of Downe to be close to Charles and Emma. Darwin loved the conviviality of family and friends, but it was also understood that he had to often retreat to his study, where he would "do his barnacles" in peace.

The following year, 1848, brought joy, sadness, and excitement: Franky, the latest addition to the family, came along in August, but just a few months later Darwin's father, Dr. Robert Darwin, died in Shrewsbury. The excitement came from his most startling barnacle discoveries yet: while most barnacle species were hermaphrodites bearing both male and female sexual organs, he came across one (*Ibla cumingii*) that was not only single-sex, but had males reduced to tiny nubs adhering to the carapace of the comparatively gigantic females. The males were mere rudiments, eyeless, limbless, near-microscopic bags of sperm. This was astonishing: having separate sexes was strange enough, but there was practically nothing to compare with the radical difference in size and development of the two sexes in this species—rather as if human males were reduced to tiny testes adhering to women like ticks. He wrote excitedly to Henslow in April 1848 about the "microscopically minute" males:

> But here comes the odd fact . . . [the males] become parasitic within the sack of the female, & thus fixed & half embedded in the flesh of their wives they pass their whole lives & can never move again.

"Is it not strange," he mused, "that nature should have made this one genus unisexual [sexes separate], & yet have fixed the males on the outside of the females—the male organs in fact being thus external instead of internal."[33] It was strange, and things got even stranger when he realized that another barnacle, *Scalpellum*, was both hermaphroditic *and* had tiny "parasitic" males. Now why have these extra, or *complemental* males, as Darwin called them, when male equipment was already built-in, as it were? He wrote excitedly to Hooker: "I had observed some minute parasites adhering to [*Scalpellum*], & these parasites, I now can show, are supplemental males . . . so we have almost a polygamous animal, simple females alone being wanting."[34] He went back and scrutinized earlier specimens, rightly thinking he may have mistaken diminutive males for the odd parasite, and indeed found several new cases of complemental males. Recall how he had been mistaken about the double penis of what turned out to be Arthrobalanus females. Now he actually found Arthrobalanus males, which turn out to be the minutest of minute complementals but equipped with a penis nearly nine times the length of the male's body.

It dawned on him that the sexual strategies of these barnacles were of tremendous significance: no less than evidence of an evolutionary differentiation from hermaphrodites into males and females. Animals and plants were initially hermaphroditic, he began to think, then slowly differentiated over eons into the two-sex arrangement so common today. *Why* was another matter, but this explained anomalous features of anatomy like the nonfunctional nipples of male mammals or rudimentary stamens of female flowers, and it dovetailed with a growing insight from his studies of reproduction and pollination (see Chapters 6 and 7); namely, that nature abhors perpetual self-fertilization. How and why the sexes formed was yet another exciting line of investigation. Something like a grand unified theory of life's evolution driven by natural selection was taking shape in his mind. "I never [should] have made this out," he exclaimed to Hooker, "had not my species theory convinced me, that an hermaphrodite species must pass into a bisexual species by insensibly small stages, & here we have it, for the male organs in the hermaphrodite are beginning to fail, & independent males ready formed." This was the kind of evidence he needed. The odd sex lives of his barnacles may not be a smoking gun for transmutation, but they were not far from it: how else to explain such strange arrangements? It certainly wasn't consistent with the conventional view of perfect design in nature. He half-joked in the same letter that Hooker would "perhaps wish my Barnacles & Species theory al Diabolo [to the devil] together. But I don't care what you say, my species theory is all gospel."[35]

And so it went as he was drawn in deeper and deeper, studying barnacles from every corner of the world. He traced homologies and reconstructed what he thought was the "archetypal" barnacle, showing how its 17 body parts derived from the 21-body-part archetype that the French zoologist Henri Milne-Edwards had identified for Crustacea as a whole. He was astounded by the variation he found, at multiple levels: abundant individual variation, but also unexpected structural and life history variations on a theme, from those puzzling complemental males to groups with elaborate cirri to yet others lacking appendages altogether. Those he classified in their own suborder: "Apoda," the footless barnacles. An "apodal cirripede" may be a contradiction in terms, but as he marveled to Lyell, when it came to barnacles "truly the schemes & wonders of nature are illimitable."[36]

By the time he was through, Darwin had produced the most comprehensive treatment of living and fossil cirripedes yet, culminating in a four-volume set of monographs, two published in 1851 and two in 1854.[37] It was a huge accomplishment, earning him the Royal Society's highest honor, the Royal Medal, in 1853, before the study was even fully published. Some of his interpretations of barnacle anatomy and development did not stand the test of time, but they give us insight into Darwin's early evolutionary thinking and working method. Janet Browne pointed out that the barnacle period marked a sea change in Darwin's awareness of the ubiquity of individual variation. Prior to poring over barnacles he seemed to have had a conception of variation mainly induced by environmental change and other factors acting on the reproductive system. As a result of scrutinizing not just individuals, but *populations* of individuals of many species (albeit populations occupying museum drawers and jars), Darwin gained a new appreciation for the sheer abundance of variability in virtually all traits, great and small. "You ask what effect studying species has had on my variation theories," he wrote Hooker from Malvern, where he was trying to combat his latest bout of illness with the water cure. "I have been struck . . . with the variability of every part in some slight degree of every species: when the same organ is rigorously compared in many individuals I always find some slight variability."[38] He joked that working on systematics would be easy were it not for "this confounded variation," but acknowledged it was a boon for his species theory. It was, after all, the very raw material that selection acted upon, and for selection to shape species as completely as he envisioned, variation must abound in even the most trivial of traits.

In this period, too, Darwin refined a research method that stood him well through all of his subsequent investigations: first, systematic and thorough study, sometimes more experimental and sometimes more observational, but always marked by noticing unnoticed phenomena, asking unasked questions, and connecting dots where others could not see the patterns. Crucially, his method included immersion in the available literature and cultivation of a worldwide network of correspondents: academics and amateurs, aristocrats and humble museum men, ship's captains and army officers— anyone and everyone expert in an area of interest to Darwin or otherwise able to help him was approached to bounce ideas off of and to ask for specimens, observations, and information.

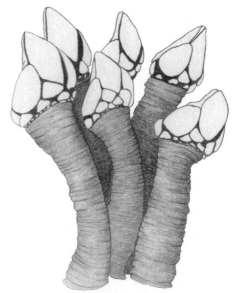

Gooseneck or pedunculated barnacles. Drawing by Leslie C. Costa.

By the time Darwin finished his 8-year barnacle study, his children had gotten used to his scientific work at home. He had seven kids ranging from a teenager of 15 years to a toddler of 3 (there had been eight, but the couple suffered the devastating loss of their eldest daughter, Annie, in 1851). His barnacle work was going on for most to all of the children's lives, depending on the child. Sir John Lubbock, a family friend and neighbor, told an anecdote about this period: when visiting a friend who lived nearby, one of the Darwin boys looked around and, seeing no microscope or dissecting equipment, asked of his friend's dad, "Then where does he do his barnacles?" For all the kids knew, *everyone's* dad worked on barnacles.[39]

Message by Pigeon

Throughout his cirripede odyssey Darwin kept up a number of parallel investigations bearing on his species theory from different angles, "to see how far they favour or are opposed to the notion that wild species are mutable or immutable," he explained to his cousin Fox in March 1855. Central to his researches were, he said, "a number of people helping me in every way & giving me most valuable assistance."[40] Barnacles packed away, he now hoped that Fox would be among the first to assist with his newly rekindled interest in pigeons.

Darwin had long recognized the significance of domestic pigeons for his theory. Back in the late 1830s he even commented in one of his notebooks that pigeons are proof that variation in animals is not limited or constrained: "Analogy will certainly allow variation as much as the difference between species—for instance pidgeons [*sic*]," he wrote.[41] Barnacles may have taken center stage as a diverse and little studied *natural* group of species that he could use to document variation and trace adaptation and descent, but those studies set the stage for the study of a complementary group, one that would serve as an *artificial* case study with its own expression of variation and demonstrated the power of selection. But what to study? Dogs or barnyard animals would not do: too big, and their development too long. His friend William

Yarrell urged him to give pigeons a go. Of course! A group with diverse breeds, yet smallish animals easily kept inexpensively and in quantity, pigeons also grew fairly quickly and had the added bonus of being edible. And when I say diverse breeds, I mean *diverse*: breeds so divergent from one another that they would be easily placed in separate genera, let alone species, were they to be collected and classified by some alien naturalist. There were barbs and fantails, pouters and runts, trumpeters and carriers: pigeons with absurdly, sometimes grotesquely, modified heads, beaks, tail feathers, crops, feet—even behavioral variants like tumblers, with their neurological loose wire that makes them suddenly somersault in midair. Why were these strange breeds developed? Some, like carriers, are "working" pigeons, but pigeon breeding is called "the fancy" for an obvious reason: breeders take a fancy to this or that odd variant, and work diligently to develop it. Back in the mists of time this was probably not deliberate, but in more recent centuries it surely was. He knew pigeons had something to teach us: the often-absurd varieties and breeds carried a message about the power of selection.

It was conventional wisdom among zoologists that all pigeon breeds descended from just one wild species: *Columba livia*, the ubiquitous rock dove of city parks and country farmyards. Or did they? He was surprised to find that, unlike the zoologists, most fanciers believed that each breed was derived from a distinct wild ancestor. He recognized that this view, if widely held among the general public, would be one of many obstacles he would face in making a convincing case for common descent and the power of selection. But even the zoologists had little concept of *how* breeds were made. They often assumed new varieties appeared more or less fully developed, anomalous sports like the odd yellow rose popping up on a red rose bush. Their view is expressed by the prolific naturalist Daniel Jay Browne in his 1850 book *The American Bird Fancier*: "It is from the wild rock pigeon, (*C. livia*) that all those numerous varieties . . . of the common inhabitants of the dove cot have descended," and that "the greater part of them owe their existence to the interference and art of man; for, by separating them from the wild rock pigeon, such accidental varieties as have occasionally occurred . . . and by assorting and pairing them together, as fancy or caprice suggested, he has, at intervals, generated all the various races."[42] Darwin would have read "variations" for "varieties," variations that are more minute and abundant than Browne might have imagined. Subjected to Sebright's "judicious selection," variation upon variation leads to cumulative change over time.

Darwin was sure that the many breeds did not simply represent the domesticated descendants of a like number of ancestral wild species. Certainly the fossil record did not support a profusion of wild canines, bovines, and equines, for example, but maybe at most a few wild progenitors. (Indeed, recent DNA analysis shows that modern cattle, sheep, pigs, and goats likely descend from only two or a few ancestral species or subspecies each.) Darwin needed to prove this for pigeons. Luckily, by this time the use of embryology and development to inform classification was well appreciated, largely thanks to Henri Milne-Edwards's important *Essay on Classification* of 1844. Darwin studied Milne-Edwards while working on his barnacles. Now he realized that by comparing the anatomy of certain domestic varieties from birth through maturation he should be able to show that breeds were identical in early stages and then diverged from one another at different points in development.

Darwin threw himself into the world of fancy pigeons with charac-

Four of the 16 pigeon breeds kept by Darwin (L to R): the pouter, with enormously inflated crop and foot feathers resembling spats; the English carrier, a sleek and strong-flying homing breed used to carry messages; the barb, with flattened, broad beak and large featherless eye-ring; and the fantail, with permanently fanned tail feathers reminiscent of displaying turkeys. From Darwin (1868), vol. I, figs. 18–21.

teristic enthusiasm, but he didn't go so far as to try to selectively breed them himself. Rather like connoisseurs or knowledgeable critics of fine art who are themselves no artists, Darwin studied and observed and immersed himself in the culture, getting to know fanciers at their meetings and visiting their aviaries and shows. And like art aficionados he became a collector, over the next two years procuring every pigeon breed available in Britain and more from further afield—some 16 breeds in all, over 90 birds cooing away in a veritable pigeon condo erected in the garden at Down House. He joined leading London pigeon clubs— the Philoperisteron and Southwark Columbarian—and gamely reported in a letter to Willy, then age 14 and away at boarding school, his adventures with the "strange set of odd men" that frequented them:

> I want to attend a meeting of the Columbarian Society . . . I think I shall belong to this Society where, I fancy, I shall meet a strange set of odd men.—Mr Brent was a very queer little fish; but I suppose Mamma told you about him; after dinner he handed me a clay pipe, saying "here is your pipe" as if it was a matter of course that I should smoke. Another odd little man (N.B all Pigeons Fanciers are little men, I begin to think) & he showed me a wretched little Polish Hen, which he said he would not sell for £50 & hoped to make £200 by her, as she had a black top-knot.[43]

Bernard Brent was a leading pigeon fancier and, odd as he and the other "little men" of the Columbarian were, Darwin respected their expertise and was eager to be coached in the fine points of pigeon breeding.

The following February, 1856, found him as game as ever, writing to William on just having acquired some prize breeds: trumpeters, nuns, and turbits. "I am building a new house for my tumblers," he was excited to report, "so as to fly them in summer."[44] Emma and the four youngest children were away visiting her sisters. It may have been then that Etty's pet cat was quietly dispatched, having developed the habit of eating her father's pigeons—something Etty never quite forgave. At one point Darwin's pigeon mania led to another point of contention: the stench of curing and preparing the skeletons led to his pigeon lab being banished as far from the house as possible. Darwin, with his butler, Parslow, presided over the foul witch's cauldrons of pigeon skeletons until even they couldn't take it any longer, and he resorted to outsourcing the skeleton preparations. Owing (once again) to his network of friends and contacts, in short order he assembled an impressive set of skeletons of nearly every breed available in England and beyond.

Indeed, specimens alive and dead arrived daily at Down House, each of which he examined minutely. William Tegetmeier, sometime London writer, editor, beekeeper (as we shall see in Chapter 4), and pigeon fancier was a big help: "Many thanks for your offers about dead Pigeons," Darwin wrote him. "If Scanderoon dies please remember I should wish carcase [sic] sent per coach by enclosed address as soon as possible to arrive fresh."[45] His old college friend Thomas Eyton, who developed a sizable collection of bird skins and skeletons at his family estate, also obliged. Darwin explained to Eyton that he had taken up his pigeon project to compare the structure of pigeons at different stages of development; "I mean to try to get Domestic Pigeons from all parts of the world." It wasn't all one-way, of course: there was a gentlemanly exchange of specimens and information, as when Darwin proffered the preserved head of a Chinese breed of dog for Eyton's study of dog skeletons: "I am delighted to hear that you are at dogs; it will be splendid for my work individually, & I am sure most desirable for Science. I have somewhere, I am almost certain, the head of a Chinese Dog: would you like to have this?" (He wrote a week later that he seemed to have misplaced it: "I have been looking everywhere for the Dog's Head. . . . I am vexed at this.")[46]

A Delightful Commencement

Barely a month before Darwin first expressed his pigeon fancy to Fox back in March 1855, a naturalist little known to Darwin, one Alfred Russel Wallace, was waiting out the rainy season in a small bungalow in Sarawak, in northern Borneo. Wallace passed the time gathering his thoughts on the striking parallel between species relationships geographically, in terms of their distribution on earth, and geologically, in terms of their distribution over time in the fossil record. He dashed out an essay and at the first opportunity mailed it off to London. Published the following September, this essay concluded that *"every species has come into existence coincident both in space and time with a pre-existing closely allied species."* The principle became known as the Sarawak Law. Its evolutionary implications were clear to nearly all except Darwin: impressed, Lyell urged his friend to make haste and publish his species theory. Darwin was less impressed, maybe because he thought of Wallace as a mere collector. Expanding his interest to poultry, and adding chicken, duck, and turkey breeds to his menagerie, Darwin had evidently written Wallace to ask for specimens from southeast Asia. In late summer 1856 Wallace enclosed instructions with his latest consignment bound for London: "The domestic duck var[iety] is for Mr. Darwin & he would perhaps also like the jungle cock, which is often domesticated here & is doubtless one of the originals of the domestic breed of poultry."[47] Wallace also wrote Darwin directly, and in his reply Darwin alluded to the Sarawak Law paper, praising it in cautious terms and remarking "I can plainly see that we have thought much alike," but he was coy about his own interests in species and varieties. He reiterated his desire to obtain "any curious breed" of poultry, and mentioned that Sir James Brooke, the "White Rajah" of Borneo, had kindly sent him pigeons, fowl, and cat skins.[48]

But Wallace was more than a collector. He was a confirmed transmutationist who collected in order to fund his travels investigating the "species question." Having spent four years in Amazonia, he had just recently embarked on what was to become an 8-year odyssey crisscrossing the vast Malay Archipelago stretching from Singapore and Malaysia in the west to Papua New Guinea in the east. Little did Darwin know that in a short couple of years Wallace would succeed in his quest to find the mechanism of species change, natural selection. In the meantime, Darwin began to work away at his big species book, to be entitled

Natural Selection. He progressed steadily, and by coincidence he just started writing up a section on pigeons in 1858 when a bombshell of a package from Wallace arrived, with a manuscript that laid out a formulation of natural selection. The brevity of Darwin's journal entry speaks volumes: "June 14th Pigeons: (interrupted)."[49] He was devastated.

The ensuing weeks were an emotional maelstrom as Darwin appealed to his friends Lyell and Hooker for help preserving his priority and honor even as his youngest child, baby Charles Waring Darwin, fell dangerously ill; he felt self-loathing for even caring about theories and priority at such a time. His friends hastily arranged for excerpts of Darwin's unpublished outlines of the theory from the 1844 *Essay* and other sources to be read at the Linnean Society, along with Wallace's paper. The papers were read on July 1st, but neither Wallace nor Darwin were present—Wallace still halfway around the world, and Darwin attending the funeral of his son. Within a few weeks, Darwin had regained enough composure to write Wallace explaining what had transpired. He then buckled down to finish his book and cement his priority. He was relieved to hear back that Wallace was delighted with all Darwin had accomplished thus far. Darwin knew he had to get his book out, but also knew that *Natural Selection* would take too long to complete. By the fall of 1858 Darwin resolved to pare it down. His "abstract," as he called it, would become *On the Origin of Species.* Lyell recommended his own publisher, John Murray of London, and by April 1859 Murray had the manuscript in hand.

He asked a few colleagues to review it. One, the Scottish literary editor Rev. Whitwell Elwin, was singularly unimpressed with all but the section on pigeons. He recommended that Darwin get rid of the rest and produce a work on pigeons instead, smoothing the path for a later book expanding on his more unconventional views. "Even if the larger work were ready it would be the best mode of preparing the way for it. Every body is interested in pigeons," Elwin said to Murray.[50] Darwin was appalled, but was reassured that Murray didn't take the recommendation seriously. Yet Elwin did have a point, in that Darwin *had* deployed his arguments on domestication in the very first chapter as a device to smooth the reception for the ideas of common descent and natural selection in subsequent chapters. Many of Darwin's key arguments about natural selection were more speculative than well supported, and he was painfully aware that the *Origin*, which duly appeared in November of 1859, really *was* an abstract of his aborted big book, lacking the range of examples, data, and citations of authorities intended to make his original book unassailable. The remedy was to come out with more detailed treatments expanding on the main arguments of the *Origin*. In fact, his friend Thomas Henry Huxley encouraged Darwin to do precisely that in the weeks after the *Origin* appeared. Darwin was ahead of the curve: "You have hit on exact plan," he assured Huxley, "which on advice of Lyell, Murray &c I mean to follow, viz bring out separate volumes in detail & I shall begin with domestic productions."[51]

Domestication was the first line of argument in the *Origin*: chapter 1 is all about modification of breeds by artificial selection as an analogy for how natural selection works in nature. He would thus begin with a supporting volume on domestication and follow it with a second volume dedicated to the case for natural selection (corresponding to *Origin* chapters 2–4), and then a third reflecting the rest of the book, tackling difficulties and detailing the empirical patterns evident in nature, from fossils to behavior to geographical distribution to comparative anatomy and more, a galaxy of dots neatly connected and facts explained by descent with modification by natural selection.

Darwin only produced the first of this projected series, and even that took nearly a decade: *The Variation of Animals and Plants Under Domestication* was published under Murray's imprint in 1868. What had been treated in a single chapter in the *Origin* was now two volumes, and Darwin's beloved pigeons, covered in a dozen pages in the *Origin*, now had two chapters of their own. Darwin's strategy in these chapters is worth noting: in the first he detailed the characteristics of breeds, highlighting their diversity and rich variation in characters such as beak and skeleton relative to the rock pigeon as "parent-form." Having made a convincing case for variation on a par with what naturalists would ordinarily associate with different genera, in the second chapter he then argued for a single origin of these breeds in all of their breathtaking diversity, culminating in a long final section entitled "Manner of formation of the chief races"—selection, both unconscious and methodical. It is here, in Darwin's treatment of pigeons, that we find the only evolutionary tree ever produced by Darwin for any group of organisms. If domestic breeds in general represented a case study or microcosm of common descent and the power of selection in nature, for Darwin pigeons were a case study of a case study.

By 1868 Darwin was able to do for pigeons what he could not or would not do for barnacles in the 1850s: namely, trace an explicit evolutionary heritage. Why? It may have been the times; in those pre-*Origin* years of barnacle work he was perhaps simply unwilling to tip his hand and publish something too obviously (and controversially) transmutational before he was ready. But, perhaps, in an important respect, barnacles may have set the stage for this pigeon family tree—and indeed helped set the course for Darwin's subsequent life's work. It is ironic that his barnacle work is sometimes portrayed as a distraction that sidetracked Darwin from publishing his species book. On the contrary it was, first, the study that clued Darwin in to the sheer abundance of variation in all points of anatomy—that all-important ingredient for selection to act upon—and the variation he subsequently sought out in pigeons. Second, barnacles were his first overtly *evolutionary* investigation, one that saw the denouement of a new working method that entailed meticulous study, understanding the oddities of barnacle biology in terms of an evolutionary history, and, very importantly, developing a worldwide network of expert contacts that he could appeal to for assistance of all kinds. Darwin's investigations of barnacles and pigeons may not have been experimental in nature, but illustrate a related working method no less important for us to understand the experimentiser.

Experimentising: Doing Your Barnacles

Of the many lessons that Darwin learned in his studies of barnacles and pigeons, perhaps the most important was gaining an appreciation for variability and how related groups of species represent variations on a theme: relationships can be traced by observing the same parts modified in different ways. To explore this idea in a hands-on way, "do your barnacles" like Darwin with these dissections.

A. Materials

- Notebook and pencil
- Barnacles (acorn and/or gooseneck)
- Limpets (optional, for comparison)
- Dissecting microscope or good hand lens
- Forceps or tweezers
- Pipet or eye dropper
- 2 in. (5 cm) C-clamp
- Shallow pan or tray for dissection (a shallow sardine tin with wax bottom works well; just melt some paraffin wax and pour enough to coat the base of a cleaned sardine can with about $^{1}/_{10}$ in. [¼ cm] of wax)
- Glass microscope slides
- Scissors (small and sharp)
- Paper towels

Note on obtaining specimens: If you don't live near the ocean, barnacles and limpets can be obtained through biological supply houses. Ward's Natural Science (www.wardsci.com), www .biologyproducts.com, and www.onlinesciencemall.com sell preserved gooseneck barnacles (generally genera *Pollicipes* or *Lepas*), and Carolina Biological Supply (www.carolina.com) offers the Carolina™ Barnacle Cluster—live acorn barnacles. Limpets, single-shelled grazing snails, can be bought online from saltwater aquarium suppliers (e.g., www.reefcleaners.org and www.liveaquaria.com).

B. Procedure

1. Select an **acorn barnacle**. Note that the barnacle's exoskeleton is shaped like a volcano, with a set of six overlapping and rigid calcareous *carinal* or *wall plates* surrounding the barnacle within. Can you identify all six plates?

2. Which way is up? Acorn barnacles are affixed to the substrate on their *dorsal* side, so the "crater" of the barnacle volcano is its *ventral* side.

3. Inside the "crater" are four movable plates called the *opercular plates* (the door is the *operculum*). The two larger kite-shaped plates of the operculum, one on each side, are called *scutal* plates. Adjacent to these are two *tergal* plates, held more or less vertically and so seen edge-on at the top. The opercular plates are opened and closed with muscles.

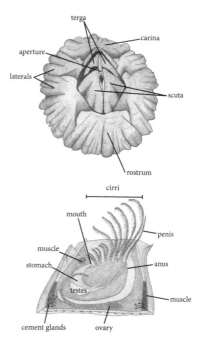

4. The slit-like opening formed when the operculum is open is the *aperture*. The upside-down barnacle extends its long modified legs (*cirri*), which are feeding organs, through the aperture when feeding.

5. Gently push the operculum open with forceps, and observe the interior with a hand lens or dissecting

External and internal anatomy of a generalized acorn barnacle. Drawing by Leslie C. Costa.

microscope. The space within the aperture is called the *mantle cavity*, a term associated with mollusk anatomy. (This is a hold-over from a time when barnacles were classified as mollusks. Note the superficial similarity of the Acorn barnacle to the **limpet**, if one is available, which is a mollusk.)

6. Use the C-clamp to loosen the articulations between the adjacent wall plates: carefully place the clamp jaws on opposite sides of the shell and apply pressure slowly until the wall plates give way and separate. Remove the clamp and repeat at intervals around the circumference until all the plates are loose.

7. Carefully remove one or two of the wall plates, and the opercular plates. Sketch the scutal and tergal plates.

8. Within the mantle cavity, note the appendages arising along the ventral (upper!) surface of the body; these are the cirri. How many pairs are present? (You should find six pairs.)

9. Carefully remove one of the cirri (singular: cirrus), and note that it has two arms (*biramous*). Place a cirrus on a glass slide and apply a drop of water to observe the fringe of hairs, or *setae*. Three of the paired cirri capture food particles in the water column, while the other three function to scrape the particles into the mouth. There are many excellent video clips online showing how the cirri function in feeding, such as this one at the Snail's Odyssey website: www.asnailsodyssey.com/VIDEOS/BARNACLE /barnacleFeed.html.

10. Select a stalked **gooseneck barnacle**. The first difference to note with respect to acorn barnacles is the *peduncle*, or stalk. This functions mainly for attachment but does house muscles and ovaries. (The peduncle of some species is covered with small round calcareous plates called *ossicles*.) Atop the peduncle is the part you think of as the barnacle proper: the *capitulum*.

11. Note that the exoskeleton is flexible; it consists of a thin membrane of chitin (the material that forms the exoskeleton of most arthropods) and protein.

12. Orient yourself to gooseneck barnacle structure starting from

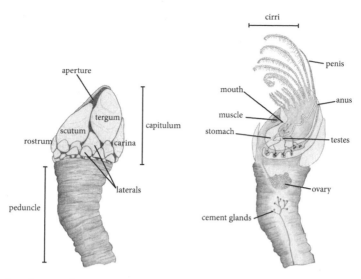

External and internal anatomy of a generalized gooseneck barnacle. Drawing by Leslie C. Costa.

the aperture at the top of the capitulum. As in acorn barnacles, the aperture is opened and closed with opercular plates. In gooseneck barnacles, however, these plates are very large and form part of the exterior body wall. The scutal plates are the largest plates, sitting more or less atop the peduncle. The tergal plates are next largest, just above and sometimes slightly to one side of the scutal plates. The largish plate just below the tergum is called the *carina*.

13. Carefully pull open the aperture with your forceps or fingers. Inside is the dark mantle cavity, and you may see the cirri within.

14. The pointy tergal side of the body is the posterior of the animal. Placing the barnacle on its side, posterior to the right and peduncle toward you, use the scissors to carefully cut the membrane *between* the large plates, taking care not to thrust the blade of the scissors more deeply than necessary to cut the membrane.

15. Lift the cut membrane with the forceps. You will find a large muscle running transversely between the scutal plates. This is

the adductor muscle, which controls the aperture. Cut this muscle to free the body wall membrane, and remove this to expose the mantle cavity within.

16. Observe the six pairs of large cirri, each two-branched (biramous). Remove a cirrus at the base and observe the two long and curled branches covered with long setae.

17. Compare the scutal and tergal plates and cirri of the acorn and gooseneck barnacles. These structures are *homologous* between the two barnacle groups, their different morphologies reflecting variation on a theme.

For further dissection, consult the barnacle section of a manual such as *Observing Marine Invertebrates* by Donald P. Abbot (Stanford, CA: Stanford Univ. Press, 1987) or *Invertebrate Zoology: A Functional Evolutionary Approach*, 7th edition, by Edward E. Ruppert, Richard S. Fox, and Robert B. Barnes (Belmont, CA: Thomson, Brooks-Cole, 2004).

See also:

M. Lowe and C. J. Boulter, "Darwin's Barnacles: Learning from Collections," in *Darwin-Inspired Learning*, ed. M. J. Reiss, C. J. Boulter, and D. L Sanders (Rotterdam: Sense Publishers, 2015), 273–284.

"Barnacles" at the Darwin Correspondence Project: www .darwinproject.ac.uk/learning/universities/getting-know-darwins-science /barnacles.

3

Untangling the Bank

One of Darwin's most enduring images from *Origin* is that of an overgrown bank teeming with life. Think how a river bank, or a country lane meandering through a rolling woodland, displays nature's exuberance in cross section: imagine low-growing creepers and fragrant herbs and shrubs growing from moist mossy cushions, here and there punctuated by mushrooms, lacy ferns, and tree trunks festooned with lichens and vines. The shrubs and trees cradle birds' nests and suspend spiders' webs; and everywhere are scurrying mice, marching ants, leaf-nibbling beetles and caterpillars, nectar-sipping bees—the whole is a living tangle. In the lovely final passage of the *Origin*, Darwin contemplated just such an "entangled bank," one "clothed with many plants of many kinds, with birds singing on the bushes, with various insects flitting about, and with worms crawling through the damp earth."[1]

Then he reflected on where the tangled bank came from. "These elaborately constructed forms, so different from each other, and dependent on each other in so complex a manner, have all been produced by laws acting around us," Darwin declared.[2] Those laws are all about reproduction, growth, variation, the struggle for existence, and natural selection. The diverse and interdependent organisms of the bank are each exquisitely adapted and honed by selection to stake their claim,

Orchis Bank, near Down House, likely inspired Darwin's enduring image of the "tangled bank." Photograph by the author.

survive, and flourish as best they can. And what's more, in a lucid ecological vision Darwin saw that the bank as a whole, its sublime messiness itself, was an outcome of selection. What seems an exuberance of vegetation with its flitting and creepy-crawly denizens, unruly enough to strike terror into the heart of gardeners whose tastes favor the manicured, has in fact a certain order or underlying structure in Darwin's eyes. "Thus," he concluded, "from the war of nature, from famine and death, the most exalted object which we are capable of conceiving, namely, the production of the higher animals, directly follows."[3]

Now, that was just it: "from the war of nature, from famine and death"—like some purifying fire, natural selection's destruction paradoxically produces beauty, adaptation, and order. Darwin knew he had a problem. The principle of natural selection had come to him in a flash of insight in the fall of 1838, and it didn't take him long to realize the implications: Malthusian population pressure, the "struggle for existence," could only mean death and destruction on a literally unimaginable scale. Exponential population growth was straightforward enough, and with it the idea of populations outstripping their means of sustenance. But he knew that the logical conclusion of Malthus's doctrine as applied to nature flew in the face of experience and received wisdom. "It is difficult to believe," he wrote in his notebook in March 1839, "in the dreadful but quiet war of organic beings going on in the peaceful woods and smiling fields."[4] It *is* difficult to believe; to our eye nature often appears bucolic, peaceful. Yes, predators catch prey, trees topple, and the occasional insect outbreak strips fields bare, but think about your experience walking in the woods some fine day, or picnicking on a sunny knoll: such scenes are the very picture of tranquility and beauty. Although we know that these places teem with life that can and does reproduce at a prodigious rate, nature seems unchanging, in balance.

Well, it *is* in a kind of balance, or perhaps stalemate is a better way to put it; each species can only grow in population to a point before it butts up against the similarly growing populations of other such species vying for the same space, food, light, or other vital resources, or it comes up against ever-ready predators, pests or pathogens trying to use *it* as a resource and expand their own population, or it hits the resource limits of its environment. Through any or all of these factors, the population of each species is held more or less in check. Individuals, Darwin realized, strive to the utmost to reproduce, but even as reproductive output is prodigious, mortality is equally so. It was not a happy thought, but reality nonetheless. Musing on an entangled bank led Darwin to think about what sorts of evidence would help make his case. This brought out the experimental side of Darwin, and in the process helped give rise to the field of ecology. But first he contemplated how to make a case for the struggle going on all around us, if only we could learn to see it.

Struggling with Struggle

Darwin did not have to read Malthus to learn about the struggle for existence. Charles Lyell wrote about it in the *Principles of Geology*, quoting at length the French naturalist Augustin de Candolle's *Géo-*

graphie botanique ("all the plants of a given country are at war one with another"), and giving the phrase "struggle for existence" more than once. But the image goes back further than that, to an earlier generation: his own grandfather, Erasmus Darwin, wrote of "one great Slaughter-house the warring world" in his epic poem *The Temple of Nature* (1803), in turn citing Carl Linnaeus's description of nature as *Bellum omnium in omnes*—the war of all against all. It's important to keep in mind, however, that by Charles Darwin's time, nature was viewed through the lens of natural theology, a philosophy that de-emphasized epic struggle and instead saw inherent harmony and goodness in nature, a view more consonant with a wise, loving, and benevolent Creator. Thus Lyell, though describing the struggle for existence in the *Principles*, took pains to stress a prevailing balance and harmony among mostly peaceably coexisting species. Darwin would have imbibed that religious tradition largely through William Paley, a famous son of Darwin's own Cambridge College, Christ's. (Darwin is thought to have occupied Paley's old rooms at Christ's, ironically). Paley was renowned for his lucid theological writings, including *Natural Theology* and *Principles of Moral and Political Philosophy*; the latter was even required reading when Darwin was a student, and he also devoured *Natural Theology*. The bucolic and pastoral English landscape seemed to almost compel a view of nature as smiling and glad—as surely as the wretched conditions endured by the urban poor compelled the Malthusian view of nature that Dickens captured so well.

When *Vestiges of the Natural History of Creation* was anonymously published in 1844, the outrage it provoked had as much to do with its vision of an uncaring pitiless nature, in which species constantly supplant other species in the history of life, as with the transmutationism it promoted. Extinction was much discussed in *Vestiges*, both of species and individuals, and it was that unsparing vision of the history of life that led the grieving Alfred Lord Tennyson to decry a "Nature red in tooth and claw" in *In Memoriam*, the poet's elegy for his friend Arthur Hallam. Despite Darwin's admiration for Lyell, and the lessons from Paley, he would have none of that natural-theological "balance and harmony." He took an unsparing view too: not even Malthus's "ener-getic language" adequately conveyed the magnitude of the "warring of the species" that constantly goes on, as he put it in one notebook. He cast about for ways to express it himself, and hit upon a power-ful metaphor: "One may say there is a force like a hundred thousand wedges trying [to] force into . . . the gaps in the oeconomy of Nature, or rather forming gaps by thrusting out weaker ones."[5] Species or vari-eties are like wedges struck repeatedly by the mighty hammer of nat-ural selection—the deeper and more securely they are driven into the "economy of nature" the better adapted they are to their niche or hab-itat, forcing out "weaker" ones in the process. The gaps are soon filled by other forms, and the competition continues. He later inserted a note between the lines: "The final cause of all this wedgings [sic], must be to sort out proper structure & adapt it to change."

It was a metaphor he stuck with for many years, varying the num-ber of imagined wedges: in the brief 1842 *Sketch* of his theory it was "a thousand wedges" forced into the economy of nature, and then "ten thousand sharp wedges" in the long *Essay* of 1844.[6] He stuck with ten thousand in the version that made it into the *Origin*: "The face of Nature may be compared to a yielding surface, with ten thousand sharp wedges packed close together and driven inwards by incessant blows, sometimes one wedge being struck, and then another with greater force."[7] Curiously, after using this metaphor from 1838 right through the publication of the *Origin*, he dropped it altogether from subsequent editions. Perhaps it was just too much. Only long and hard reflection would bring the reality of this truth home, he realized, but how to get people to even try? He struggled to express the concept—how to get the point across that the "war of nature" is incessantly rag-ing all around us, though most of us are completely oblivious? In the *Origin* he wrote that "Nothing is easier than to admit in words the truth of the universal struggle for life, or more difficult . . . than con-stantly to bear this conclusion in mind." That took imagination. Think, he urged, how the birds that may seem to be idly singing away in the trees must eat to survive. Reflect on the fact that to live and raise their young each and every one must consume prodigious quantities of seeds and insects. Or reflect on the realities of population growth: "There is no exception to the rule that every organic being naturally increases at so high a rate, that if not destroyed, the earth would soon be covered by the progeny of a single pair."[8] Even the slowest-reproducing species will outpace space and resources in short order.

We saw in the last chapter how Darwin immersed himself in the world of barnacles for an 8-year period beginning in 1846. But all the while he never stopped contemplating the nuts and bolts of his transmutation theory—struggle, extinction, selection, favored variants, diversification, competition, and so on. Not only did he contemplate all these things, but more importantly he also contemplated how they interrelated. In addition he struggled with struggle very personally with the traumatic loss of his 10-year-old daughter Annie, in 1851, a shock that may have extinguished in him any remaining sense of a purposeful and benevolent Creator. The reality of an indifferent and uncaring nature was brought into unbearably sharp focus by her death; grieving, he and Emma soldiered on, as he immersed himself in his work. They were buoyed, too, by their other children. 1850 saw the birth of Leonard, called Lenny, and Horace came along the year they lost Annie—the Darwins had seven children by 1851, all under the age of 13.

Divergent Thinking

The second of Darwin's monographs on living barnacles appeared in 1854, bearing a heartfelt dedication: "To Professor H. Milne Edwards this work is dedicated, with the most sincere respect, as the only, though very inadequate acknowledgement which the author can make of his great and continued obligations to the *Histoire Naturelle des Crustacés*,' and to the other memoirs and works on natural history published by this illustrious naturalist." Up to the publication of Darwin's monographs on the subject, the remarkable French naturalist Henri Milne-Edwards (1800–1885) had provided the definitive treatment of barnacles. Of the many great insights Darwin gained from

Milne-Edwards, one in particular stands out—an insight with special relevance to Darwin's anxiety. In 1852 Darwin read key sections of Milne-Edwards's *Introduction à la Zoologie générale*, a masterful overview of anatomy and physiology throughout the animal kingdom. He was struck by Milne-Edwards's interconnected laws of diversity and economy: "Nature is prodigal in the variety of her creations, yet parsimonious in the means of diversifying her works."[9] The beetles provide a good example, perhaps the single largest taxon in the animal kingdom with nearly 300,000 named species, all different yet in another sense all carbon copies of one another with the same beetle-defining characteristics. Naturalists (including Milne-Edwards and Darwin) struggled with the concepts of so-called "highness" and "lowness" of organisms as they tried to map out diversity and classify species—a holdover from Aristotelian ladder thinking where species were arrayed on a scale of supposed perfection (always culminating in humans, needless to say). Milne-Edwards maintained that highness or lowness in the scale of nature was best expressed by structural complexity, measured by the differentiation of cells and tissues. He called this the "physiological division of labor," a concept that captured Darwin's imagination.

The physiological division of labor was an industrial analogy from economics and manufacturing. In chapter 3 of his *Introduction*, in a line that Darwin scored, Milne-Edwards explained how "in the creations of nature, as in the manufacture of men, it is mostly by the division of labor that perfectibility is obtained." The key word there is "perfectibility." In economics that might mean increased production efficiency, per capita productivity, or perhaps overall economic prosperity. In the analogy of Milne-Edwards it refers to increased complexity and improved functionality of the organism—as tissues and organs divide the "labor" of physiology, the organism as a whole prospers. Coelenterates like sea jellies, say, with a simple sac for a stomach that also functions for respiration are less efficient at feeding *and* breathing than animals with separate gut and respiratory organs, the implication being that by doing double-duty, tissues and organs by definition cannot specialize and so cannot achieve a high degree of efficiency in any one thing. That limits the adaptive possibilities for the organism. Similarly, having a single type of muscle is more limiting than the adaptive possibilities inherent in having multiple muscle types (e.g., skeletal, smooth, and cardiac), each specialized for a different function.

Darwin transferred Milne-Edwards's idea from the tissues and organs of organisms to different organisms in the economy of nature: in an analogy of an analogy he conceived of an *ecological* division of labor. It was a remarkable idea: just as the overall productivity of the organism as a whole increases with increasing physiological division of labor, so too, Darwin thought, can more species live and prosper together in any one area if they are specialized in slightly but sufficiently different ways. This hypothesis resonated with and helped crystallize ideas on natural selection, extinction, adaptation, and diversity swirling around in Darwin's head. Before reading Milne-Edwards, Darwin had a general conception of diversification of species over time through isolation and selection. He studied variation with an eye to intermediate forms as evidence of a slow diversification; his barnacle studies were conducted with an eye to that, as well as to the evolution of separate sexes from a hermaphroditic ancestral state. After Milne-Edwards, Darwin saw diversification more in the context of division of labor, and his idea of an ecological division of labor was key. As he labored to finish up his barnacle work this idea was never far from his mind—not to mention his perennial back-burner researches into variation and domestic animals, cross-pollination, and other subjects.

In November 1854 Darwin scribbled down a thought: "It is indispensable to show that in small & uniform areas there are many Families & genera. For otherwise we cannot show that there is a tendency to diverge (if it may be so expressed) in offspring of every class, & so to give diverging tree-like appearance to the natural genealogy of the organised world."[10] What he was saying was that if in a small area, homogeneous in its conditions (of soil characteristics, say, and temperature, rainfall, etc.), you were to find great diversity, then that would suggest the existence of a tendency or force (like natural selection) promoting diversification and coexistence. Otherwise, a small homogeneous area would be expected to support one or a few species. Darwin thinks that was likely the case ancestrally, but over time through competition and selection the more divergent offspring variants tended to persist while those too similar in their ecological needs were rendered extinct. It's a dynamic that leads to more and more divergent varieties and species occupying that small and homogeneous area. He soon came to realize that this dynamic is not simply an interesting by-product of selection, but is, in fact, key to how selection drives the evolution of the diversity of life. But what evidence could he bring to bear on this idea?

Darwin thought back to the monotonous heathlands of Hartfield, in Sussex, home of Emma's sister Sarah Wedgwood and Charles's sister Charlotte, and of Leith Hill, in Surrey, where his family had just visited his sister Caroline and her husband Joe Wedgwood. (Like Charles, Caroline had married a Wedgwood cousin—Emma's brother Josiah III, who went by Joe.) Contrast the heathland monoculture with the abundant flora of fertile meadows; both are crowded with plants, but one cannot doubt, Darwin mused, that overall more plant (and thus animal) life is supported in the latter. Maybe this could be proved experimentally: "I think [the] amount of chemical change should if possible be taken as measure of life, viz amount of carbonic acid [carbon dioxide] expired or oxygen in plants." Easier said than done, however— "Gas" Darwin's chemistry days were a distant memory now.

That winter and spring, 1855, Darwin's preoccupation with diversity overlapped with his growing interest in the related problem of the geographical distribution of species and varieties. In fact, the two interests dovetailed nicely: it was all a matter of spatial scale. Diversification and diversity in small areas, where individuals experience the direct effects of competition and selection, ultimately bears on geographical distribution on the scale of continents. Relevant, too, is the question of how species become distributed as we see them on local, regional, and global scales. He began a series of experiments looking at seed dispersal: it was astounding to him that no one had ever considered doing so before. All through the severe winter of 1855, Darwin and the children diligently followed the progress of seeds they set floating in artificial saltwater, tipping in snow to simulate cold seawater, chalking up floaters versus sinkers, and testing germination after progressively longer periods of immersion. We'll explore that rich line of experimental investigation fully in Chapter 5; for now just note that through that winter Darwin was dreaming of field experiments to come.

A Taste for Botany

Nineteen-year-old Catherine Thorley of Tarporley, Cheshire, joined the family as governess in 1848. She worked at Down House for nearly 10 years, through the dark days of the loss of Annie and the happy ones that saw the birth of Frank, Lenny, and Horace. And although she had her work cut out for her with Etty, then an adolescent with an attitude, all in all she was well loved by the family. She taught the children French, dancing, music, and etiquette, and also had a love of plants and a curiosity about her employer's work. So, it's perhaps no surprise that June 1855 found Miss Thorley gamely assisting Darwin in his latest hobbyhorse: a flowering plant survey of Great Pucklands Meadow and environs, the 13-acre field just over the hedgerow from Darwin's sandwalk, owned by his friend and neighbor Sir John Lubbock. Darwin wrote to Hooker about it on June 5th:

> Miss Thorley & I are doing a little Botanical work (!) for our amusement, & it does amuse me very much, viz making a collection of all the plants, which grow in a field, which has been allowed to run waste for 15 years, but which before was cultivated from time immemorial; & we are also collecting all the plants in an adjoining & similar but cultivated field; just for the fun of seeing what plants have arrived or dyed out. Hereafter we shall want a bit of help in naming puzzlers.—How dreadfully difficult it is to name plants.[11]

Before posting the letter he surprised himself with his first grass identification, excitedly appending a P.S.: "I have just made out my first

Grass, hurrah! hurrah! I must confess that Fortune favours the bold, for as good luck would have it, it was the easy Anthoxanthum odoratum: nevertheless it is a great discovery; I never expected to make out a grass in all my life. So Hurrah. It has done my stomach surprising good." Darwin may have exaggerated his inability to identify plants, but it's true that *A. odoratum*, pleasantly scented sweet vernal grass common throughout Europe and Eurasia, is a rather easy one to identify. Others he shipped off to Hooker. Ten days later he sent his thanks: "You cannot imagine what amusement you have given me by naming those 3 grasses: I have just got paper to dry & collect all grasses.— If ever you catch quite a beginner, & want to give him a taste for Botany tell him to make perfect list of some little field or wood. Both Miss Thorley & I agree that it gives a really uncommon interest to the work, having a nice little definite world to work on, instead of the awful abyss & immensity of all British Plants."[12]

The survey continued over the course of the year, often with the kids helping out. Their approach to surveying was more casual than systematic, randomly searching for anything new rather than sweeping along transects or scrutinizing quadrats the way a modern botanist would do it. But it suited his purpose: he may have said the survey was done for his amusement, but there was more to it than that. He saw the field and wood as a "small uniform area"; how much diversity was found there? In the end a whopping 142 plant species falling into 108 genera and 32 orders were identified in Great Pucklands alone. One hundred fifty years later, in June 2005, three generations of Darwin's descendants (ranging from 21 months to 78 years of age) gathered at Great Pucklands to help kick off a repeat of their ancestor's survey along with botanists from London's Natural History Museum.[13] The NHM team surveyed in the summers of 2005–2007, taking samples in the last year for DNA barcoding. They found some new species and many of Darwin's old friends (including his triumphantly identified sweet vernal grass)—overall nearly the same numbers of species and genera as Darwin and Miss Thorley had found, but rather fewer orders. Why? It largely reflects changes in classification: in Darwin's day British flowering plants were classified into about 86 plant orders, while the latest classification recognizes 62 orders. That means some of Darwin and Miss Thorley's plant orders have since been combined with others.

Darwin would have found these results interesting indeed—but perhaps unsurprising. Those 140-something plants are perpetually vying to expand, yet they are kept in mutual check generation after generation. In Darwin's thinking, what we are seeing is a showdown in slow motion. The reason we see rich diversity in the meadow, the reason that highly varied species manage to live closely together for a long period of time in rather small and uniform areas, is that competition and selection have over the eons led to that very diversification and promoted their coexistence—each of the plants is adapted to ever-so-slightly different conditions, just different enough on a microgeographical scale that for all intents and purposes they compete little though they grow side by side. If you think about it, this is an *ecological* vision: the community itself emerges, evolves even, as constituent species become coadapted. Coadaptation does not have to mean adapted for direct interaction like pollinator and flower. Instead, it is a simple coexistence, with the species partitioning resources in such a way that they co-adapt to living closely together.

The one thing missing from Darwin and Miss Thorley's botanical survey was the *temporal* dimension. Such assemblages of species in a given locale change over long periods of time, with species coming and going, some persisting long periods and others winking out quickly. These ideas began to resonate more deeply with his interest in both dispersal (on a global as well as a local scale) and seed viability—again, taken up in Chapter 5. Darwin's musings on dispersal and seed viability in connection with his interest in divergence led him to discover the seed bank concept: the idea that the soil of a given area is chock full of seeds deposited there by birds, wind, or water, or dropped by plants that once occupied the spot. In large part the stock of seeds in the ground represent ghosts of plant communities past. Often they are very much alive, even after decades or centuries. Think of lotus seeds found in ancient Egyptian tombs germinating after eons. Maybe the earth is saturated with seeds similarly entombed, and that if the seeds become exposed to the right conditions, they would germinate and flourish anew.

In another ecological vision Darwin realized that as assemblages develop, they alter the very space they occupy. The concept of ecological succession was articulated by the American plant ecologist Frederic Clements a half century after Darwin and Miss Thorley botanized in Great Pucklands Meadow, but we can see early glimmers of this concept in Darwin's experiments that began in the summer of 1855. That July, for example, he noticed that where he had pulled up some thorny shrubs the previous spring, he now saw weedy yellow-flowered charlock (*Sinapis arvensis*), a mustard, coming up. Charlock often gets into cultivated fields, thriving in open conditions, but cannot survive being shaded out by overtopping shrubs and trees. He thought it was curious that they should appear. Maybe the plant is a fast colonist of favorable sites, or maybe it had been there for many years, dating to when that area was under cultivation. He did an experiment to find out, and recorded it in his notes as follows:

> July 21. dug 3 places about 2 by nearly 3 ft square: cut off turf of weeds, & dug up ground one spit deep. [a "spit" was the length of a spade blade]
>
> Aug 1st numerous seedlings came up. some of them crucifers from taste
>
> Aug 19th. Young Plants now recognizable, but marked by sticks to make quite sure: (a few rather younger) six in one place, 10 in another, 5 in the third = 21 young plants.—Grass fields all round, thick hedges—no charlock even at end at present, from there in clover 120 yd. distant.[14]

Note his simple but effective technique of marking the seedlings as they come up with sticks—we'll see him put that to good use again. The 21 young charlock plants that sprung up in his little test plots (with none in sight in nearby fields) pointed to the seeds being there the whole time, just awaiting exposure under the right conditions. He was beginning to realize that the earth was thick with the propagules of flowering plants, mosses, ferns, fungi, just waiting. It reinforced his growing "ecological" sense of an interconnectedness that extends in both space and time. Interconnectedness at varying *geographical* scales, with diverse assemblages of species coexisting, an entangled bank with continued competition as well as mutual interdependencies; and varying *temporal* scales, with the makeup and structure of these species assemblages of the entangled bank dynamically changing, morphing according to the vagaries of inexorable changes in climate, geology, and selection. The geographical and temporal sense of ecological interconnectedness resonated deeply with Darwin's overarching revelatory insight back in 1837, on how species are related in space and time.

Now he was beginning to see how diversity could emerge from an area even the size of the palm of one's hand. That August he made a telling comment in his notes:

> When many individuals crowded together some will die, so will forms. creations cause ‹death› extinction—like birth of young causes death of old.—All classification follows from more distinct forms being supported on same area.[15]

Two insights from this brief note are worth pointing out. First, this statement reveals that Darwin saw that the process of close competition forces divergence (which involves the extinction of intermediate forms) and gives rise to the classification system used by naturalists. "All classification follows" because the dynamic process of diversification, along with its simultaneous extinction and gap-creation, explains the nested hierarchy we see in taxonomic classification. Several strands of thought were gelling here for Darwin: selection, diversification, geographical distribution, extinction, classification. It was a tremendously exciting synthesis.

Earlier in this note, Darwin restates his hypothesis that far more "forms" can be supported in an area when they are diverse, and adds a

note to himself to give examples. Later he inserted which case studies
to use to support his thesis: "Trifolium at Lands end. Larch wood. Coral
[island]. Hooker facts." There is so much behind even that one line: the
first example comes from an 1847 paper on the plants of remote Land's
End, at the tip of the Penwith Peninsula of Cornwall, in which C. A.
Johns reported finding nine species of legumes in a very small space—
an area no bigger than his hat. Darwin copied out Johns' comment that
"Had the rim been a little wider, I might have included *Genista tinc-
toria* & *Lotus corniculatus*." These too are legumes, so Johns found
nearly a dozen legume species coexisting in an area a foot across or
less. The second case, "Larch wood" refers to the larch woodland at
one end of Great Pucklands Meadow, which was included in Darwin's
botanical study with Miss Thorley. "Coral island" refers to data he collected on the Cocos (Keeling) Islands, atolls in the Indian Ocean that
Darwin visited on the *Beagle*. The plants he collected there, described
a few years later by Henslow, were now called into service to support
his idea of small and uniform areas supporting great taxonomic diversity. As for "Hooker facts," Darwin meant the information furnished by
his friend Joseph Hooker, in answer to Darwin's incessant inquiries for
information.

Think Globally, Act Locally

Darwin may have been inspired by his little palm-sized testing plots
to transfer that approach to his diversity studies. The following March,
of 1856, he fenced off a 3 × 4-foot plot of lawn in his backyard and
set orders for it not to be mown. It was another "small and uniform"
area, an unremarkable patch that like the rest of the lawn had been
mown regularly probably since the house was built. It ended up being
the Great Pucklands Meadow in miniature. "I found the species 20
in number, & as these belonged [to] 18 genera & these to 8 orders &
they were clearly much diversified," he reported in *Natural Selection*.[16]
The results of his Lawn Plot Experiment, as it came to be called, were
duly reported in the *Origin* as well. Thinking big, he extrapolated

the results of Great Pucklands and his lawn plot to entire regions and continents—his own version of the environmental activists' mantra to "think globally, act locally." "We see on a great scale, the same general law in the natural distribution of organic beings," he declared, pointing to data gleaned from the literature on diversity found co-existing in heaths, mountaintops, salt- and freshwater marshes, and lakes and rivers. He began to see that Johns's observation at Land's End was something of an exception: where Johns found nearly a dozen species of the same family in a small area, usually Darwin was finding cases of multiple families and orders. The preponderance of cases led him to believe that greater diversity in uniform areas was the general rule.

Why should this be the case so often? Darwin was convinced that it was the result of selection, of course, but it is important to keep in mind that he doesn't just mean that selection has led to a more diverse roll call of species in a given area. His "ecological division of labor" idea goes beyond that. To him, so many species packed in a small area is an indication of greater overall *productivity*. This is one expression of "progress," that slippery concept that dominated (and often still dominates) social and evolutionary thinking from the Enlightenment onward. The idea of progress is linked with improved productivity, efficiency, increased resources. It is inherent in the division of labor promoted by Adam Smith, the physiological division of labor promoted by Milne-Edwards, and it, too, is inherent in the ecological division of labor promoted by Darwin. He was sure that "a greater absolute amount of life can be supported in any country or on the globe; when life is developed under many & widely different forms, than when under a few & allied forms."[17]

Darwin's plant diversity surveys and Lawn Plot Experiment helped him demonstrate that more times than not small and uniform areas really do support surprisingly diverse forms. The logical next step was to demonstrate that such areas were disproportionately more productive of life, say, or biomass than comparably sized areas dominated by just a few forms.

In the course of his researches, Darwin came across an account of a most remarkable experiment that had been conducted some decades before at Woburn Abbey, Bedfordshire, just north of London. The Scotsman George Sinclair was gardener to the 6th Duke of Bedford at Woburn Abbey from 1807 until 1825, during which time he oversaw various experiments in agricultural improvement, the most significant of which became known as the Grass Garden Experiment. It was aimed at determining how grass monocultures compared with mixed grass plantings in terms of productivity (reckoned by yield, herbage produced). Some 242 2 × 2-square foot plots were planted with different grasses singly or in various combinations, and with different experimental soil treatments overseen by the renowned Sir Humphrey Davy. It may not have met modern standards of replication and other hallmarks of statistical rigor, but for its time this experiment was cutting-edge scientific agriculture—nothing like it had ever been attempted before.

Sinclair had published the results of the experiment several times, but Darwin saw Sincair's paper of 1826 that was published in *Loudon's Gardeners Magazine* describing the plan of the garden experiment and its results. In it, Sinclair wrote of how the Woburn Abbey experiments "clearly prove that any certain soil will maintain a greater, and produce more nutritious produce, if cropped with a number of different species of grasses, than it will maintain and produce if cropped with only one or two species." Sinclair went on to note that "This is a curious and important fact" generally neglected in practice. Then he gave the data: "If an acre of good land is sown with three pecks of rye-grass, and one peck of the clovers . . . 470 plants only will be maintained on the square foot of such land." However, "if, instead of two species of grasses, from eight to twenty different sorts are sown on the same soil . . . a thousand plants will be maintained on the same space, and the weight of produce in herbage and in hay increased in proportion." Sinclair concluded that this is why "every variety of soil and situation, from the alpine rock to water itself, is provided with its appropriate grasses, destined for the support of animal life, and for covering the soil with the colour most pleasing to man."[18]

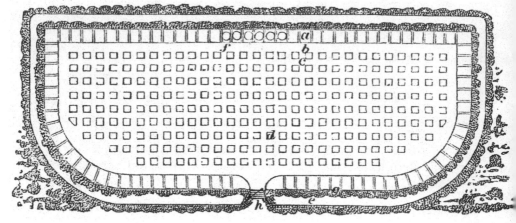

Plan of George Sinclair's *Hortus Gramineus*, or grass garden, at Woburn Abbey: "The spaces allotted to the proper grasses are in number two hundred and forty-two, of two square feet each, inclosed by cast-iron frames. Paths of gravel two feet nine inches wide separate the spaces on every side; these are surrounded by a path three feet wide, with a border for the herbage plants . . . a hedge of hornbeam separates the compartment from the rest of the grounds, and an outside border of roses completes the grass-garden." Each of the 2 × 2-foot plots was either transplanted with natural turf or planted with a single grass species or various combinations of multiple grass species. From Sinclair (1826), pp. 114–115.

Sinclair's experiment found that when between 8 and 20 species were simultaneously sown, the number of individual stems supported more than doubled to 1000 per square foot. Monoculture, or near-monoculture, gives a poorer yield than, well, "polyculture" (to perhaps coin a term)—precisely what is to be expected from the ecological division of labor, and precisely, Darwin was sure, what lies behind the surprising diversity found in even the smallest and most uniform of small, uniform lawn plots. Selection had made it so.

Botanical Arithmetic

The Lawn Plot experiment was a project that Darwin kept up for 2 years. It dovetailed with a number of other seemingly unrelated interests in the 1850s, the most significant of which could not seem more different in approach: a foray into the ratios and distribution of species and varieties gleaned from stacks of botanical manuals. We saw in the previous chapter how Darwin's barnacle work gave him a new appreciation for the abundance of natural variation. Darwin always saw this variation as the raw material for selection to act upon, but beginning around the summer of 1855 he began to consider that variation in a statistical way, trying to quantify it to somehow find selection's signature. He had an idea sparked by an analogy in an article by Swedish botanist Elias Magnus Fries. Darwin had come across a passage where Fries remarked that in genera containing lots of species, "the individual species stand much closer together" than they do in species-poor genera. "Hence," recommended Fries, "it is well in the former case to collect them around certain types or principal species, about which, as round a centre, the others arrange themselves as satellites."[19] Darwin was struck by this. He later told Hooker that he was led to undertake his analysis of varieties in large versus small genera by Fries' remark "that the species in large genera were more closely related to each other than in small genera; and if this were so, seeing that varieties and species are so hardly distinguishable, I concluded that I should find more varieties in the large genera than in the small."[20] What Darwin came to realize was that the inverse was also true: small genera with few species, limited in range, comparatively depauperate in variation, are on their way out. Speciation and extinction go hand in hand, with groups diversifying even as others wither. In a telling note to himself, he declared: "We can look far into future by looking to the larger groups."[21] In other words, the large groups of today are likely to be larger still in the future. The reason, he thought, was because the ancestors of species in large, widespread genera were favored by selection in the past, multiplying into the large number of the present. In all likelihood current species in large genera continue to be favored and selection is actively promoting their growth, competitive success, and diversification. Hence you should see more varieties associated with species in such genera. Since varieties are "little species," with time they become bona fide species that produce more varieties in turn, and so on.

One way to test this "success breeds success" way of looking at things is to quantify the variation found in different sized groups—an idea that led him to tabulate and compare the amount of variation in genera large and small. Are distinct varieties—which can be seen as products of selection acting upon past variation—more commonly found in genera with large numbers of species? Botanical manuals, he realized, are handy compendia of such information and are available for just about any region or country thanks to the labors of ardent botanists everywhere. Hooker recommended he look at three for British plants: Babington's *Manual of British Botany*, Henslow's *Catalogue of British Plants*, Watson's and Syme's *London Catalogue of British Plants*. Darwin's approach was his own version of a by-then well-established line of investigation called "botanical arithmetic." Coined and pioneered by the polymath Alexander von Humboldt and pursued by other naturalists like Alphonse de Candolle in France and Robert Brown in England, botanical arithmetic was an indispensable tool of early to mid-nineteenth century geographical botany. With a simplicity that belies its significance at the time, botanical arithmetic involved enumeration of species, genera, etc. of different plant groups according to variables such as elevation, latitude, and continental area in the pursuit of divining underlying relationships that may give insight into the workings of nature. How does the proportion of mosses, ferns, or grasses compare to that of other groups on this or that continent? Does it vary according to elevation? Does the average number of species per genus vary between regions or between continental areas and oceanic islands? The ledgers and tables of botanical diversity worldwide patiently assembled by explorer-naturalists since the seventeenth century held the key.

From Henslow's tutelage, Darwin was long familiar with botanical arithmetic, and his transmutation notebooks show that he used the technique in different ways. In the B notebook, for example, in an entry probably made in early 1838, we find Darwin tabulating genus-to-species ratios for the plants of north Africa, St. Helena, and the Canary Islands. He noticed that the "poorness" of the flora of the islands was "in exact proportion to distance" from the African mainland—something he soon came to understand in the context of chance voyagers landing at oceanic islands from the nearest mainland. Now, in the 1850s, Darwin realized that just as facts in the distribution of plant species and genera could be used to shed light on underlying principles governing biogeography, so too could facts in the number, ratio, and distribution of varieties in relation to species and genera be used to test his ideas about the action of selection. If the current geographical distribution of species represents a clue to—or even a diagnostic marker of—geological, climatological, or evolutionary history, Darwin was sure that the frequency and distribution of varieties was diagnostic too. He worked his way through botanical manual after manual doing his own brand of botanical arithmetic. Besides the three on British flora, he looked at about nine more on the flora of the United States, Russia, Germany, France, and Holland.

In the Weeds

Even as Darwin was poring over his botanical manuals, he was hard at work on *Natural Selection*—and experimentising. By mid-October 1856 he finished a chapter on domestication, and by mid-December, just a week after the birth of his 10th child, Charles Waring, he completed another on the crucial importance of intercrossing—two great strands of his thinking that we explored in the last chapter. As 1856 became 1857 he jumped into the all-important chapters on variation (completed near the end of January), the struggle for existence (finished up in early March), and natural selection (done by the end of March). They got him thinking about how struggle and natural selection gave rise to the close coexistence of divergent plant groups he had found in those small lawn plots. That means struggle must be evident even—or maybe especially—in that small spatial scale. He decided to find out with another kind of experimental plot.

In January he cleared an old planting bed about 2 × 3 feet square that had been used for strawberries for a year or two. He called it his "weed garden." Given the myriad pressures that organisms face, by dynamically changing over time with age, size, reproductive condition, complex interactions with competitors, predators, and more, he needed to simplify to get at the basic question of mortality at one stage of life. He chose a single kind of organism—plants—at one life stage—seedling—in one small plot where conditions were uniform and no crowding was experienced. The plan was to check the bed daily and mark the position of each weed seedling as it naturally sprouted by sticking a piece of wire in the ground next to it. Then over time he could census the wires and see which weeds were still alive and which had died. He recorded his progress: "Early in March seeds began to spring up: marked each daily. March 31st About 55 marked, of which about 25 Killed already. April 10th. Pulled up 59 wires marking where seedlings before development of two leaves had been devoured, I suppose by slugs . . . April 20th Pulled up 28 wires, dead (I think dry weather is beginning to tell against some) . . . May 8th Pulled up 95 wires (I suspect that some seedlings are Killed by drought) . . . June 1st Pulled up 70 wires."[22] By the beginning of August 1857 he found that just 62 of the 357 plants that came up survived. Slugs and insects had taken their toll. These results were eventually given in *Origin* as an illustration of the struggle going on all around us that we fail to notice. Tracking seedling mortality over 8 months, he found that of 357 seedlings that appeared, 83 percent of them were destroyed.[23]

"I can often gaze for a long [time] at a square yard of turf," he mused, "and reflect with astonishment at the play of forces which determine the presence & relative number of the 30 or 40 plants which may be counted in it."[24] He saw epic struggle in a tiny plot of land and wanted to find other ways to get this point across. He soon realized that the abundant heathlands and commons he encountered on his countryside rambles told the story too, in a different way. All the better that such spaces were familiar to everyone. Who didn't know the stark and beautiful heaths, those wide open spaces clothed in grasses and heather? Some might say they know the heathlands like the back of their hand, yet who really knows the back of their hand? The seemingly familiar is really unfamiliar, once you learn to see. Darwin made observations in Surrey, near pine-covered Crooksbury Hill. A scan of the hundreds of acres where cattle grazed revealed not one pine. Yet a closer look, low

A replication of Darwin's "Weed Garden" experiment at Down House. Photograph by the author.

to the ground, showed trees were everywhere. Free-ranging cattle kept them cropped. These were pygmy trees, he thought, and he estimated the age of one at 26 years. Such is the power of one species—cattle—to keep others in check and shape the landscape. He observed another heath in Staffordshire where pines abounded in a fenced area enclosed 25 years prior. The whole ecology was altered by the exclusion of cattle: he counted 12 plant species in the enclosure not found outside of it, and six insectivorous bird species. These enclosures became a central part of Darwin's ecological thinking: species influencing species, great and small, in a web of interconnectedness. It was all there, plain as day, evident in a woodlot, heath, or backyard weed garden.

Disgraceful Blunder Averted

The distinguished philosopher of science Michael Ruse once referred to Darwin's chapters on variation, struggle for existence, and natural selection as the "deductive core" of *On the Origin of Species*. They hang together, building an argument almost syllogistically. In them Darwin almost seems to be saying that if there is abundant variation in nature that can be passed down to descendants, along with constant and intense population pressure and struggle, then survival and successful reproduction based on those variations must follow. He dubbed that process "natural selection," and it is well known that he first conceived of the idea in 1838. But it is often unappreciated that Darwin came to see his initial concept of natural selection as incomplete: natural selection as such might explain how groups supplant groups, or how species slowly change one into another, but how does it generate the *diversity* of life?

A crucial element of these chapters was his case for the mechanism of divergence, and so all the while he was also (among many other things!) making steady progress with his botanical arithmetic. He went through the manuals one after another, counting up varieties that the authors had listed for larger and smaller plant genera. As he expected, larger genera (i.e., those with more species) also had more varieties. On March 31st he recorded in his diary having completed his "Natural Selection" chapter—which included his ideas to date about divergence, or diversification, and the evidence from larger genera. He was writing confidently but soon needed a break. Exhausted, he headed to Dr. Lane's establishment at Moor Park, in Surrey, the hydropathic spa he now favored. After the death of Annie, the Malvern facility was too fraught with painful memories. "My health has been very poor of late," he wrote to Lyell in April. Two weeks later he was home—"Returned May 6th Did me astonishing good."[25]

He would be back soon. Fourteen-year-old Etty had been afflicted with a mysterious ailment for months. Emma had first taken her to the seaside town of Hastings. When that failed, they went to Moor Park. In a long letter to Hooker, Darwin mentioned that he'd be tag-teaming in June with Emma: "My wife Emma & Etty have just started for Moor Park; she [Emma] will stay a fortnight, & then I shall relieve guard for another fortnight."[26] Etty ended up staying through October except for a brief trip home in August, the delicate state of her health a chronic source of worry for her parents.

Darwin worked away at his "everlasting species-book" as best he could. By the end of September 1857, he had finished chapters seven and eight. But he was practically derailed in mid-July 1857 when he received a letter from his neighbor John Lubbock, informing him that his botanical arithmetic calculations were all wrong. Lubbock pointed out that Darwin had been rather arbitrarily labeling genera as "big" or "small," but a more rigorous approach would have been to predetermine the size categories and then look at the frequency of varieties found in each. This way he would not be using relative "bigness" or "smallness" of genera, but an absolute size measure that would be used to categorize all genera for each region in the study. Of course, deciding on "large" and "small" categories is itself arbitrary, but at least there is then a consistent standard. Darwin was initially shocked by Lubbock's revelation. Still, he knew Lubbock was right and he thanked him profusely: "You have done me the greatest possible service in helping me to clarify my brains. If I am as muzzy on all subjects as I am on proportion and chance— what a book I shall produce! . . . I am quite shocked to find how easily I am muddled, for I had before thought over the subject much, and concluded my way was fair. It is dreadfully erroneous. What a disgraceful blunder you have saved me from."[27] He couldn't resist a frustrated P.S.: "It is enough to make me tear up all my M.S. & give up in despair—It will take me several weeks to go over all my materials. But oh if you knew how thankful I am to you." He immediately wrote to Hooker too, kicking himself in the pants and crying that he was the "most miserable, bemuddled, stupid dog in all England."[28]

It was bad, but not disastrous. He followed Lubbock's suggestion
with a test case, the flora of New Zealand, dividing the plants into two
groups: genera with four or more species ("large" genera), and those
with one, two, or three species ("small" genera). He found 339 species
in the large category and 323 in the small one. Of the 339 species, 51
presented varieties. Proportionally speaking, if the incidence of varieties was the same in the two groups then by proportion he should find
48.5 varieties in the small category. In fact, there were only 37 varieties
within the smaller genera. That was completely consistent with Darwin's
overarching argument, and then some: he realized that by this method
of proportional analysis larger genera could be shown to not only have
more varieties, but *disproportionately* more. This was the smoking gun
he was seeking. "The case goes as I want it," he wrote cautiously to Lubbock, "but not strong enough, without it be general, for me to have much
confidence in."[29] Prevailing upon Hooker to resend all of those botanical books he had long since returned, he threw himself into redoing
his calculations. He generated some 200 pages of tables by the time he
was done.[30] Actually, he didn't generate all those tables himself: Darwin
employed the local school master, Ebenezer Norman, to do much of the
laborious tabulating for him. He emerged from his growing mountain
of tables with not only more robust evidence for divergence of species,
but also a *principle* of divergence. He first used the term that August,
reporting his results to Hooker and asking for still more botanical manuals: "If it will all hold good it is very important for me; for it explains,
as I think, all classification, ie the quasi-branching & sub-branching of
forms, as if from one root, big genera increasing & splitting up &c &c,
as you will perceive.— But then comes, also, in what I call a principle of
divergence, which I think I can explain, but which is too long & perhaps
you would not care to hear."[31] It may have been too long to explain to
Hooker, but a month later he sent Asa Gray, in Massachusetts, a long
but concise explanation of his now all-important principle after defining
natural selection. It's worth quoting in full here, especially as he has
worked in several lines of evidence such as the Woburn Abbey experiment and his biodiversity surveys:

One other principle, which may be called the principle of diver-gence plays, I believe, an important part in the origin of species. The same spot will support more life if occupied by very diverse forms: we see this in the many generic forms in a square yard of turf (I have counted 20 species belonging to 18 genera),—or in the plants and insects, on any little uniform islet, belonging almost to as many genera and families as to species.— We can understand this with the higher, animals whose habits we best understand. We know that it has been experimentally shown that a plot of land will yield a greater weight, if cropped with several species of grasses than with 2 or 3 species. Now every single organic being, by prop-agating so rapidly, may be said to be striving its utmost to increase in numbers. So it will be with the offspring of any species after it has broken into varieties or sub-species or true species. And it follows, I think, from the foregoing facts that the varying offspring of each species will try (only few will succeed) to seize on as many and as diverse places in the economy of nature, as possible. Each new variety or species, when formed will generally take the places of and so exterminate its less well-fitted parent. This, I believe, to be the origin of the classification or arrangement of all organic beings at all times. These always seem to branch and sub-branch like a tree from a common trunk; the flourishing twigs destroying the less vigorous,—the dead and lost branches rudely representing extinct genera and families.[32]

It was a grand ecological and evolutionary vision, and his new Principle of Divergence was the lynchpin.

Analysis of manual after botanical manual all through the remain-der of 1857 yielded, with a few exceptions, the same result: species of larger genera had disproportionately more varieties than their counter-parts in smaller genera. There was only one explanation, he thought: size of genus really is a direct indicator of evolutionary success in the recent past. Abundant and widespread, these species are even now rich in variation and yielding varieties, just as he thought. Thus Darwin's

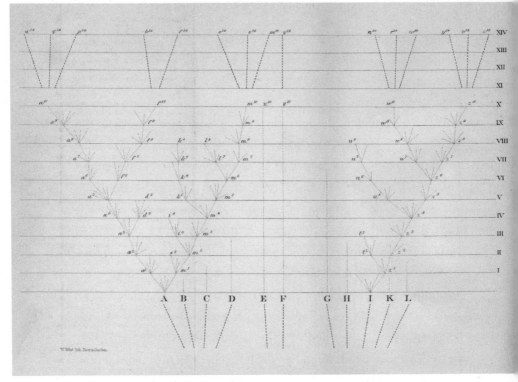

The sole figure in *On the Origin of Species*, illustrating the outcome of Darwin's Principle of Divergence. Time, reckoned in thousands or tens of thousands of generations, is represented on the vertical axis, and morphological and ecological divergence is represented on the horizontal. The diagram illustrates how descendant species become increasingly more different (divergent) from ancestral parental forms (capital letters along the bottom) over time. Darwin envisioned this divergence process giving rise to branching and re-branching lineages, forming the Tree of Life. From Darwin (1859), p. 117.

"disgraceful blunder" had a silver lining. His Principle of Divergence brought together several key strands of thought: variation, struggle, ecological division of labor, extinction, classification, and diversity in small uniform areas. Selection favors *distinctness* among variant forms of a given species, since by virtue of that distinctness those variant forms can more stably coexist. It is a *de facto* ecological division of labor, stemming from a selection dynamic that promotes branching and rebranching of lineages over time. That is, the Tree of Life grows and spreads not simply through natural selection, but through natural selection's divergence dynamic that forces branching.

Darwin's Principle of Divergence is a concept as sobering as it is enthralling. The reality of it is that destruction and extinction are the inevitable flip sides to growth, adaptation, and diversification. It must have been jarring to him though, realizing that it wasn't simply that larger, growing groups take the place of smaller, declining ones. No, more poignantly the growing groups *displace* the declining groups— the offspring varieties and species drive their parents to extinction. He well knew how *that* would be received. But it is what it is, he recognized; there was no going back or softening the blow. Most people then, as now, could only "behold the face of nature bright with gladness," as he put it in *Origin*. "Nothing is easier than to admit in words the truth of the universal struggle for life, or more difficult . . . than constantly to bear this conclusion in mind." Failing this, he cautioned, "the whole economy of nature, with every fact on distribution, rarity, abundance, extinction, and variation, will be dimly seen or quite misunderstood."[33]

Paley, Darwin's sometime hero and fellow son of Christ's College, would have been appalled. But in the spirit of Cardinal Cesare Baronio's admission to Galileo that "The Bible teaches us how to go to heaven, not how the heavens go," Darwin held true to his conviction that scientific truths must be followed however unwelcome or uncomfortable they may be to received wisdom, or presumed scriptural truth. So it was that Darwin's taste for field botany and little experimental plots helped introduce principles that eventually became central concepts of the science of ecology, namely, competition, competitive exclusion, niche partitioning, and even character displacement (mutual adaptation by divergence in morphology between two species with similar ecological needs that overlap in their ranges). In a case of giants standing on giants' shoulders, these concepts were elaborated and refined by several twentieth-century pioneers of mathematical ecology, including Charles Elton, G. Evelyn Hutchinson, Robert MacArthur, and others.

British-born Yale ecologist Hutchinson in particular brought fresh perspectives both empirical and theoretical to Darwin's dual interests in competition and diversification. In his classic 1959 paper "Homage to Santa Rosalia, or why are there so many kinds of animals?"—given, fittingly enough, on the occasion of his presidential address to the American Society of Naturalists 100 years after the reading of the Darwin and Wallace papers at the Linnean Society—Hutchinson refined the niche concept, now called the "Hutchinsonian niche," and stimulated interest in the central role of energy flow in food chains and food webs, and how these relate to community stability and the maintenance of biodiversity. Darwin would have appreciated how Hutchinson's insights, like his own, had come from extrapolating the local to the global, in this case observations made at Lindley Pond, not far from the Yale campus, and a small pond on Mount Pellegrino in Sicily where the Shrine of Santa Rosalia is found. And, given Darwin's sense of humor, he would have been tickled to know that this renowned ecologist who helped extend and formalize Darwin's nascent "ecological" ideas had had an inauspicious first encounter with the Darwin family as a youth growing up in Cambridge, where his father was a lecturer in mineralogy. Hutchinson, then 11 years old, and his 9-year-old brother Leslie played a prank on Darwin's son George (by then Sir George) and his wife when they came for a dinner party one evening. The impish boys locked their parents and distinguished guests in the dining room, tossing the keys into the garden and cutting off the dining room lights from the main breaker. If that was an "experiment" to see what would happen the results were likely less than happy ("The subsequent parts of the story," Hutchinson concluded, "are repressed"), and the same was true of at least one other ill conceived "experiment" of Hutchinson's youth: he pushed his friend Christina into a pond to see if she would float ("an early interest in the hydrodynamics of organisms of which I am not proud."[34]). Thank goodness Hutchinson turned his energies to pursuing the power of Darwinian experimentising to untangle the bewilderingly tangled bank of nature, building upon Darwin to found the modern field of evolutionary ecology in the process.

Experimentising: A Taste for Botany

"I believe a leaf of grass is no less than the journey-work of the stars." Those immortal words in *Leaves of Grass* by Walt Whitman were published just a year before Darwin established his Lawn Plot experiment. Darwin would have agreed, and saw even greater significance in a whole patch of grass blades. To him a little plot of lawn might as well have been all of teeming Amazonia: the same principles of competition, selection, and adaptation applied no matter how small the land area.

I. Darwin's Lawn Plot

We'll make similar observations in our own little plots, whether with lawn or some other ground cover. It may be difficult to fully replicate Darwin's experiment, insofar as a history of mowing was important for his lawn plots. This is because he thought that if plant growth is limited by grazing or mowing, a greater diversity will result, whereas unchecked growth will eventually result in some plants outcompeting the others, lowering overall diversity. If you know your plot's mowing history, your experiment will be more comparable to Darwin's. Regardless, you can learn much by looking closely at the number and frequency of species found in a small area of uniform habitat.

A. Materials
- Notebook and pencil
- Tape measure
- String or cord
- Garden stakes or sticks (4 per plot, to mark corners)
- Basic plant identification book, such as the *Weeds* and *Wildflowers* Golden Guides (St. Martin's Press, New York), the *Peterson's Field Guide to Wildflowers*, or online plant identification guides like eNature, searchable by geographical area (www.enature.com/fieldguides).

B. Procedure

1. Measure a 3 × 4-ft (0.9 × 1.2-m) plot in a grassy or weedy area, ideally untreated with pesticides or fertilizer, marking the corner of each with a garden stake and fencing off with the string or cord (attaching or looping around the stakes). You now have a "small and uniform" plot, just like the one Darwin established in his back yard in March of 1856. (Make sure you contact anyone who mows this area to let them know to avoid your plot.)

2. Every 2–3 weeks, census your plot by identifying its plants and counting how many of each species there are. Identify plants as best you can to genus or species. This can be difficult for plants lacking flowers, and for some groups like grasses, even when you do have the flowers. In a pinch simply identify to "morphospecies," or "kinds" based on obviously different morphology. If you are able to identify to genus or species it will be easy to determine the taxonomic family or order to which the plant belongs. For example, sites such as Wikipedia (en.wikipedia.org), Tropicos (www.tropicos.org), and efloras (www.efloras.org) list the full classification for plant genera or species. It might be best to tag or label your plants to help keep track of what you have identified and what's new at your next census.

3. Compare your results with Darwin's: over the course of the growing season he found 20 species belonging to 18 genera and 8 orders in his lawn plot. Is your plot as diverse as his, or more or less so? Your findings will likely differ from Darwin's, maybe

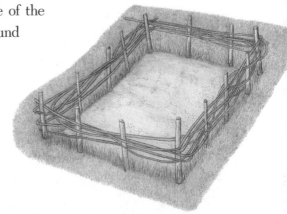

Sketch of Darwin's "Lawn Plot."

dramatically. Since many factors can influence local diversity (soil characteristics, local climate, time since last site disturbance, etc.) it would be difficult to say *why* your results are similar or different from Darwin's. In fact, even replications at Down House, where Darwin did the experiment, give varying results (see McLauchlin below). Such are the vagaries of ecology.

II. Darwin's Weed Garden Experiment

Darwin recognized the difficulty in conveying to the average person the organisms' ceaseless struggle for existence and the creative role this plays through selection. Comprehending this point is all the more difficult because struggle is experienced in different ways for different organisms, seasons, times of day, life stages, location, and many other factors. When we observe nature we often miss the struggle, seeing only peace and harmony, and mistake this for the natural condition of the living world. Darwin's aim in the Weed Garden Experiment was to focus on one group of organisms (plants) at just one life stage (seedling). He thought if he could remove one set of pressures, namely crowding by other plants, he could identify the other destructive forces acting on them (e.g., frost, drought, insects, slugs) and measure their cumulative effects over time. His experiment—a pioneering investigation in population ecology—is easily replicated. See if you can measure seedling mortality and compare your results with his.

A. Materials
- Notebook and pencil
- Tape measure and ruler
- String or cord
- Garden stakes or sticks (4 per plot)
- Chicken wire or aluminum flashing, to erect a barrier around the plot 1 ft (~ 0.3 m) high
- Wood dowels 0.125 in. (~3 mm) in diameter, galvanized wire, or twist ties in 4 in. (~10 cm) lengths

- Permanent marker
- Scissors
- Paper, cut into ~½ in. (1.3 cm) squares

B. Procedure

1. In late winter measure one or more 3 × 2-ft (0.9 × 0.6-m) plots, marking the corners with a garden stake or stick and marking the perimeter with string or cord (attaching the cord to the stakes or looping around them).
2. Clear the plot by removing all existing plants; dig the soil as necessary to remove the roots or rhizomes of any perennials.
3. Place the chicken wire around the plot, embedding it into the ground as best you can, as a barrier to protect the area from animals digging or trampling the emerging seedlings.
4. Check the plot on a regular schedule, at least every 2 or 3 days. As seedlings first appear, mark their location by inserting a dowel or wire bearing a number into the ground adjacent to the plant, taking care not to contact the seedling, and record the number in the notebook. Label each seedling individually using the paper squares numbered with indelible marker; use scissors or a sharp pencil to make a small slit or hole on opposite sides of each numbered square, and insert the dowel or wire through these to affix the square as above.

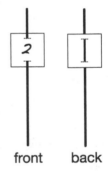

front   back

5. While you continue marking the position of each seedling as it appears, observe and note any evidence of activity or presence of slugs, snails, insects, birds, etc. that may harm the seedlings. At appropriate intervals, perhaps once or twice weekly, keep a log of the climate and soil conditions, noting average temperature, rainfall and moisture of soil. If, as the plants grow, it is possible to identify any of them, note that, too.
6. Census days: Once or twice monthly, approximately 2–4 weeks

apart depending on the schedule you settle upon, remove the numbered wires *without* a seedling, indicating that the seedling has been destroyed. Count and record the number of numbered wires. Record any notes about climate or other factors that may explain the numbers of lost and surviving plants.

C. Analysis

1. Following Darwin's lead, for each census day record (1) the number of seedlings killed since the previous count, (2) the number of surviving seedlings, and (3) the total number of seedlings that have emerged to date.

2. Use your data to produce a survivorship table, showing the percentage of seedlings surviving out of the cumulative number emerging over time. For example, on March 31, 1857, Darwin recorded that out of 55 seedlings total to that date, 30 survived and 25 had died. That gave a survivorship of 30/55 = 54 percent. At his June 1, 1857, census Darwin found just 80 surviving and 277 that had died, giving 80/357 = 22 percent survivorship. Survivorship in his experiment dropped to 19 percent by July 1st, and 17 percent by August 1st.

3. Graph a survivorship curve for all the seedlings in your sample, plotting percent surviving on the *y*-axis against census date on the *x*-axis.

See also:

K. E. James, "DNA Barcoding Darwin's Meadow," in *Darwin-Inspired Learning*, ed. M. J. Reiss, C. J. Boulter, and D. L. Sanders (Rotterdam: Sense Publishers, 2015), 257–270.

J. McLauchlin, "Charles Darwin's Lawn Plot Experiment," *The London Naturalist* 88 (2009), 107–113.

4

Buzzing Places

Insects always had a special fascination and charm for Darwin, especially beetles, his first love in natural history collecting. "Whenever I hear of the capture of rare beetles," he wrote to his friend and protégé John Lubbock, "I feel like an old war-horse at the sound of a trumpet."[1] But in the development of Darwin's evolutionary thinking, beetles played second fiddle to the Hymenoptera: ants, bees, wasps, and their ilk. Not for their diversity—this group is handily outdone in that department by beetles, the single largest taxon in the animal kingdom—but for their astonishing behaviors and even more astonishing life histories. Bumblebees first caught his attention, a group he knew as "humblebees." Who is not familiar with the buzz and hum of bumblebees as they make their floral rounds, so evocative of lazy summer days in garden and meadow? In September 1854—the very month he wrote about feeling like an old warhorse—these bees caught his son Georgy's attention while they were out one day. As Georgy, later Sir George, by then a distinguished astronomer, recalled many years later:

> I was once the unconscious means of making a discovery in natural history . . . When I was about 8 or 9 I was one day in August or September in the "Sand-walk" (so-called because at one time the path had been covered with red sand), & as my father paced

round the walk I waited by the old oak tree at the end. On his coming round I told him that there was a humble bee's nest in the tree. This he declared to be impossible, but I stuck to it that there was, & that the bees were going & coming from it. Accordingly we waited there & presently a bee came & buzzed about & went away, and then another and another. That there was no nest was obvious but the fact excited his curiosity & he determined to investigate it.[2]

Georgy had discovered a bumblebee "buzzing place." His father was delighted and puzzled in equal measure—he loved a natural history mystery. The "buzzing place" was a way station of sorts that the bees would briefly visit, dozens of them one after another. Darwin determined that the bees were male *Bombus hortorum*, the common garden bumblebee of Europe and Eurasia. He and Georgy started watching them, and soon discovered that this wasn't the only buzzing place—the bees flew well-established routes along the hedgerows and paths of the garden, stopping to buzz at certain spots along the way. Just where they came from, where they were going, and what their way-station buzzing was all about were puzzles. They also found that keeping up with bees on the wing is not easy. Field assistants were needed, and the rest of the kids were soon recruited—besides Georgy there were Willy (15), Etty (11), Bessy (7), Franky (6), and even little Lenny, just 4.

Over time they mapped 11 buzzing places over a distance of about 300 yards. In one investigation their father arranged them at intervals along the bees' flight path, each near a buzzing place, so they could follow them as far as possible and also observe the buzzing places closely. "The routes," he wrote later, "remain the same for a considerable time, and the buzzing places are fixed within an inch. I was able to prove this by stationing five or six of my children each close to a buzzing place, and telling the one farthest away to shout out 'here is a bee' as soon as one was buzzing around. The others followed this up, so that the same cry of 'here is a bee' was passed on from child to child without interruption until the bees reached the buzzing place where I myself was standing."[3] The insects were too fast to follow for long. To improve visibility Darwin tied a flour-dredger to the end of a stick, stationing himself at a buzzing place at one end of the bees' route, and giving each bee a liberal sprinkling as it appeared. As far as they could tell the bees were unfazed by being flour-coated. George remembered that they "could see the bee much further when he was whitened all over. We traced the bees over a line of 300 yds or so, & then they disappeared like lightening [*sic*] over the corner of the kitchen garden wall." Another time Darwin followed the bees into a dry ditch thick with brush, where he could just see the insects fly slowly along the ground between the thorn bushes. Too overgrown for an adult to get through, he marshaled the troops: "I could only follow them along this ditch by making several of my children crawl in, and lie on their tummies, but in this way I was able to track the bees for about twenty-five yards."[4] The kids didn't seem to mind; in fact, they found it a great adventure. To Lenny their buzzing place study seemed part game and part scientific investigation, and he recalled how, insofar as they were playing, "my father was like a boy amongst other boys."[5]

The mysteries were many: How did the bees recognize the buzzing places? And, since only new bumblebee queens overwinter, how did the bees of successive generations return to much the same buzzing places as the previous years' bees? Darwin kept field notes, recording his and the kids' observations over the course of three years. He noticed that the bumblebees seemed to prefer to fly along hedges and paths, and to buzz around the bases of trees, concluding that the same routes and the buzzing places must have some kind of attraction for them. He was at a loss to identify anything special about the buzzing places, though. He experimented with changing the appearance of some buzzing places by removing vegetation and sprinkling flour on the spot, yet still the bees knew where they were.

This suggested that the bees did not locate the buzzing places visually. Darwin and his young field assistants confirmed this with another experiment, this time by hiding a buzzing place with a net. Still, the bees stopped there: "Hoop-net . . . did not prevent bees coming . . . so not guided by vision," he jotted in his notes.[6]

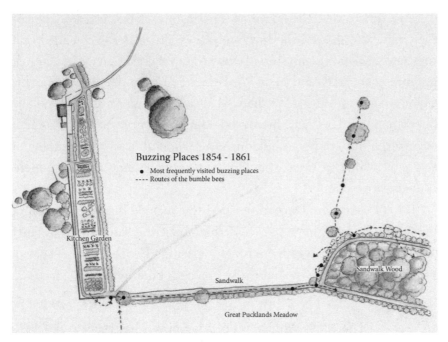

Sketch of the routes of male bumblebees and their buzzing places at Down House, 1854–1861. Adapted from Freeman (1968), p. 180 by Leslie C. Costa.

What was going on? Darwin never did figure it out. The bees were using markers that naturalists only really cottoned onto in the twentieth century: pheromones, airborne chemical cues secreted from specialized glands. He didn't realize how close he was to putting his finger on it, though. In his field notes from July 1856 Darwin wondered: "How on earth do bees coming separately out of nest discover same place, is it like dogs at corner-stones?" Dogs, of course, mark their territory with urine—using scent to say something to other dogs. Later study showed that the bees, too, use chemical cues, marking their buzzing places with secretions from their mandibular glands. The bees are not being territorial, however. The fact that the patrolling and buzzing bees were all males might have been a clue to Darwin that this had something to do with mate-finding. The amorous males scent-mark prominent objects or sites (to them, anyway) along set paths they patrol, hoping to attract new queens. Different species establish routes at different heights. Darwin and his children were fortunate that the common garden bumblebee is a low flier or they may have missed the buzzing places altogether. No matter that they never solved the mystery. Lenny (then Major Leonard, sometime military man, MP, and economist) recalled how deeply he and his siblings were impressed by their father's keenness to understand the phenomenon. "The fact," he mused, "that it never even for a moment occurred to him to try to hide his entire ignorance of the meaning of the habit which we were watching was probably not without beneficial educative effects."[7]

It's not clear why Darwin decided to conclude his investigation of bumblebee buzzing places, except that about the same time, in early 1858, his attention was increasingly consumed with *another* sort of buzzing place: bee hives, and the waxen cells built by bees.

The Most Wonderful of All Known Instincts

In early 1858 Darwin was poring over experiments, diagrams, and calculations relating to the cells constructed by honeybees and related bees and wasps. It was his latest hobbyhorse, and he was utterly consumed. Why honeybees? Insofar as honeybees were thought to construct their cells and comb with the accuracy of an architect equipped with the latest precision tools, they had long been celebrated as the height of instinctive intelligence in animals. They thus posed a special challenge, because to many they defied naturalistic explanation. These insects were already animals with a powerful mystique and great symbolic significance. They were exemplars of perfect monarchy or commonwealth to political philosophers, models of industriousness in folklore, an irresistible mascot for communities emphasizing the collective spirit over the individual. Beyond that they were hailed as living proof of divine design.

Rev. William Kirby, vicar and entomologist of Kirby and Spence fame (the authoritative textbook of entomology Darwin knew backward and forward), described honeybees as "those Heaven-instructed mathematicians, who before any geometer could calculate under what form a cell would occupy the least space without diminishing its capacity, and before any chemist existed to discover how wax might be elaborated from vegetable sweets, instructed by the Fountain of Wisdom, had built their hexagonal cells of that pure material."[8] A similar thought came from the pen of Henry Lord Brougham, barrister, politician, and intellectual, who wrote *Dissertations on Subjects of Science Connected with Natural Theology* in 1839. He made an argument about bees reminiscent of Paley's watch and watchmaker: the kind of "disciplined labor" seen in bees is only found in people under the direction of a "superintendent with a design." The bees, too, must therefore be guided by such a superintendent. Darwin made many marginal notes in his copy of *Dissertations*, scribbling on one page that bees' cells were "very wonderful—it is as wonderful in the mind as certain adaptations in the body—the eye for instance, if my theory explains one it may explain other." Indeed, to natural theologians the constructions of bees were the behavioral equivalent of the eye. Darwin well knew that William Paley had written that the "examination of the eye [is] a cure for atheism," quoting Saint Sturm, a medieval missionary. If he could not explain such wonderful structures as the eye or honeybee comb in light of natural selection, Darwin knew his theory would be seen as fatally flawed.

Honeybees (*Apis mellifera*) and their relatives are remarkable in their abilities, including in ways that Darwin and his contemporaries never knew. These insects exhibit one of the most complex forms of social behavior found in nature, each colony essentially a large family headed by the queen. Only she reproduces, her myriad offspring functionally sterile and laboring to assist their mother to rear more siblings. The family is mainly female: males are transient and are produced infrequently. The life of the honeybee revolves around the honeycomb, each consisting of two layers of cells arranged back-to-back, opening on opposite sides. Fashioned largely of wax secreted by special glands on the underside of the female bees' abdomen, honeycomb is both nursery and larder: some cells are used to rear brood while others become honeypots. An average hive can easily consist of 100,000 cells or more, often distributed among multiple combs. A hive of that size can take well over

2 pounds (1 kilogram) of wax to build, providing space enough to rear thousands of bees and store 44 pounds (20 kilograms) or more of honey (a substance that, as entomologists like to point out, may be considered the world's first stored-food product). Honey is the colony's winter food supply, made from flower nectar modified by the bees' salivary secretions and evaporation—but it's best not to think about eating bee spit as you enjoy a dollop with toast or tea. Protein-rich pollen is mainly used to rear the brood, fashioned into "bee bread" by mixing it with honey.

It is difficult to overestimate the powerful allure of these bees and the irresistibility of seeing them through a human lens, like engineers or architects setting out to solve a building problem. But the bees have no abstract concept of a design or a problem, let alone an ability to cogitate on solutions.

A Recondite Problem

In the lead-up to the events of 1858, Darwin was busy as a bee. Home improvements were on the agenda in 1857, building a new dining room with a bedroom above to accommodate the needs of a growing family. Worries continued, too, about Etty's health. She was periodically whisked off to the coast where her fretful parents hoped the sea air would do her some good. In research, Darwin oversaw his Lawn Plot and Weed Garden Experiments that year, observed the diversity sheltered within heathland enclosures, and computed (and frantically recomputed) his botanical arithmetic. But of course there was more—he also experimented with crossing chicken breeds to study their coloration patterns, documented the striping patterns of horses, and continued experiments with floating dried plants in saltwater. And through it all, he chipped steadily away at his big species book, *Natural Selection*—six chapters were completed in 1857, with the last of these, "Hybridism," finished on December 29th. The family had no sooner rung in the new year when Darwin had already jumped into the next chapter, on the "Mental powers and instincts of animals." That was when he became infatuated with bees.

In his earliest notebook jottings on honeybees Darwin only remarked on how bees must have an instinct for measurement, given the cells' geometrical properties. That instinct now became an urgent issue for him. All through 1857 he had collected random bits of information on cells built by different kinds of bee, but with the initiation of his chapter on instinct in 1858 he started to think more deeply about cells and comb—perfect networks of waxen amber-hued hexagons. These insects were said to "have practically solved a recondite problem," he later wrote. How did bees achieve that regularity, coordinate their building, and come to use hexagonal cells, a shape that happened to be *the* best possible one to maximize cell number per unit area? Hexagons are optimal for cell-packing, as geometricians since classical antiquity appreciated.[9] The problem for Darwin and his contemporaries was, how did the bees "know" this, if that's the right word? The natural theologians claimed that they were divinely inspired.

Darwin sought to demystify the cell-building behavior of these remarkable insects, knocking them off of their apian pedestal (or at least bring them down a notch or two—he was a great admirer of bees, after all). He soon came to a twofold approach. First, he sought to show that teamwork was not necessary per se to produce beehive cells. Instead, the action of multiple independent individuals, each attuned to its own work, resulted in multiple hexagonal cells (a process reminiscent of Adam Smith's "invisible hand"). Second, he had a hunch that by stepping back and looking at honey bees' structures in relation to those of other bees, he would find less complex nests that would look like transitional steps along the evolutionary pathway to the honeybees' perfect honeycomb (although not necessarily an actual evolutionary series). Darwin was sure that highly complex structures were not pinnacles of perfection, isolated and unreachable from the rest of nature, as natural theologians believed (and as their intellectual descendants, today's intelligent design fans, still believe). A transitional series pointed to a stepwise process, moving from the simple to the more complex.

When grooming a new "hobbyhorse" Darwin would read up on the subject, exhaustively scouring the literature and pumping any-

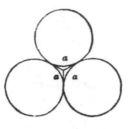 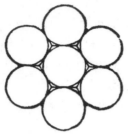

George Waterhouse's diagrams illustrating his hypothesis for the origin of hexagonal bees' cells. (Left) The cells start out cylindrical (a shape easily built by pivoting around a fixed point as they build). The bees then excavate the empty spaces between touching cylinders and flatten the walls where cylinders meet (a, a, and a). (Right) As this proceeds with other surrounding cells, the flat walls surrounding each cell collectively take on a hexagonal shape. From Waterhouse (1835), figs. 7 and 16.

one he knew for relevant information. He soon came across an 1835 encyclopedia article on bees by George Waterhouse, a naturalist and friend at the British Museum who had described Darwin's mammal and insect specimens from the *Beagle* voyage. Waterhouse gave him the seed of an idea: he thought that hexagonal bees' cells started out cylindrical, circular structures being easily made merely by pivoting around a fixed point. This is why circular or cylindrical structures are so common in the animal world, from birds' nests to burrows to the leaf discs excised by leaf-cutter bees and ants. "We observe," Waterhouse wrote, "that the works of almost all insects (perhaps we may say almost all animals) proceed in circles or segments of circles. The cells of almost all the various species of bees are of this construction."[10] And so, in working a mass of wax honey bees would "most probably" fashion cylinders. "Supposing," he continued, "the circumferences of three cylinders were to touch, the bees working in each of these cylinders would cut away the wax at the point where they meet." Because each cylinder would tend to be surrounded by six others (the only number of equal-sized circles that could fit symmetrically around one of the same diameter), as wax is removed from the spaces where each pair meets and flat walls are built between them, the cylinders would become hexagons. This is the simplified version: there was much more detailed math to the honeycomb's construction, and Waterhouse gave an elaborate discussion of how the rhombic bases of cells develop and how cells are fashioned in back-to-back layers into comb—far more detail than we need to get into here.

Darwin was intrigued. He wrote to Waterhouse for more information, including details of cell building by wasps and solitary bees. He thought his friend was basically correct, but that there was more to it. How did cell-building by the honeybee's relatives compare? What were the steps in cell construction? Did cells in the middle of the comb, surrounded by other cells, have the same shape as those on the edge? Was each cell built by just one bee? He had so much work to do, but could not get bees' cells out of his mind. In one almost stream-of-consciousness letter written to Hooker in late February 1858, he dipped into topic after topic like one of the foraging bees he loved to watch in the garden. Does Hooker know what bees are pollinators in New Zealand? Will he kindly answer the enclosed questions on classification? Can he spare volumes 10, 11, and 12 of Candolle's flora again? Are all the species of a genus classified together because they really are very similar, or because there are fewer differences between them and those of other genera? But he closed the letter with a comment revealing what was really on his mind: "Farewell, I have partly written this note to drive Bees-cells out of my head; for I am half mad on subject to try to make out some simple steps from which all the wondrous angles may result. Forgive your intolerable but affectionate friend." A week later he wrote Hooker again, thanking him for the botanical books and other information, and declaring that lately he had been "beyond measure interested on the construction instinct of the Hive bee."[11]

He wasn't the only one. William Tegetmeier (1816–1912), editor of *The Field, the Farm, the Garden: The Country Gentleman's Newspaper*, was an authority on poultry and fancy pigeons, subjects on which he and Darwin had corresponded for a few years. Darwin was surprised to discover that his correspondent was also something of a bee fancier. Tegetmeier regularly exhibited new bee hive designs at Entomological Society meetings, and came up with the idea of giving bees blocks of wax in various configurations to better understand their building behavior. Tegetmeier's ingenuity would prove to be helpful to Darwin, but for now he was still struggling to master the geometry of bees' cells. His brother tried to help—for all of his idleness Erasmus certainly had a flair for mathematics. But Charles, who candidly lamented that his "noddle is not capacious enough to retain or comprehend Mathematics," may have found that 'Ras obscured rather than clarified the matter by sending him calculations like this: "The obtuse $\angle$ at c is in the plane of the paper [therefore sign] Kc = radius of $\odot$ passing thro' obtuse $\angle$s Kb = 12 distance of centers = distance from center of sphere to rhomb."[12] Whew! Darwin may have been feeling more obtuse than the angles here.

He fared better with his old friend William Hallowes Miller, a crystallographer and professor of mineralogy at Cambridge. Who better to help him understand bee geometry, "the most difficult of all instincts to comprehend on my theory," as he lamented to Hooker, than someone whose stock-in-trade is planes and angles? Darwin left a long list of questions with Miller, and after a false start his friend got back to him with encouraging results. Miller even sent him instructions for constructing a paper model of a bees' cell based on a rhombic dodecahedron—a 12-sided polygon where each facet is a 4-sided rhomb. If certain axes of this dodecahedron were elongated, Miller said, the structure would take the familiar form of a bee's cell.

Darwin soldiered on. In the weeks before Wallace's Ternate essay arrived at Down House he oscillated between triumph and despair. In early May he declared to his eldest son Willy, then staying at his tutor's to prepare for his entrance exams for Cambridge, that he thought he had "settled the theory of Bees-cells." A couple of weeks later, however, he wrote again saying that he had "come to heavy grief" over bees' cells. The cause was Swiss naturalist François Huber, author of *Nouvelles Observations sur les Abeilles* (*New Observations on Bees*), first published in 1792. In a later edition, Darwin read that Huber, who made significant contributions to the study of bees despite being blind from an early age, suggested that the walls of certain cells were already angular before other adjoining cells were constructed. Alarmed, Darwin wrote in the margin "fatal to my theory." His only hope was that Huber was wrong.

Darwin would have to see for himself, using "a grand observatory Hive" lent by a neighbor. "I am going to watch," he declared to Willy, "for I cannot bear throwing up all my work." That's the kind of stick-to-itiveness so typical of Darwin. His friend John Innes, the local vicar, helped out, and barely a week later he had thrown himself into apiculture with gusto, making quite an impression on Emma: "Only think how bold Papa has become," she wrote Willy, marveling how "he hived a swarm of bees all himself . . . he & Mr Innes go about wonderful figures in their bee dresses with white veils on their hats."

This was one area of investigation where Darwin was playing catch-up, unlike so many research topics where he was in the vanguard from the get-go. Beekeeping was an ancient art form that was then experiencing much innovation in hive designs, methods of honey extraction, and other husbandry techniques. It was difficult to become an apiculturist overnight, and though Darwin tried in his usual ebullient hands-on way he also relied heavily upon the assistance and advice of friends more expert than himself, especially Waterhouse, Innes, and Tegetmeier. About the same time he was donning bee suits with Innes in early June 1858, Tegetmeier suggested giving the bees unworked wax as a way to observe the earliest stages of cell construction. If these start out as concavities, little bowl-shaped impressions like you would get if you pressed the curved bottom of a test tube into soft wax, that would be consistent with his and Darwin's conviction that the bees initially construct *cylindrical* cells that are subsequently modified into hexagons. Darwin put it to the test.

It was a success! Darwin wrote Tegetmeier with the news: "I have got some excavated hemispherical bases in artificial wax—hurrah! I thank you cordially for this capital suggestion." Weeks later he was caught up in the turmoil over his sick child, 2-year-old Charles Waring, who had contracted and would die from scarlet fever, and Wallace's manuscript, which looked like it would scoop his life's work. Thinking about his bees may have helped keep him anchored. Tegetmeier too was always experimentising, and reported his latest results in a paper at the July 5, 1858, meeting of the Entomological Society of London. He also displayed his latest observation hive, consisting of three glass plates per side. When he gave the bees a bar of wax to work, they fashioned cylindrical cells with concave bases. As new cells were added at the comb's edge, he found that the outer edges of the original ones, now surrounded by other cells, were transformed into hexagons.

One would think that this settled the matter, but some naturalists apparently had much invested in the belief that bees constructed perfect hexagons from the start, and, more to the point, that this "fact" proved the bees were divinely inspired. John Edward Gray, the Keeper of Zoology at the British Museum and president of the Society, spoke up to express his impatience with that view: "the attempt made by Natural Theologians to prove that the formation of an hexagonal rather than a cylindrical cell indicated the possession of a greater degree of Divine wisdom bestowed upon the insect, was the greatest piece of humbug they had ever brought forward."[13] Hymenopterist Frederick Smith of the British Museum, about whom we will learn more later, said he was not prepared to argue one way or another, but pointed out that when he tried to convert paper cylinders to hexagons he was not successful, and that some wasps construct hexagons from scratch, rather like building up a hexagonal chimney bottom to top. That was when Tegetmeier announced his next round of experiments, this time using colored wax to trace precisely how and where the bees excavate and what they do with the excavated wax.

It was an inspired idea, and inspiring: Darwin soon seized on the technique. Alkanet was the coloring agent of choice, a popular source of red dye since antiquity. Made from plant roots, it puts red in rouge, lipstick, and other cosmetics; wood stains ("rosewood"); and even foods like the curry Rogan Josh, which takes its name from the Indian term for alkanet, *Ratan Jot*. That summer Darwin and Tegetmeier gave the bees blocks of wax with a thin red layer, a marker to see what the bees were doing as they built their cells.

Accidental Architects

In early September 1858 Tegetmeier visited Darwin to talk bees and compare notes. Darwin was happy to let his friend take the lead on reporting their findings, rather than his usual approach of publishing himself in "letter to the editor" fashion. He was distracted, grieving the loss of his son in late June and in the throes of deciding whether to abandon his "big species book" *Natural Selection* and, if he did, the best format to make his theory public and preserve his priority in the wake of Wallace's unwelcome essay. Later that month Tegetmeier gave a paper at the British Association meeting in Leeds, describing his and Darwin's experiments with red-colored wax. Their experiments, repeated with various modifications, showed "that the excavations were in all cases hemisphaerical, that the wax excavated was always used to raise the walls of the cells, and that the cells themselves, before others were formed in contact with them, were always cylindrical."

Tegetmeier's idea of using red-tinted wax as a tracer of the bees' activity was just the kind of simple and effective experimental approach that Darwin loved. He put the red-wax technique to good use in several experiments. In one he gave the bees a thick wax block coated with just a thin layer of red wax. This revealed that the first thing the bees do as they make new comb is to excavate small circular pits. As they deepen and widen these pits they turn into shallow bowls or cup-shaped depressions about the diameter of a honeycomb cell. When the excavations by adjacent bees intersected, the bees began to erect a flat wall between the growing cells. But then Darwin noticed something really interesting: in this single-layer comb, hexagonal walls went atop the *circular* edge of the depressions, with no sign of the usual three-sided pyramidal bases of double-sided comb. This was because the wax block was so thick it was not possible for the bees to build back-to-back comb in their usual way. However, when he put a thinner strip of red wax in the hive the bees started excavating pits on both sides. They do not line up the cells exactly opposite one another, however. Rather, the cells are offset, such that a given cell on one side sits at the intersection of three cells on the other side. Each cell shares a flat wall with each of three cells opposite—thus terminating at the base in a three-sided pyramid rather than a bowl. This makes the honeycomb stronger than an arrangement with cells directly back-to-back, since each cell is backed by walls of three opposite cells.

In this experiment, however, the strip of wax Darwin gave the bees was so thin that they only worked it to a limited extent. The bees were on their way to producing pyramidal bottoms of cells, and different degrees of this development were in evidence at the points where they ceased to excavate, before they could break into another cell. In yet another experiment, Darwin covered edges of cells with a thin layer of red-colored wax. "I invariably found," he wrote, "that the colour was most delicately diffused by the bees by atoms of the coloured wax having been taken from the spot on which it had been placed, and worked into the growing edges of cells all around." In other words, the bees were continually working and reworking cells, taking a bit of wax here and adding a bit there as the comb grew. Trial and error, constant improvement. In arguing that the bees' cells are not perfect, Darwin would have been intrigued to learn that some 80 years later mathematicians discovered a hexagonal cell shape with a prismatic four-sided base that is more efficient at cell-packing than the honeybee's pyramidal-ended cells.[14]

By the time Tegetmeier's paper was published early in 1859 Darwin had settled on the format he would use for explaining his theory in full: a concise book—concise relative to the length *Natural Selection* would have been, anyway. After some deliberation he dubbed it *On the Origin of Species*. There, in chapter 7, he summarized his experiments with Tegetmeier, satisfied that the hexagonal cells of honeybees emerged from properties of the placement and elaboration of initially circular cells. "Grant whatever instincts you please," he wrote in the *Origin*, "and it seems at first quite inconceivable how they can make all the necessary angles and planes, or even perceive when they are correctly made. But the difficulty is not nearly so great as it at first appears: all this beautiful work can be shown, I think, to follow from a few very simple instincts." Modern biologists would agree. Honeycomb likely

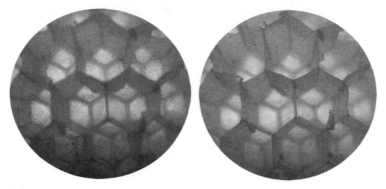

Detail showing the three-dimensional structure of honeycomb. (Left) Three-sided pyramidal bases of cells, each face of which is also one face of the base of each of three opposite cells. (Right) Hexagonal cells, showing 120-degree angles between walls. Photographs by the author.

takes on its distinctive form through context-dependent rules followed by the bees during construction, in combination with the physical properties of wax and cell shape. The intriguing hypothesis of "stigmergy" postulates that properties of the nest structure *itself* triggers certain construction rules in insect builders, a form of self-organization where each step in construction, simple behaviors in and of themselves, stimulates the next step. This fascinating idea, applied most successfully to the construction of wasp and termite nests, is one expression of the "extended phenotype" concept articulated by the Oxford evolutionary biologist Richard Dawkins. Here the comb is an extension of the bees themselves. Insofar as the physical structure affects survival and reproductive success, and is constructed in whole or in part through heritable genetic traits, natural selection can act upon it. And, the simplest way for selection to alter nest architecture is by altering those simple behavioral rules that worker bees follow in cell construction—in the same way that a slight deviation from a path of travel, like a simple fork in the road, can lead to dramatically different destinations, small changes in construction rules early on can lead to dramatically different structural outcomes. That's where the bigger picture comes in, as Darwin appreciated. He wondered if there was evidence that modification of simple behavioral rules of construction can lead to this "most wonderful of all known instincts?" He thought so.

Ask the Relatives

Even before throwing himself into honeycomb geometry Darwin started collecting as many examples of bees' handiwork as he could—characteristically, by writing to contacts near and far imploring them to send him specimens. Just as individual bees built their hexagonal cells in stages beginning with circles and cylinders, so too, Darwin thought, did this stepwise process reflect the likely evolutionary trajectory leading to the exquisitely complex honeybees. In something of a behavioral version of von Baer's "developmental law," where earlier developmental stages were thought to reflect ancestral states of a species, he was sure that the incipient, cylindrical, stage of honeybees' hexagonal cells meant that ancestors of these bees built solely spherical or cylindrical cells. Honeybees thus represented a more recent, advanced, state of cell development. Why would selection have favored this? The idea was that in the struggle for existence, selection acting on the ancestors of honeybees to improve reproductive success led to maximizing the number of cells per unit area while also economizing on the quantity of wax needed for those cells. More cells potentially mean more brood that can be reared in a given period of time, and more storage space for honey (vital for getting the colony through the winter). And, crucially, the hexagonal shape requires less wax.

He asked Tegetmeier what wax "costs" the bees to produce, in terms of amount of energy. Their currency is food, mainly sugar in the form of the honey they produce. His friend calculated that 15 pounds (6.8 kg) of sugar was needed for every pound (0.45 kg) of wax, probably a rough calculation based on how much he fed his colonies per season and how much they grew on average in that time. Our current estimate is that it takes about 0.2 ounce (6 g) of honey for the bees to synthesize 0.04 ounce (1 g) of wax. To try to understand what that means for the energy budget of a colony, here's how much effort it takes for one bee to make 6 g of honey. Each flower yields a tiny amount of nectar. On average a bee's crop can store about 30 mg of nectar at a time, so besides the multiple floral visits to fill up its crop, it takes more than 200 crops-full of nectar just to condense down and process to 6 g of honey. The hexagonal cell means less wax needed and more energy saved—energy that could be put into brood production and nest hygiene, and that much more food put up for the winter. It's worth pointing out that overwintering as honeybees do is unusual for insects: rather than going dormant, these insects create a warm microenvironment by metabolically burning honey for heat production. It's an expensive lifestyle (rather like our own).

Darwin was sure that natural selection acted on honeybee ancestors to favor the evolution of hexagonal cells to save the bees energy. He thus needed to show that simpler, less efficient, transitional forms were possible. He looked to "collateral descendants"—related species—to see if gradations are found. He began with his old friends the bumblebees.

Bumblebees are taxonomically of a different genus (*Bombus*) than honeybees, but are from the same tribe (Apini). Like honeybees, they are social, but differ in important ways. Bumblebee colonies last only a year, and the bees typically nest in the ground. And, as Darwin knew, they build more or less spheroidal waxen cells of various sizes that are randomly clumped together, in which they raise their brood and store honey. Bumblebees are the perfect contrast with their geometrician relatives, but Darwin needed to find species intermediate between the two. He was excited to learn about the so-called "Mexican bee," *Melipona domestica,* which make a rudimentary comb of spherical to cylindrical cells, with flat walls where they intersect. Now dubbed *M. beecheii* and placed in the same family (Apidae) as honeybees and bumblebees but a different subfamily (the Meliponinae), this stingless bee was semidomesticated centuries if not thousands of years ago by the indigenous people of Mexico, where it is known as *Xunan-kab,* or Royal Mayan bee. Darwin came across an account of this bee in a paper by Pierre Huber, son of the Swiss apidologist François Huber, whom we met earlier in this chapter. The younger Huber, a distinguished entomologist like his father, published a paper with wonderfully detailed plates illustrating the bees' nests and the unique "hives" created by the indigenous Mexicans to maintain colonies. They basically took a hollowed log or branch, stoppered both ends, and hung it from trees or

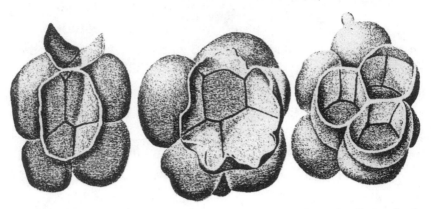

Three examples of *Melipona beecheii* cells dissected to show faceted walls that develop where cells intersect. From Plate III of Pierre Huber's paper "Notice sur la Mélipone domestique, abeille domestique Mexicaine," published in 1836.

the eaves of their dwellings. Darwin was more excited by the cutaway drawings of the cells, showing spheroid cells stuck together in a rudimentary comb, with curved walls except where they intersect other such cells. There the walls were made flat by the bees, inevitably taking on a multifaceted form where several came into contact. It was the very intermediate arrangement that Darwin sought: placing their cells next to one another allowed wall-sharing between adjacent cells, saving wax relative to the independent cells made by bumblebees.

He argued as much in the *Origin*: "it occurred to me that if the Melipona had made its spheres at some given distance from each other, and had made them of equal sizes and had arranged them symmetrically in a double layer, the resulting structure would probably have been as perfect as the comb of the hive-bee." To back this up he sent some calculations to Prof. Miller, his mathematician friend, who concurred. In the *Origin* Darwin describes his and Tegetmeier's experiments, and finishes with a stirring summary of hexagonal cell evolution: "Thus, as I believe, the most wonderful of all known instincts, that of the hive-bee, can be explained by natural selection having taken advantage of numerous, successive, slight modifications of simpler instincts." Natural selection worked by "slow degrees, more and more perfectly" to lead the bees to build more regularly:

That individual swarm which wasted least honey in the secretion of wax, having succeeded best, and having transmitted by inheritance its newly acquired economical instinct to new swarms, which in their turn will have had the best chance of succeeding in the struggle for existence.[15]

Darwin didn't intend his bumblebee-to-*Melipona*-to-honeybee series to be taken literally. That is, he did not mean that bumblebees evolved into *Melipona* bees, which in turn gave rise to honeybees. Rather, the bumblebee and *Melipona* style of cell construction sheds light on what the cell-building behavior of the honeybee's ancestors must have been like, showing the evolutionary pathway to modern honeybees' cells. His point is that simplified intermediate stages must exist for even the most complex, intractable structures or behavioral adaptations. "A series," as he put it in regard to a different but related subject, "impresses the mind with the idea of an actual passage."[16]

Time-Out for Bad Behavior

Honeybee cell-building was just one of three cases Darwin explored as "difficulties" in understanding instinct. The other two, slave-making ants and cuckoos—both bizarre cases of brood parasitism—seemed to cry out for explanation. He found himself distracted from his honeybee work thinking about them, probably because they, too, held deeper implications. He would go about his work on them similarly to the way he found relatives of honeybees that had transitional behaviors. Call it Darwin's time-out to study some bad behavior.

The case of the cuckoos (*Cuculus canorus*) had been known for centuries. These birds, widespread in Europe, are obligate brood parasites incapable of rearing their own young, not even able to build a nest. (Their penchant for getting others to rear their young gave rise to the term "cuckold.") How could such an extreme form of parasitic behavior get its start? Piecing together information on the behavior of other

species in the cuckoo family (Cuculidae) in Europe and North America, Darwin found variation in degree and form of brood parasitism,
consistent with the idea of an evolutionary series. On one end of the
spectrum Darwin found opportunistic parasites occasionally practicing
"egg dumping" (laying eggs surreptitiously in the nests of other birds)—
having eggs in more than one basket, as it were, helps spread around
the effort and risk of brood-rearing, and boosts the parasite's reproductive success. This is termed "facultative brood parasitism" today, and
is found in North American yellow-billed cuckoos. These birds usually
rear their own young, but given a chance will also lay some of their
eggs in the nests of other yellow-billed cuckoos, or even of other species. Darwin noted that the European cuckoo is found at the other end
of the spectrum: it is a parasite not only wholly dependent upon hosts
for brood-rearing, but whose young also invariably kill the young of
the hosts and monopolize their care efforts. This is an extreme form
of obligate brood parasitism. Darwin suggested a scenario where distant ancestors of today's cuckoos first benefitted from the accidental
laying of eggs in the wrong nest. Selection may have then favored the
evolution of facultative brood parasitism, a kind of opportunism. This
in turn could have set the stage for the evolution of the extreme form
of brood parasitism found in European cuckoos, where the birds avoid
expending the time and energy of building their own nest and focus
completely on locating the nests of other birds to parasitize.

The other form of "bad behavior" that caught Darwin's attention
was the recently discovered phenomenon of slave-making ants. Like
the cuckoos, in its extreme form this behavior entails a parasite that is
completely dependent upon a host to rear young. Again, how to explain
the evolution of such obligate parasitic behavior as a gradual process?
Once again Darwin looked to "collateral relatives," making a case for
transitions based on what related species do. From the accidental parasite to the occasional or opportunistic (facultative parasitism) to the
extreme of complete dependency (obligate parasitism), he sought to
find natural variation in the behavior among related species as a possible evolutionary pathway.

Darwin knew something about slave-making ants from Kirby and Spence's *Introduction to Entomology*, first published in 1818. Many a person had witnessed the predatory raids of certain ants, assuming they were contests for territory or food stores, not comprehending what was actually happening. Henry David Thoreau, in 1846, witnessed a battle near

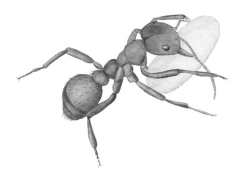

Formica sanguinea, a dulotic (slave-making) ant species studied by Darwin. Drawing by Leslie C. Costa.

his cabin on Walden Pond: "The legions of these Myrmidons covered all the hills and vales in my woodyard, and the ground was already strewn with the dead and dying," he wrote in *Walden*. Thoreau saw it as "internecine war, the red republicans on the one hand, and the black imperialists on the other," evoking the carnage of human conflict.[17]

Given Thoreau's strong antislavery sentiments he might have viewed the conflict in a different light had he read Kirby and Spence. These authors anticipated readers' incredulity at an astonishing form of insect behavior. What would the reader say when told that "certain ants are affirmed to sally forth from their nests on predatory expeditions, for the singular purpose of procuring *slaves* to employ in their domestic business?", they asked. Kirby and Spence credited Pierre Huber as "the discoverer of this almost incredible deviation of nature." Huber found that two ant species engage in raids on the nests of other ants in their vicinity: *Formica rufescens* and the appropriately named *F. sanguinea* (whose name means "bloody ant"—for their reddish coloration, but resonant with their behavior). These ants invade the nests of their quarry, killing off any adult ants that resist them and "ant-napping" the young, carrying the brood back to their own nest like booty-bearing pirates making for their lair. The fate of the brood is not, however, to become so many ant snacks; the raiders permit them to mature, and having imprinted (in the scientific sense) on their captors they believe themselves part of the group, and go about performing all of the work of the colony as if it were their own—and I mean *all* the work: nest building, cleaning, foraging, raising the brood of their captives, even defending to the death should the colony be threatened. Huber experimentally showed that *F. rufescens* was completely dependent upon the "slave" ants, so utterly incapable of even feeding themselves that they would die if deprived of their care. *F. sanguinea*, in contrast, could do some work, but not much. It appears to be an opportunistic (or facultative) slave-maker, engaging in the practice as opportunity knocks in contrast to the "obligate" slaver *F. rufescens*, which can't survive on its own.

The fact that early observers referred to this phenomenon as "slavery" speaks volumes: they could not help comparing it with human slavery, but the comparison is a poor one in several respects—not least the fact that these ant interactions are interspecific, while the human institution of slavery involves members of the same species. Nonetheless, the label stuck even in the scientific literature, although the preferred scientific term is *dulosis*. The inevitable comparison with human slavery resonated deeply with Darwin—he and his family were vehemently abolitionist, heirs to a proud family tradition of antislavery activism going back to his maternal grandfather Josiah Wedgwood I, whose famous abolitionist cameo brooch raised funds for the cause with its striking image of a manacled African and the resonant plea "Am I not a man and a brother?" Slavery was only abolished completely in Britain and its territories in 1833, but Darwin witnessed its brutality firsthand in Brazil while on the *Beagle* voyage. Darwin was not one to anthropomorphize nature, yet he could not resist referring to this predatory instinct of *Formica* as "odious."

One of the two species that Huber discovered as slave-makers was not found in Britain and the other one was rare, so naturalists there had little opportunity to study the phenomenon. Then Darwin read a paper by Frederick Smith, his entomological correspondent in the British Museum. Smith mentioned in his paper that *F. sanguinea* was locally abundant in Hampshire, a coastal county in southern England, where he had witnessed them launch a raid. Darwin wrote Smith in late February 1858 with a list of questions: in nonslavemaking ants, has Smith ever seen stray workers of one species inhabiting the nest of another species? No, Smith replied. Does *Formica sanguinea* "always & invariably make slaves," Darwin wondered, or only occasionally? Smith thought they always did—"I never myself met with a community without numbers of other species" in the nest. In regard to ants that tend to be "enslaved," do they feed their own queens in their own nests, or do the queens feed themselves with food brought by the workers? Smith was unsure about this.[18] Darwin also asked for pointers on identifying this species. He wanted to see a raid himself. Smith advised him that morning or evening was the best time to see a raid, especially in the summertime when vulnerable pupae were in the nest for the taking.

That April of 1858 Darwin headed south to Moor Park for a couple of weeks' treatment with Dr. Lane—perfect for relaxation. He wrote Emma that he fell asleep on the grass after one long walk, awaking to "a chorus of birds singing around me, & squirrels running up the trees & some Woodpeckers laughing, & it was as pleasant a rural scene as ever I saw, & I did not care one penny how any of the beasts or birds had been formed."[19] An idyll, but the question of the origin of the beasts and birds was never really far from his mind. The park was perfect for ant-watching, and Darwin described in another letter to Emma how he would "loiter for hours in the Park, & amuse myself by watching the Ants: I have great hopes I have found the rare Slave-making species."[20]

He sent specimens to Smith for confirmation. Darwin got to wondering if and how ants of different colonies recognized one another, and tried a simple transplantation experiment: what happens if an ant from one colony is plunked down among ants of another colony? Well, it didn't end well for the stranger, as his notes reveal: "I took several times some hill-ants (*F. rufa*) from their own nest & placed them on another; they were always extremely much agitated & were instantaneously attacked by the inhabitants: whereas when I returned several of the same lots to their own nest, they seemed immediately to recognise their comrades & be recognised by them."[21] In a letter to Willy from Moor Park he commented on this experiment, reporting that although thousands of ants inhabit each ant mound, "each seems to know all its comrades, for they pitch unmercifully into a stranger brought from another ant-hill."[22] Darwin did not do much more than this with his ants, but John Lubbock (who combined a facility for experimental entomology with the more traditional Lubbock family pursuits of banking and politics) took up the question, experimenting extensively with recognition and other forms of communication in ants in the 1880s (see Chapter 10). The key to understanding what's going on here, like the mystery of bumblebee buzzing places Darwin also never grasped, was chemical markers.

Darwin returned home in early May and got back to working on his bees. With all the ups and downs of bees' cells, ants were out of mind for a while. But following the tumultuous events of late June and early July 1858, he got another chance to look for slave-making ants when the family headed south for some rest. They stopped for a week in Hartfield to visit with Emma's sister, and then went on to Sandown on the Isle of Wight. From Hartfield he wrote a letter to his old friend Hooker, first thanking him for his part in what has been called the "delicate arrangement" over Wallace's manuscript, but also commenting that when he got to the Isle of Wight he would begin working on an abstract of his species book. The plan at that time was to write a long review paper, an unhappy prospect. "I will set to work at abstract," he wrote, "though how on earth I shall make anything of an abstract in 30 pages of Journal I know not, but will try my best." He also revealed a more interesting diversion, telling Hooker that he "had some fun here in watching a slave-making ant, for I could not help rather doubting the wonderful stories, but I have now seen a defeated marauding party, & I have seen a migration from one nest to another of the slave-makers, carrying their slaves . . . in their mouths."[23]

The family soon arrived on the Isle of Wight. He began on the abstract, but before long gave up the idea of a journal paper and eventually decided instead on writing what was to become *On the Origin of Species*. Exploring the sandy heaths and woodlands, he was rewarded with witnessing another myrmecine slave raid. He noted that "only the largest & medium sized were carrying pupae."[24] He continued to make observations at every opportunity over the next year, observations that were later reported in the *Origin*. How to understand the evolution of so extreme a lifestyle, one in which in some cases the marauding ants are so dependent on their quarry that they cannot even feed themselves? One key question was whether slave-making behavior varies between and within species. Over the course of two field seasons he observed that it did have a range. That meant selection could act upon the behavior, and as with bees' cells a transitional series was at least possible. "By what steps the instinct of *F. sanguinea* originated I will not pretend to conjecture," he wrote in the *Origin*.[25] But then he does: if a few ant pupae captured for food escaped this fate and, once emerged as adults, helped the colony instead, the host colony could benefit just enough relative to their competitors that selection might favor raiding for the sake of adding to their labor force and not their larder. Once this instinct is acquired, Darwin reasoned, natural selection could further modify it, step by useful step, even to the point of the ants being "abjectly dependent" on slaves. Thus he met his objective of coming up with a plausible scenario of a gradual evolutionary sequence for slave-making in ants, just as he did for the parasitic cuckoos. Modern biologists agree that Darwin's scenario for the evolution of dulosis is basically correct—it is an elaboration of "normal" foraging and predatory activity that involved capturing the young for food. But Darwin saw further lessons: "Finally, it may not be a logical deduction, but to my imagination it is far more satisfactory to look at such instincts as the young cuckoo ejecting its foster-brothers,—ants making slaves,—the larvae of ichneumonidae feeding within the live bodies of caterpillars,—not as specially endowed or created instincts, but as small consequences of one general law, leading to the advancement of all organic beings, namely, multiply, vary, let the strongest live and the weakest die."[26]

That last statement may have seemed jarring to readers, but writing as one who had felt the deep, unspeakable pain of losing a cherished child to illness, he would rather not attribute life's miseries—including those inflicted by parasitic species—to a benevolent deity, however much his vicar friends urged that there was a plan behind it all, that all was for the best. He couldn't accept that. An important implication, too, was that there was no slavery by divine design, running counter to

pro-slavery advocates who saw the occurrence of the "peculiar institution" in nature as justification.[27]

The fact that behavioral traits can be bred into domesticated varieties is an indication that some instincts or behaviors can be shaped by
selection in nature. Darwin was sure it was true throughout the animal
kingdom, including both the most "wonderful" and most "odious" of
behaviors, from the busy bees to the noxious cuckoos and slave-making
ants. He wanted that fact about species to be acknowledged widely: the
good, the bad, and the ugly alike arise through a gradual process of
evolution by natural selection.

Experimentising: Bees' Cells and Bubbles

Unless you're a beekeeper, or good friends with one, it won't be very
easy to experiment with cell building à la Darwin, but the next best
thing is dissecting some honeycomb. First, check out some hexagonal
cell-building online: in the BBC program *The Code* (www.bbc.co.uk/
code), Oxford mathematician Marcus du Sautoy explains why the hexagonal cell-building of bees is more space-efficient than the other two
polygons (squares and equilateral triangles) that fit together without
leaving gaps: www.youtube.com/watch?v=F5rWmGe0HBI.

I. Honeycomb Dissection

A. Materials

- Honeycomb (can be purchased online, and can be found packed
  with honey at some health foods stores and farmers' markets)
- Razor blade, craft knife, or sharp paring knife
- Tray, plate, or shallow dish

B. Procedure

1. Drain the honeycomb of honey as best you can over a bowl, to save as much honey as possible for enjoyment later. The remainder can be dissolved by letting the honeycomb sit in a bowl of luke-warm water, changing the water every 10 minutes, until most of the honey is gone. The comb is very delicate—handle with care.

2. Honeycomb is best studied in chunks, so you can see how multiple cells relate to one another. The comb consists of two back-to-back layers of cells. Note that the cells are not perfectly horizontal, but angle upwards slightly.

3. Hold the comb up to the light and note how each of the three facets of the pyramidal cell base is shared by one of the three cells opposite—in other words, how each cell on one side of the comb is centered on the intersection of three cells on the opposite side of the comb.

4. Try to dissect out a single cell or a pair of cells by cutting away cells on either side of the one or two cells you selected. This

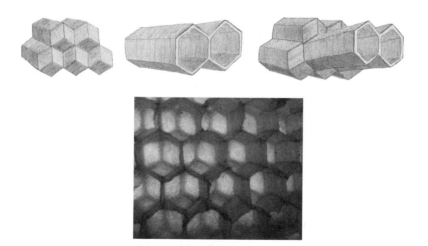

Hexagonal prism structure of honeybee cells, each of which terminates in a three-sided pyramid at the base as seen at upper left and the photograph. Note that cells on each side of the comb align with one facet of each of three cells on the side opposite as shown at top center and right. Drawing and photograph by Leslie C. Costa.

won't be easy, but isolating a cell or two will help you better discern cell shape: note the hexagonal prism shape along the long axis, and the three-sided rhombic pyramidal base. The shape of the base is important; each facet of the three-sided base serves as one of the facets for three cells opposite in the intact comb.

II. Exploring the Geometry of Bees' Cells by Analogy

Mathematician Joel Hass and colleagues may have experienced toil and trouble of Shakespearean proportions as they labored to provide a formal proof of the Double Bubble Conjecture, which held that two joined bubbles enclose volumes separated by the least possible surface area (see: en.wikipedia.org/wiki/Double_bubble_conjecture). The minimal area of the shared wall means the least amount of soap solution goes into it. When honeybees build a common wall between two adjacent cells, they too are conserving building material. The most economical state is a flat wall, hence the tendency of initially circular or cylindrical cells surrounded by six such cells to morph into a regular hexagon. The wall-sharing principle can be demonstrated with bubbles—a fun if messy way to explore bee cell geometry.

A. Materials
- Plastic drinking straw
- Plastic plate or lid of a large margarine container
- For an all-purpose bubble solution, mix together the following ingredients and let the solution sit at room temperature for a couple of hours:
  1. 2 cups (500 ml) water
  2. 2 cups (500 ml) Johnson's® baby shampoo
  3. 2 teaspoons (10 ml) glycerine*

  *Many recipes call for the addition of small amounts of glycerine or corn starch for added strength, but don't overdo it. Note, too, that hard (high mineral content) water makes bubble-making difficult. Try distilled water, available at many grocery stores, if you live in an area with hard water.

B. Procedure

1. Let's practice making bubbles. Pour a small amount of bubble solution onto the plastic plate or a plastic lid from a margarine container.

2. Gently blow into the straw as it touches the bubble solution on the plate to make bubbles. Remove the straw and continue to blow bubbles to form clustered bubbles.

3. Observe how bubbles in a cluster intersect, in particular the angles that form between shared surfaces.

4. Let's do this more methodically: starting again with a clear bubble-free surface, form two bubbles roughly the same size and gently bring them together and attempt to link them. The resulting double bubble should look something like the left-hand bubble figure (A) below, sharing a common flat wall. The closer the two bubbles are in size, the flatter the wall they share—that is the minimal surface area.

5. Next, attempt to merge a third bubble, about the same size as

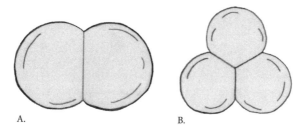

A.   B.

the first two, with the joined double bubble. The resulting triple bubble should look something like (B). The shared walls between the three bubbles will form a Y shape, and the angles between the bubble walls will be 120° (360° divided by 3). It may be easiest to measure them by photographing the triple bubble from the side and taking measurements of the angle of the shared walls from a printout of the photo. Why do the triple bubbles form an angle of 120° and not others? This is a result of each bubble exerting equal pressure such that the walls between them are rendered flat.

6. Try to add more bubbles to a cluster on a flat surface. They will rearrange themselves to create multiple 120° angles in the shape of a Y. A network of 120° angles creates a pattern of hexagons, like bees' cells—and for the same reason: this is the configuration that produces the most cells with the least amount of building material (in this case bubble solution) to form the walls.

5

A Grand Game of Chess

I t was early May of 1855, and Darwin was having a bad day. He had been at the Zoological Gardens in London, feeding a nice sampler of soaked millet, lettuce, cabbage, linseed, barley, and onion seeds to the goldfish. Things began promisingly enough: "they took them in mouth & kept them for some seconds," he jotted in his notebook. But it quickly became clear that these fish were not going to be willing participants in his study; after half a minute or so the fish forcibly spat out the seeds, and he could not tempt them to nibble further. Glumly, he wrote a woebegone letter to his cousin William Darwin Fox: "I am rather low today about all my experiments . . . all nature is perverse & will not do as I wish it, & just at present I wish I had the old Barnacles to work at & nothing new."[1]

What led to this curious scene, with the respectable middle-aged naturalist trying to force-feed goldfish? As usual there was a method to the apparent madness. Darwin was convinced that plants and animals were far more mobile than many of his colleagues realized, and he was devising all sorts of experiments to prove his point. His seed-eating fish scheme was just one of many.[2] He imagined that birds eating fish that in turn had just eaten seeds might turn out to be an unexpectedly common way that plants became dispersed far and wide. It seemed perfectly plausible that once a bird voided the remains of the fish with its (hopefully mostly undigested) seed cargo in some far-flung place—the distant shores of a remote island, or in a valley beyond a lofty mountain range—the bird would have performed a useful seed dispersal service: carried far and conveniently fertilized, to boot, the seeds just might germinate and establish a new population. The first step was to establish that fish would indeed eat seeds, but those pesky fish would have none of it.

Fortunately, not all of Darwin's dispersal investigations went awry. In the years leading up to the publication of the *Origin* he carried out an astonishing range of experiments and related studies bearing on the dispersal and movement of plants and animals. Explaining the geographical distribution of species was key to Darwin's evolutionary ideas, after all, and he puzzled over every aspect from modes of seed and snail dispersal to the ebb and flow of plant and animal populations as climate cycles over the eons. Ensconced in his study at Down House he nonetheless had an exhilarating global perspective—it was a "grand game of chess," he enthused, "with the world for a board."[3]

A Grand Subject

The question of geographical distribution was of the highest importance in Darwin's day. Naturalists knew that species distribution was not random—there were obvious differences among species on continents and even more in regions, but just what gave rise to that pattern? Beginning in the late eighteenth and early nineteenth centuries, naturalists had first wondered if the earth could be divided up into distinct centers of creation, judging by the richness of their species. Whatever the source of species, whether by special creation or some mysterious natural process, tropical regions and remote islands clearly seemed to be special—islands more than continents, because while the absolute number of species may be greater on continents, islands often have far greater endemism (unique species) and greater diversity at higher taxonomic levels like genus, family, and even order.

A related concern was the question of single versus multiple centers of creation; that is, did *all* species originate from *one* geographic locale from which they dispersed the world over, or could species arise in different areas, whether simultaneously or over time? Some early biogeographers trying to reconcile the puzzle of species distribution with their commitment to scripture (the Noachian flood in particular) favored the single-location idea: the Swedish naturalist Linnaeus, for example, suggested that Noah's ark came to rest on the peak of a towering mountain. When the waters receded and the area was completely exposed, there would be a range of climatic zones available for the Noachian menagerie, from alpine tundra and boreal forest through grasslands and lowland forest. Never mind the difficulties of the frigid conditions on the summit to begin with—that's only the tip of the iceberg of difficulties in Linnaeus' scheme. The important point is that all the species were imagined to have migrated worldwide from a single area.

This and similar models for a single source and subsequent spread of all species raised as many questions as it answered, of course, like how multitudes of land species subsequently colonized remote oceanic islands, why related groups of species tended to be found in adjacent geographical areas, and, later, how to explain why the species of a given continent are related to the now extinct, fossil species of that same continent? Most naturalists rejected religion-based scenarios—even those otherwise committed to a divine origin for species. The alternative was multiple centers of creation, and here there are important distinctions to be made. Different groups of species might arise in different parts of the world, resulting in the obvious regionalism we see: marsupials largely confined to Australia, say, bears and cats primarily in northern hemisphere areas, and remote islands having their own sets of species found nowhere else. Charles Lyell preferred the idea of multiple centers of origin to the extent that it neatly explained island endemics. But could the *same* species arise in more than one geographic locale, maybe even at different times? Explaining regional differences in species is one side of the coin; the other is regional similarity. Widespread groups pose their own puzzle; how to explain the occurrence of the very same species on mountaintops in Europe and the Appalachians of eastern North America, for example, or continental Europe and islands like Madeira, far out to sea?

In the late eighteenth and early nineteenth centuries, some naturalists became explorers precisely to obtain ever more detailed information about the distribution of species. Some, like the Prussian polymath Alexander von Humboldt and his botanist friend Aimé Bonpland, paid special attention to the distribution of plants of the American tropics and tried to quantify their similarities and differences along altitudinal and latitudinal gradients. Humboldt was trying to gain a glimpse into the plan of the Creator by uncovering the underlying pattern of species distributions.

As a young undergraduate at Cambridge in the late 1820s under the tutelage of John Stevens Henslow, Darwin was exposed to a great many pressing philosophical questions, but perhaps none was so pressing as the "centers of creation" issue and its relationship to the mys-

Alexander von Humboldt's profile of the Andean volcano Chimborazo, first published in his 1807 *Essai sur la géographie des plantes*. Pioneering the cross-sectional representation of altitudinal zonation, this map shows the transition of plant communities from the lowland tropics through the frozen arctic zone at the summit. Humboldt inspired Darwin's determination to travel abroad.

tery of species origins. His vicar-naturalist professors impressed upon him the conviction that a detailed comparative study of geographical distribution held the solution. Perhaps this is why so many naturalists considered travel a rite of passage of sorts, as it was the only way to obtain firsthand knowledge of different species and make the careful observations and collections required to discern patterns. This was what led Humboldt abroad to the New World, and his rapturous writings in turn inspired a generation of eager young naturalists. Darwin was no exception.

The peculiarities of geographical distribution and its meaning were very much on Darwin's mind throughout his voyage on HMS *Beagle*. Recall from Chapter 1, for example, how in the final year of the voyage Darwin mused on this very topic amid the strange fauna of Australia: "I had been lying on a sunny bank & was reflecting on the strange character of the Animals of this country as compared to the rest of the world," he wrote in his diary in January 1836. "An unbeliever in everything beyond his own reason, might exclaim 'Surely two distinct Creators must have been [at] work." But the ant lion he spied, a species distinct from those he knew back home but clearly related, got him thinking about commonalities rather than differences around the globe. "Would any two workmen ever hit on so beautiful, so simple and yet so artificial a contrivance? It cannot be thought so. The one hand has surely worked throughout the universe."[4]

Darwin's "one hand"—a single Creator—does not necessarily entail that each species *arose* at a single location, but this is very likely what he believed. Certainly Lyell, whose watershed work *Principles of Geology* Darwin was reading on the *Beagle* voyage, thought so. In volume II of *Principles* Lyell suggested that "each species may have had its origin in a single pair . . . and species may have been created in succession at such times and in such places as to enable them to multiply and endure for an appointed period, and occupy an appointed space on the globe."[5] Darwin received this eagerly awaited second volume of the *Principles* in November 1832, when the *Beagle* stopped at Monte Video in South America. Having enthusiastically embraced Lyell's arguments of the first volume for what would later be called a uniformitarian view of earth evolution—understanding the earth in terms of geological forces now in operation—Darwin was now intrigued by the long discussion of species in Lyell's second volume. It may seem odd that a geological work would devote so much space to subjects like species variability and hybridization, habitats and biogeography, migration and dispersal, but it only underscores the central tenet of naturalists like Lyell and others that earth and the life upon it must be studied together to gain insight into the ultimate philosophical and theological question of origins: origins of earth, life, species, *us*. It was in this spirit that Lyell thoroughly reviewed contemporary understanding of geographical distribution and theories of the "original introduction of species"—all utterly compelling to the young Darwin. One later observer, the geologist John Wesley Judd, even suggested that Darwin's reading the second *Principles* volume was the key event putting him on the path to solving the mystery of species origins. If the *Principles* did not spark Darwin's interest in the origin of species, it certainly crystallized the issues for him and provided a research agenda.

The second *Principles* volume probably also sparked Darwin's interest in dispersal. The single-origin idea for species requires high dispersal ability. How else do species come to occupy often vast and sometimes even separated ranges, especially when the same mainland species is found on distant islands? In characteristic form, Lyell thoroughly reviewed the means of plant and animal movement on the globe by wind, water, and other agencies, from animal transport (within and without: seeds in stomachs and those adhering to fur) to accidental transport by rafting on floating vegetation or debris, whirlwinds, and icebergs. This all seemed very plausible to Darwin, though as we shall see next it didn't deter other naturalists from postulating virtually any mechanism *but* simple dispersal to explain these patterns. There ensued a genteel debate, one that pitted Darwin against his sage mentor Lyell for a time. Few students today would scarcely believe that questions of the geographical distribution of species were once heady stuff. To Darwin, it was "that grand subject, that almost keystone of the laws of creation."[6]

Atlantis Rising

Lyell's vision of earth's history was one of fluctuating levels of land and sea, and an equally fluctuating climate. The continents were thought to be incapable of moving, but there seemed ample evidence to show that the level of the land could change dramatically, if gradually, over vast stretches of geological time. The land is in a state of continual flux, and the geographical ranges of species ebb and flow in response. As Lyell expressed it in the *Principles*,

> Every flood and landslip, every wave which a hurricane or earthquake throws upon the shore, every shower of volcanic dust and ashes which buries a country far and wide to the depth of many feet, every advance of the sand-flood, every conversion of salt-water into fresh when rivers alter their main channel of discharge, every permanent variation in the rise or fall of tides in an estuary—these and countless other causes displace in the course of a few centuries certain plants and animals from stations which they had previously occupied.[7]

It was clear to Lyell that God endowed the "numerous contrivances" of plant and animal dispersal to harmonize with the fluctuations of the inanimate world, fluctuations that he also ordained. Otherwise, we would not find species in the most unlikely, far-flung, and often environmentally extreme locales where we find them today. In his reference to harmonic balance of the animate and inanimate worlds, Lyell was reinforcing the idea that earth and the life upon it are intimately bound together.

Lyell also saw dispersal, whether by chance or design, as augmented by shifting landscapes and climates. Corridors of movement open and close, volcanic islands like stepping stones across the seas form and eventually erode back into the depths, mountain chains erupt on land, forming barriers for some species and ridgelines of migration for others. The sea encroaches on the land and then withdraws, and the climate cycles through glacial and warm periods. Lyell's vision of the earth was one of great areas of balanced uplift and subsidence: as mountains are raised on the west coast of South America, for example, the east coast slowly subsides into the sea. It was a grand vision, one that profoundly influenced Darwin and many other naturalists. Yet some were not convinced that species were quite so adept at getting themselves dispersed far and wide. They thought that most species needed help, at least when it comes to getting across the wide ocean basins. Island stepping stones might seem to fit the bill, but there arose—no pun intended—a school of thought that took that idea to an extreme, maintaining that now-sunken continental areas of great extent were once found in the ocean basins.

The chief exponent of this view was the Manx naturalist Edward Forbes (1815–1854), a brilliant and creative thinker who is now best known, somewhat unfairly, as someone who was consistently wrong in his theorizing. Perhaps the biggest of Forbes's blunders was his advocacy of the existence of vast former continents or continental bridges that have disappeared without a trace. To be fair, there are elements of truth in the idea. Forbes published a lengthy memoir in 1846 aiming to explain striking disjunctions in species' ranges, like the similarities in the arctic and alpine flora of Scotland, Scandinavia, and Switzerland. How did plant species of the arctic regions, certain saxifrages, say, also end up high in the Alps, separated by vast stretches of inhospitable habitat? His explanation was recent subsidence of large expanses of land during the recent glacial period, followed by uplift. During subsidence the ocean covers continental areas, and any higher regions—the tops of ancient mountains of Wales and Scotland as well as the younger but still very old Alps—become islands in a chill sea. In such a cold climate the northern species easily disperse southward on island stepping stones. As the climate warmed, Forbes suggested, and the seas receded by uplift of the land, we are left with disjunct populations: the arctic species can only persist at the highest elevations in the Alps, so they are isolated there far from the main range of the species which occurs at lower elevations much farther north.

This model was far from unreasonable, and proved extremely influential. In spirit it is not so far off the modern explanation—the cool climate of glacial periods did indeed play a role in creating such disjunctions, but not in the context of a subsided and flooded European continent. Darwin put his finger on this idea as early as 1842, in the brief sketch of his species theory. He opened his discussion of geographical distribution stressing the importance of barriers but then immediately turned to alpine plants, proposing precisely the same kind of climatic shifts as Forbes, explaining the ebb and flow of high-latitude plants according to cycles of climate. Darwin's model did not involve wholesale subsidence and uplift of the continent, however, but simply the natural climactic push and pull of species according to the dictates of their environmental needs. During cold periods high-latitude species are pushed south and high-elevation species pushed down. As climate warms and alpine ice shrinks, "arctic fauna would take place of ice, and an inundation of plants from different temperate countries [would] seize the lowlands, leaving islands of arctic forms."[8] Darwin elaborated on this in his longer species *Essay* of 1844, and dedicated a whole section of the *Origin* to the idea. There he gave Forbes full credit, though he was disappointed that he hadn't published it first: "I was forestalled in only one important point, which my vanity has always made me regret," he wrote in his *Autobiography* years later. "Namely, the explanation by means of the Glacial period of the presence of the same species of plants and of some few animals on distant mountain summits and in the arctic regions. This view pleased me so much that I wrote it out *in extenso*, and I believe that it was read by Hooker some years before E. Forbes published his celebrated memoir on the subject. In the very few points in which we differed, I still think that I was in the right."[9]

Some aspects of Forbes' model stand today, but not his idea of uplift and subsidence. That aspect of his model is problematic enough, but Forbes' real blunder comes in a related part of his 1846 memoir. Having nicely explained the alpine and arctic disjunctions, Forbes turned to the problem of continental-island relationships: How did some species of the mainland also end up far out to sea on islands like Madeira or the Azores? "There are three modes in which an isolated area may become peopled by animals and plants," he wrote. "1st. By special creation within that area. 2nd. By transport to it. 3rd. By migration before isolation."[10] Forbes reasoned against the first by pointing out that with a few exceptions the flora and fauna of the isolated areas he was considering are identical with continental species, implying that special creation of the same species in several locations did not make sense. He dismissed the second as insufficient, arguing that a few groups of high mobility might be accounted for in this way, but if dispersal by wind and water explains the sizable number of identical species in continental and island areas, shouldn't this have made these areas *completely* similar? He thought it most plausible that colonization from "neighbouring lands" prior to isolation of the islands made sense for his model of movement of arctic species, and Forbes reasoned that there must also have been some sort of intervening land, or at least a scattering of island stepping stones, between the continent and these distant islands.

From a modern perspective we would agree with Forbes' commitment to the idea of single rather than multiple centers of origin for species. We also recognize that sea level rises and falls with glacial cycles, exposing continental shelves whenever it falls low enough and thereby unites the mainland with any outlying islands sitting on the shelf. When sea level rises, the lowlands are flooded and the higher areas once again become islands—termed, appropriately enough, landbridge islands. (Note that seemingly isolated habitat islands like mountaintops are basically land-bridge islands too—climatic cycles open and close overland migration corridors.) Flora and fauna are free to move over the continuous land areas, and when sea level rises lots of those same species become marooned on those portions that become islands.

This is a very different dynamic from truly oceanic islands that never had any connection with a continent. In those cases, colonization is largely a matter of chance as well as factors like distance from the nearest mainland, and prevailing winds and currents. We expect lots of shared species between land-bridge islands and the mainland since continental species repeatedly migrate there as intervening land comes and goes. The last time that happened was during the last glacial maximum a mere 12,000–15,000 years ago. Forbes was right to be struck by the similarities he was seeing, and to argue that chance dispersal was unlikely given that degree of similarity. However, he took the idea of connecting continental land bridges to an extreme, proposing vast extensions of the European continent stretching far out to sea, in some cases across the entire Atlantic. Forbes saw the Atlantic islands like the Azores, Canaries, Madeira, Iceland, and others as remnants of a former continental mass, and inevitably this idea of an ancient continent in the Atlantic connecting western Europe, north Africa, and its outlying islands with North America, became associated with the mythical lost continent of Atlantis. To Darwin's shock quite a few naturalists were taken with this idea, and some began postulating land bridges and disintegrated continents with abandon, which drove Darwin to distraction.

He, too, argued against multiple centers of origin, and agreed with Forbes about the idea of climatic fluctuations influencing the movement of species. This became known as the "dual flux" idea—climate and species shift in concert as a result of geological processes. But Darwin saw little evidence that the sea had flooded the European continent so far inland as to make islands of the Alps, and no evidence that vast continental areas once occupied the abyssal sea—certainly not in recent geological time. Climate-induced shifts in the distribution of species, not uplift, explained how species got to mountaintops, Darwin maintained. And as for those islands, he was convinced that the powers of dispersal were far underrated by Forbes and his followers.

Getting There

Although Darwin was a supporter of the Lyellian idea of slow uplift and subsidence of large areas of earth's crust, he felt strongly that continental extensions of the kind that Forbes and others were proposing was unnecessary. Even worse than unnecessary—he felt the idea was deeply flawed, maybe spurred by his own growing commitment to Forbes' self-rejected hypothesis that an isolated area might become "peopled" by animals and plants "by transport to it." Very soon after becoming convinced of evolution in spring of 1837, Darwin realized the importance of barriers, and in the 1842 *Sketch* reflected on how "new forms" could arise on islands. "Here the geologist calls in creationists," he concluded—clearly not an explanation he found satisfactory. His preferred explanation is clear: "Discuss one or more centres of creation: allude strongly to facilities of dispersal and amount of geological change."[11]

While the *Sketch* says little about dispersal abilities beyond a mention of icebergs, currents, and storms, his 1844 *Essay* goes into great detail. Let us not be too hasty in concluding that island species had to be created *in situ*, Darwin urges, "Considering our ignorance of the many strange chances of diffusion."[12] Or, for that matter, that they required land bridges or stepping stones to get there. Yes, stepping stones, which he called "intermediate spots," could play a role, and must have in some cases, but his imagination was already running wild with scenarios for long-distance transport by floods and currents, whirlwinds and hurricanes: dispersal by birds, rafting quadrupeds carrying seeds in their stomachs or adhering to their fur, floating trees with seeds wedged in root masses, insects with seeds or eggs stuck to their legs, icebergs, and more.

Forbes published his continental extension idea in 1846, just two years after Darwin's essay. Fellow naturalists proceeded to discover evidence for ancient continents and continental bridges in virtually every ocean basin. Darwin's friend Joseph Dalton Hooker invoked a vast land mass in the southern ocean as the most likely explanation for the distribution of what we now call Gondwanan flora, distributed among the southern continents and circumantarctic islands. Hooker was not wrong outright—there *was* a southern continent, Gondwanaland, but it consisted of a union of *existing* continents, not a different landmass that had since sunk into the ocean depths. An appreciation of this would not come for more than a century, so in the meantime Darwin got more and more exasperated, especially with talk of Atlantis. Chagrined, he fired off a testy note to Hooker:

> *I got so wrath about the Atlantic continent, more especially from a note from Woodward . . . who does not seem to doubt that* every *island in Pacific & Atlantic are the remains of continents, submerged within period of existing species; that I fairly exploded & wrote to Lyell to protest & summed up all the continents created of late years by Forbes, (the head sinner!) yourself, Wollaston, & Woodward & a pretty nice little extension of land they make altogether! I am fairly* rabid *on the question & therefore, if not wrong already, am pretty sure to become so.*[13] (emphases Darwin's)

As "rabid" as he was on the subject, he concluded the letter on a characteristically humorous note: "I must try & cease being rabid & try to feel humble, & allow you all to make continents, as easily as a Cook does pancakes." That same humor, only thinly masking his frustration with the wanton raising and sinking of continents to explain every biogeographical puzzle, is on display in a letter to Lyell, the geological authority of the day. Darwin took Lyell to task for encouraging the continental extension nonsense. He declared that his "blood gets hot with passion & runs cold alternately at the geological strides which many of your disciples are taking."

> *Here, poor Forbes* [recently deceased] *made a continent to N. America & another (or the same) to the Gulf weed [Sargasso Sea].— Hooker makes one from New Zealand to S. America & round the world to Kerguelen Land. Here is Wollaston speaking of Madeira & P. Santo "as the sure & certain witnesses" of a former continent . . . why not extend a continent to every island in the Pacific and Atlantic Oceans! And all this within the existence of recent species! If you do not stop this, if there be a lower region for the punishment of geologists, I believe, my great master, you will go there. Why your disciples in a slow & creeping*

*manner beat all the old Catastrophists who ever lived.— You
will live to be the great chief of the Catastrophists!*[14]

"Don't answer this, I did it to ease myself," he concluded. Darwin may
have been half-joking with Lyell, but only half. How much simpler was
long-distance dispersal of individuals, or their seeds, eggs, or spores!
Improbable as it was that, aided by wing or wave, propagules from a
mainland could make it to distant islands, mere specks in the ocean,
in the fullness of geological time even the odd and highly improbable
event is bound to happen. All it takes is one successful colonization. To
Darwin such an idea was not only a simpler explanation, but it was also
testable.

In fact, by this time Darwin had already embarked upon an ambitious program of experiments. As with Darwin's earlier barnacle studies, these experiments were prompted by Hooker, who threw down the
gauntlet some years before by maintaining that mere dispersal could
not explain the similarities among the flora of Tierra del Fuego, Tasmania, and New Zealand: "I fear that they neither belong to transportable
species or orders or present any facilities for transport,"[15] he wrote to
Darwin in June of 1847, but reassured him with a promise of "an hon-
est investigation." Darwin was to be disappointed.

Hooker had served as assistant surgeon and naturalist on the Ross
expedition to Antarctica and the subantarctic islands aboard HMS
Erebus between 1839 and 1843 (the same HMS *Erebus* later lost on
the ill-fated Franklin arctic expedition—the wreckage of which was
only discovered in 2014 by the Canadian government). Over the next
15 years he produced his monumental three-volume *Flora Antarctica:
The Botany of the Antarctic Voyage*. By 1853, when he completed the
third volume—the *Flora Novae-Zealandiae*—Hooker had carefully
weighed the evidence for and against continental extensionism, and
the extensionists won his support. The obstacles to Darwinian dispersal were many: these scattered plant species are hardly abundant
enough to produce many seeds for chance transport, he thought, and
were unsuited for transoceanic travel, having "feeble vitality" and "soft

or brittle integuments." And what were the chances of seeds germinating anyway after floating in frigid saltwater for days on end? Take the Kerguelen Islands, an archipelago of the remote Antarctic ocean: "seedlings are extremely rare," Hooker pointed out. "The seeds, if not eaten by birds, either rot on the ground or are washed away . . . if such mortality attends them in their own island, the chances [of survival] must be small indeed for a solitary individual, after being transported perhaps thousands of miles, to some spot where the available soil is pre-occupied."[16] No, he concluded, the chances of successful oceanic dispersal are remote; we must look elsewhere for an explanation for dispersal. The idea of movement across long-sunken land bridges seemed like the only reasonable alternative.

Darwin and Hooker had argued back and forth over the matter for a couple of years when Darwin decided to try to show Hooker the error of his views. In March 1855 he started what he called his "seed-salting experiments," testing seed flotation and viability with homemade saltwater. He put seeds of various species—cabbage, cress, radishes, lettuce, carrots, onions—into a cellar tank with saltwater made with melted snow, to ensure the water was ocean cold. He then tried a variant using small open bottles kept in the shade outdoors. Both were successful: "These after immersion for exactly one week, have all germinated, which I did not in the least expect (& thought how you would sneer at me)," he crowed.[17] Darwin impishly told Hooker that he initially held back on telling him of the success of this experiment in hopes that the skeptical Hooker would say he would eat all the plants Darwin could raise after immersion. Hooker would have had quite a nice salad—everything but the cabbage seeds did well after saltwater immersion. Darwin was elated that his data showed that long-distance dispersal by ocean currents is indeed possible. He then planted the seeds to prove they not only could travel many miles by sea, but would succeed in growing once they came upon land. Hooker was impressed.

About the same time that Darwin published the pictured request for information in the weekly *Gardeners' Chronicle*, he followed up with a short report of his own results. This caught the attention of his

Darwin's first letter on the effects of salt water on seeds, published in the *Gardeners' Chronicle* for April 14, 1855 (no. 15, p. 242). He published several open queries in this magazine and others over the years, asking readers for information or urging them to replicate his experiments and share their results—a nineteenth-century form of crowd-sourcing.

long-time botanical correspondent Rev. Miles Berkeley, who offered to try his hand at the seed immersion experiments. Between the two of them they collected lots of valuable data, with Berkeley reporting suc-cessfully floating and germinating seeds of 53 species.

Meanwhile, annoyance with his friend Thomas Wollaston's notion of a continental bridge linking Europe with the Azores, out in the Atlantic, undoubtedly led Darwin to undertake a set of "seed-salting" experiments especially with Azoran species. In the summer of 1855 he wrote to enlist the help of Henslow and his kids in gathering up the seeds of plants near their home in Hitcham that were also found in the Azores so he could test their germination rate after immersion in saltwater. He would reward the children with a few shillings for their help. "The experiment seems to me worth trying; what do you think?"[18] he ventured to Henslow. By that fall Henslow and daughters provided seeds from 22 Azoran species, and Darwin delightedly paid up, send-ing the money on to Henslow's "good little Botanists."

Darwin's sense of humor and humility often come through in the many letters he exchanged with Henslow, Hooker, and others over his "seed-salting" experiments. But as he tested more and more species, he couldn't hide his pleasure at proving Hooker wrong time and again. It became a sport with his kids, who delighted in every experiment that rebutted Hooker. Then came Hooker's observation that sent Darwin back to the drawing board: many seeds did not float on the saltwater. He overlooked the fact that most sank to the tank's bottom, taking with them his spirits. "As for showing your satisfaction in confounding my experiments," he wrote Hooker, "I assure you I am quite enough confounded—those horrid seeds, which, as you truly observe if they sink they won't float. . . . The bore is if the confounded seeds will sink, I have been taking all this trouble in salting the ungrateful rascals for nothing."[19]

Darwin came up with a new experiment, testing whether seeds, flowers, and even whole plants bearing flowers and seed pods and other fruits would float in seawater and germinate. These experiments ran all through the summer of 1855 and into the fall. By October his enthusiasm seemed to be waning, as he wrote Hooker: "I am sick of the job . . . NB. capsicum & celery seed have come up after 137 days immersion."[20] Hooker became as fascinated as he was amused by Darwin's experiments, and duly provided him with all sorts of seeds and fruits from Kew Gardens. While the problem of the sinking seeds seemed at first to ruin Darwin's argument for long-distance dispersal by ocean currents, he kept at it and the eventual results surprised him as much as Hooker. "I have *almost* finished my floating experiments on salt-water," he wrote Hooker two years later, in April 1857. "72/94 sunk under 10 days—seven plants, however, floated *on average* 67 days each.— I then dried all these (with in each case, with pods, bits of twig & few leaves) & 62/94 sunk under 10 days, so that generally the drying had no great effect, but . . . sometimes it had great effect."[21] Dried asparagus was one, floating about 23 days when fresh but up to 86 days dried, and still "germinated excellently." By his calculations, ocean currents could have carried that asparagus over 2800 miles. He concluded that about a tenth of all plants would float an average of 30 days dried, and with an average current of 33 miles per day they could easily make it across the ocean. By now Hooker received Darwin's latest results as eagerly as did the Darwin children. But although Darwin's seed-salting experiments were a great success, it turned out he was being overly optimistic about plant flotation in that letter to Hooker. He was eventually forced to acknowledge that, on balance, being ensconced in fruits is a hindrance, not a help, to seed flotation.

What he wasn't ready to share yet was the results of his varied spin-off experiments, some of which were suggested by his children. One son came up with an imaginative experiment recounted in yet another letter to Hooker: "I must tell you another of my *profound* experiments!" Eight-year-old Franky suggested that if a seed-eating bird was killed over the ocean (by, say, lightning or hail), it could drift for some time. "No sooner said, than done," says Darwin. An unlucky pigeon from his dovecote "floated for 30 days in salt water with seeds in crop & they have grown splendidly." Anticipating the ever-skeptical Hooker's reply, Darwin headed him off: "You will say gulls & dog-fish &c would eat up the [carcass], & so they would 999 out of a thousand, but one might escape: I have seen dead land bird in sea-drift."[22] Fish, however, were not so obliging. Darwin envisioned that fish might eat seeds washed out to sea, then either swim great distances or themselves be eaten by a bird that would then carry the fish, seeds in stomach, to distant shores where they would be voided and grow wonderfully. The letter to his cousin Fox quoted at the beginning of this chapter gives you an idea of how the seed-eating fish experiments went: not well.

He persevered, and as with so many of his research projects, recruited all manner of assistants to aid him in his seed-dispersal obses-sion, from friends and family to colleagues, acquaintances, and fellow naturalists around the world. In early 1857, at Darwin's suggestion, the keeper of the aquarium at the Zoological Society's gardens tried feed-ing wheat seeds to minnows. It was a success, the keeper reported, and what's more: "a Fellow of the Society who is a great Angler told me, that he has taken Barbel and dissected them and found a quantity of Wheat in them . . . he says he has caught them near the water mills where the wheat has been spilt into the river."[23] So some fish *would* eat floating seeds after all. Darwin's nephew Edmund Langton, son of Emma's sister Charlotte and her husband Charles Langton, obligingly tried this experiment at the family's home in Sussex, but without the zookeeper's success. He planned on doing it again on a warmer day, thinking the weather might have been the problem, he promised his uncle. On another front, discovering that he could get dozens of plants to germinate from as little as 2 tablespoons of pond mud, Darwin thought that herons, ducks, and other water birds would carry seeds far and wide on their muddy feet. He implored his college friend Thomas Eyton, who studied ducks and other waterfowl, to wash the feet of water birds at his Shropshire estate and send him the dirty water. Eyton was ever happy to oblige Darwin, despite disagreeing with his friend's evolutionary ideas.

Nature the Careful Gardener

Darwin had been thinking about the implications of experiments such as these for a long time. A decade earlier, in the mid-1840s, he recorded several dispersal-related studies in his Questions and Experiments (Q & E) notebook: "shoot tame duck on pond with Duck-weed—coots—waterhens—examine dog, which has swum . . . every kind of seed must be distributed." And: "Spread sheets of Paper covered with some sticky stuff in flat places & see whether wind, on windy day, will drift many seeds . . . Have paper ruled in squares to facilitate investigation—Capital in middle of ploughed field—on hills."[24] It's not clear if Darwin followed up on these ideas, but there is another experiment in this notebook that he definitely pursued later: one looking at whether predators could disperse seeds within the prey they consumed. In October of 1856 he took some birds with seeds in their crops to the Zoological Society and fed them to eagles and owls. The zoo keeper retrieved the resulting pellets for him and kindly separated out the seeds. Darwin then planted them and reported the results in his experiment book: "Nov. 13: Pellet from Snowy owl from Bird with seeds 18 hours in stomach: ([December]. Germinated 5. Oats. 1 Wheat. 1. Hemp 2. Millets)."[25] So birds of prey could indeed disperse seeds carried within the birds they had eaten.

The experiment book is the most thorough record of Darwin's experiments in that period, and the entries make for fascinating reading, full, as it is, with his research on fish, rats, pigeons, pelicans, storks, eagles, and even guinea pigs. Darwin was nothing if not thorough in his investigations. These notes were never intended for others to read, but we are exceedingly fortunate to have them, as they are a portal into Darwin's working method.

His creativity wasn't limited to the varied ways that seeds are transported, either. Imagining ponds and lakes as islands on the land, he assumed they were colonized by aquatic plants and animals through accidental transport of one kind or another. As with so many natural processes, he imagined that something mundane, regular, and dependable would do the trick. What about waterfowl as a dispersal agent? There are untold millions of ducks and geese, and in the course of their lives they must visit innumerable ponds and lakes, inadvertently carrying all manner of seeds, algae, insects, eggs, snails, and more. Little by little, over the decades and centuries, pond dwellers surely hopscotch from pond to pond by the collective activity of waterfowl. Logical, but he needed to find evidence. It turned out that little was actually known about the propensity for waterfowl to accidentally carry stowaways. So, besides checking the muddy feet of herons and other wading birds for seeds, in June of 1856 Darwin turned his eye to the most obvious candidate for carriage by waterfowl: ubiquitous duckweed, the minute plant often mistaken for algae, found the world over covering the surface of lakes and ponds like tiny green confetti. Minute duckweed (the most common being genus *Lemna*) is the smallest of the flowering plants, lacking stems and leaves. If any aquatic species is going to get carried by ducks, surely this is the one. In what was surely a comical sight, he simply dunked a couple of ducks upside-down in a pan of duckweed-covered water and carefully counted how many stuck. And indeed, duckweed lived up to its name: "June 17th it does stick to feathers of Ducks,"[26] he reported in his experiment book. We might have expected as much, but nothing should be taken for granted in science.

That was an easy experiment with a positive result. But Darwin needed to prove that more than plants and seeds could be carried from one place to another. He set up experiments aimed at determining how well aquatic snails survive out of freshwater and immersed in saltwater, "tormented," he said, by the question of how land snails got to remote oceanic islands. In one experiment he immersed snails in saltwater for a week. To his relief and delight, he wrote Hooker, in each of two trials the snails revived. "I feel as if a thousand pound weight was taken off my back."[27] Darwin only needed to show that survival was possible in conditions similar to natural ones. He didn't need to have a large percentage of the mollusks survive, just show that some *could* survive an extended period exposed to saltwater.

Darwin did not fare as well with frog eggs—nor did Emma. In one experiment, he laid sheets of damp paper covered with frog eggs up and down the hallways adjacent to his study to see how long they would remain viable. After 3 days they dried up on the paper and did not revive when placed in water. This gives some insight into the curious family life of the Darwins at Down House—and how the experiments were a perpetual presence in the household. We have no record of how Emma felt about her home being turned into a field station, but chances are she not only took it in stride, she was experimentising too.

Failure did not make Darwin think his idea was wrong. Instead, he enlisted help. Puzzling over the dispersal of fish led him to write Edinburgh surgeon and naturalist John Davy, brother of the celebrated chemist Sir Humphry Davy, to do experiments on the viability of fish eggs out of water. Davy obliged by undertaking a series of experiments exploring how long fertilized salmon eggs remained alive when exposed to air and water (fresh and brackish) for different periods of time and at different temperatures. He concluded that moisture is critical for the survival of fish eggs during transit, which, incidentally, is not the only way freshwater fish might be dispersed: Davy told Darwin of an account read before the Royal Society of Edinburgh of a fingerling char that survived for 72 hours barely covered with water, but small enough that a bird could have carried it on its feet or stuck to its feathers. Davy concluded that his results confirmed Darwin's hypothesis of freshwater fish dispersal, and would shed light on the geographical distribution of migratory fish such as salmon. Darwin was delighted, and passed on Davy's report to the Royal Society.[28] But he needed to show the possibility of dispersal for more species, if he were going to make an ironclad case.

March and April of 1857 saw Darwin doing some of the oddest experiments yet at Down House. Darwin had an aquarium that he jokingly called his "snailery," with a thriving colony of two or three species of freshwater snails. Into it, he placed a pair of dried duck's feet. He wanted to see if any snails would climb aboard, but more than that he wondered how long, once on the feet, the snails could survive out of water. The first time he tried the experiment, a small *Planorbis* (a freshwater air-breathing snail that looks a bit like a miniature ammonite) held fast, surviving some 20 hours on the mantel once the foot was removed from the snailery. He tried again, this time with *Lymnaea* snails, removing the feet to a small jar where they could be kept a bit damp. He repeated the experiment, and both times found the young snails survived 17–20 hours out of water. Freshwater snails clearly could endure a journey out of water hanging on to the feet of waterfowl, especially considering that ducks tend to close their feet

when flying and the webbing inside remains damp. He reported his duck-foot findings in the *Origin,* pointing out that in 20 hours "a duck or heron might fly at least six or seven hundred miles, and would be sure to alight on a pool or rivulet, if blown across sea to an oceanic island or to any other distant point."[29]

Nor did that end the subject for him. He was still at it a decade later, when he put a dried goose's foot on the ground in a field and counted the snails and slugs that climbed on a day or so later. (They dropped off in about 5 hours.) And over a decade after that, in 1878, Darwin published a short letter

Unio clam adhering to a duck's foot. From Darwin (1878), p. 121.

in the journal *Nature* reporting an odd account of long-distance dispersal from a correspondent in Massachusetts who had shot a teal on the wing and found a living freshwater clam firmly clamped to its middle toe. "It would have undoubtedly been transplanted to some pond or river, perhaps miles from its original home, had the bird not been shot, and might then have propagated its kind," Darwin's correspondent wrote.[30] *Nature* readers followed suit, with many people writing to Darwin with similar accounts. He compiled and published them with his own further observations in an article in *Nature* in early April 1882—his last publication to appear in his lifetime, as he died later that month.[31] Thus, Darwin accumulated many rich observations—more than he could ever witness himself—over decades without leaving home.

Unexpected Means of Migration

Improbable as any given dispersal event is, Darwin was convinced that species have a remarkable tendency to get moved around in the fullness of time, whether through habitat corridors opened and closed by cycling climate and the rise and fall of land and sea, or by means of wind, water, and wing. He wanted to be the fly in the ointment to the continental extensionists, and although he lambasted his friend Charles Lyell over encouraging such explanations, Lyell himself was intrigued by Darwin's novel approach to arguing against such views: observation and experiment. In fact, inspired by Darwin, Lyell collected dispersal information of his own. April 1856 found him visiting Down House, making notes about Darwin's latest results with his seed flotation experiments, and the next month he had news for Darwin. A friend of a friend had found a freshwater limpet attached to a water beetle. "Here is a new light as to the way by which these sedentary mollusks may get transported from one river basin to another," he wrote Darwin. "How far can an *Hydrobius* [water beetle] fly with a favourable gale?"[32] Like Darwin's ducks, this beetle carried a mollusk on the wing. Lyell followed this report of water beetle transportation services with another, this time one found bearing the egg sac of a water spider under its wings. "What unexpected means of migration will in time be found out," he marveled.

Indeed, the "chance dispersal" view that Darwin championed would eventually become dominant in the field of biogeography, though even a century later it would have its detractors, some of whom dismissed transoceanic dispersal as "a science of the improbable, the rare, the mysterious, and the miraculous" (as famously quipped by ichthyologist Gareth Nelson).[33] Such sniping was not uncommon in the 1970s, in a reaction against the difficulty of testing dispersal scenarios in natural circumstances, and the relevance of the new science of plate tectonics and continental drift to explain geographical distribution. Ironically, there is a kernel of truth behind the old idea of continental extensionism: rearranged and sometime contiguous continents, not long-sunken land bridges, explain the distribution of some groups. But chance long-distance dispersal has never gone away. Improbable and rare as such events are, they are far from mysterious, and certainly not miraculous.

A multitude of cases like Lyell's water beetles have been documented, and the even rarer—or at least harder to show—phenomenon of long-distance "rafting" has been convincingly shown as well. For example, two back-to-back hurricanes had tracked through the Caribbean just weeks apart in the fall of 1995. Immediately after, a species of iguana known not to live on a particular island were documented there and shown to be in breeding condition. Herpetologist Ellen Censky and colleagues made a strong case for the iguanas having come from an island further west, traveling to their new home via huge mats of uprooted trees and other debris.[34] But it is the unprecedented treasure trove of genetic evidence made possible through molecular techniques that has brought long-distance dispersal back to the fore,[35] showing that in many cases relationships between groups separated by even vast expanses of ocean are best explained by oceanic dispersal, not continental drift. Certainly dispersal is the rule in regard to populating remote volcanic islands never connected with land. A gourd species (*Sicyos villosus*) collected by Darwin himself provides a nice example. In 1835 Darwin collected specimens of this gourd on Floreana Island in the Galápagos (and it has never been seen there since, perhaps having been grazed to extinction by feral livestock). A DNA-based analysis conducted 175 years later by a team of German and British botanists revealed that its closest relatives are found in North America and Mexico, while a second Galápagos gourd species likely descends from a species in Peru and Ecuador.[36] Ancestors of the two species floated to the islands at different times, perhaps thousands or even millions of years apart.

Darwin's notion of the power of dispersal to explain geographical distribution was largely correct—"Nature, like a careful gardener, thus takes her seeds from a bed of a particular nature, and drops them in another equally well fitted for them," as he put it in the *Origin*.[37] Darwin's view has not always been borne out, but his tenacity and creativity in *testing* his ideas nonetheless provides a powerful lesson in the pursuit of scientific knowledge. As with so many of Darwin's pursuits, his long adventure collecting evidence of dispersal, carrying out quirky experiments in his homespun fashion, shows that as often as not all it takes to probe nature's secrets is a bit of creativity and resourcefulness.

Experimentising: Getting Around

I. Sink or Swim: Seeds in a Pickle

Darwin's "seed salting" experiments were aimed at determining, first, how long seeds might retain their vitality after exposure to saltwater, and second, how long they can float versus sink in saltwater. He performed several versions of this experiment using seeds from many species, including garden vegetables, common weeds, and tropical plants.

A. Materials

Saltwater: prepare anywhere from 1 to 10 gallons (3.8 to 38 liters) of saltwater, depending on the scale of your experiment. We're interested in approximating seawater. Like Darwin, you can readily make artificial seawater with commercially available salt and mineral preparations. You can purchase your salt mixture from any of the numerous online aquarium supply retailers (Instant Ocean® and Natural Sea Aquarium Salt Mix from Oceanic Systems, Inc. are less expensive brands), or from a local pet shop or aquarium supply store. Or, make your own with this recipe:

Into every 3 gallons (11.4 liters) of fresh water, dissolve:

- 10½ ounces (298 grams) pure (not iodized) table salt
- 1½ ounces (43 grams) magnesium chloride
- 1 ounce (28 grams) Epsom salts
- ½ ounce (14 grams) plaster of Paris

Chemical and mineral mixtures such as these give freshwater all the qualities of seawater, making it suitable for marine life. If you were setting up a saltwater aquarium to support fish, we would have to pay attention to salinity and quality of the freshwater used. Since we're simply floating seeds in our saltwater, there is no need to worry about that. By the way, even if you live close to the ocean it is still preferable to use artificial saltwater, as ocean water is full of algae and other microorganisms that will die and decay, fouling the water.

B. Other materials
- Notebook and pencil
- Seeds of at least six species (e.g., assorted vegetable and wildflower seed packets)
- 500 ml (17 fl. oz.) beakers or flasks (or even mason jars), one for each plant species. The bottom of a 1- or 2-liter (1- to 2-quart) beverage bottle will work too.
- Pipette or turkey baster
- Spoon
- Forceps or tweezers
- Plastic wrap
- Labeling tape and marker
- Potting soil

- For seed planting: Petri dishes or similar chambers, or one or more germination flats (ideally, partitioned into individual planting units) or paper cups, with potting soil

C. Procedure

1. Prepare the saltwater in an aquarium tank, carboy, or large flask, depending on volume, and keep at room temperature. Measure 300 ml (10 fl. oz.) saltwater for each beaker or flask. (If the vessel does not have a 300-ml (10 fl. oz.) measurement mark, make one with a marker or tape as a reference for maintaining the water level.) Choose a location out of direct sunlight where the flasks won't be disturbed. Place the beakers or flasks there and, once positioned, using fingers or forceps carefully place (not drop) 10 seeds of a single species onto the water surface in one vessel, labeling it with the date, time, and plant species. Repeat for each of the remaining seed species, one species per beaker. (Don't worry if your seeds sink: this situation is addressed below.) Loosely cover the mouth of each vessel with plastic wrap; some water will evaporate, and a pipette or baster can be used to replenish evaporated water. To do this, carefully dispense water along the inner wall of the vessel so as to minimize agitating the water surface. (If you have a large number of participants doing this experiment it may be desirable to set up duplicates of the experiment at several different stations, comparing the results of each group later.)

2. Record the status of the seeds daily, logging for each vessel the number of seeds still floating and the number that have sunk. This can be continued indefinitely, as Darwin did by removing seeds at intervals to see how long they would remain viable after prolonged exposure to saltwater, but it will be more practical for most people to run the experiment for a set period of time— 1, 2, or 3 weeks, or whatever you prefer. Larger groups with replicates of each seed species might harvest one replicate per species after, say, 1 week of exposure, a second replicate after

2 weeks, etc. Alternatively, all replicates can be floated for the same period of time in order to obtain descriptive statistics on seed performance among replicates.

3. At the conclusion of the time-span determined for the experiment, record the number of floating versus sunken seeds for each species. First retrieve the floating seeds from each vessel using the spoon, taking care not to sink any seeds. Then retrieve the sunken seeds, making sure to keep each species separate and the still-floating and sunken seeds of each species separate. Rinse seeds in fresh water and plant the seeds using one of the two following methods:

   Paper toweling method. Use separate dishes for floaters and sinkers of each species. Place one piece of paper towel on the bottom of the dish, saturate with water, and place seeds. Saturate the second piece of paper toweling and place over the bottom piece, covering the seeds. Place the lid on the dish to seal in moisture. Ensure that all dishes are labeled (species, floater versus sinker, date).

   Soil method. Carefully plant each set in adjacent units of the germination flat. (If the flat does not have individual planting units, use string to grid off the flat.) Plant floating and sunken seeds of each species separately. Repeat as necessary, water, and cover the flats or cups with plastic wrap.

4. When all the seeds are planted, expose dishes, germination flats, or cups to indirect sunlight (not direct sunlight) or a growth lamp. Monitor daily and record numbers of seeds of each category (species, floating versus sunken, etc.) germinating.

   Record, for each species and each saltwater exposure time (1 week, 2 weeks, etc.):

   (a) Number and percentage of floating seeds germinated versus non-germinated

   (b) Number and percentage of sunken seeds germinated versus non-germinated

5. At the conclusion of this experiment tabulate the data, noting percentage of seeds sinking and remaining afloat for the experimental period, and the percentage of each category germinating. The results can be related back to Darwin's original puzzle: How do species colonize remote islands? Two sets of conclusions can be drawn from our Darwin-inspired experiment, relating to flotation and viability in saltwater. First, the seeds of some species do remain afloat for an extended period. What was the duration of your experiment, and how far might they be carried by currents in that time? In a letter to Hooker, Darwin pointed out that "many sea-current go a mile an hour: even in a week they might be transported 168 miles: the Gulf-stream is said to go 50 & 60 miles a day." Indeed, this is likely an underestimate for some currents: with a maximum estimated surface velocity of 5.6 mph, the Gulf Stream could carry floating seeds over 130 miles per day! Try your hand at calculating this.

6. The sunken seeds might seem less informative for Darwin's purposes. However, they too provide insights, in showing to what extent seeds might remain viable after saltwater exposure whether floating or sinking. What proportion of your sunken seeds germinated as compared to the floating ones?

II. Avian Airlift: Darwin's Duck-Feet Experiment

Struck by the occurrence of freshwater snails on many remote islands, Darwin looked into transport by sea and by air: floating or rafting, or airlifted by birds. Experiments showed that freshwater snails could not long survive in saltwater, so he decided that they were given a lift instead. Can waterfowl carry aquatic snails? Snails or other creatures might climb aboard the feet of ducks and geese, especially when ducks sleep, with their feet dangling in the water. To test this Darwin turned to his "snailery"—an aquarium with freshwater snails of all ages—in which he dangled dried duck's feet. He soon found that several young snails climbed aboard. How long could they hang on? As he later wrote in the *Origin*: "These just hatched molluscs, though aquatic in their nature, survived on the duck's feet, in damp air, from twelve to twenty hours; and in this length of time a duck or heron might fly at least six or seven hundred miles, and would be sure to alight on a pool or rivulet, if blown across the sea to an oceanic island or to any other distant point." Here's a duck-foot experiment that would do Darwin proud. We'll see how much aquatic life might colonize model duck feet that we construct.

A. Materials

- ³/₁₆ in. diameter wood dowels, cut into 6 in. (~15 cm) lengths
- Carpet tacks
- Small fishing tackle weights, assorted sizes
- Ping-Pong balls or corks
- One or more water-durable materials (e.g., denim, Velcro, canvas), cut into shape of duck foot using accompanying sketch
- Shallow pan
- Squirt bottle, pipet, or turkey baster
- Magnifying glass or dissecting microscope
- Petri dishes
- Small knife, nail, or wood screw to carefully bore hole in the Ping-Pong ball or cork
- Silicon or rubber cement

B. Procedure

Making your duck foot model:

1. Using the sketch pictured here as a template or one of your own making, trace the duck-foot outline onto fabric or other materials used to simulate the webbed portion of a duck's foot.
2. Using the knife or nail, carefully bore a hole large enough to insert the dowel in opposite sides of the Ping-Pong ball. Insert the dowel through the ball, with approximately 1 in. (2 cm) protruding on one end of the ball ("top").

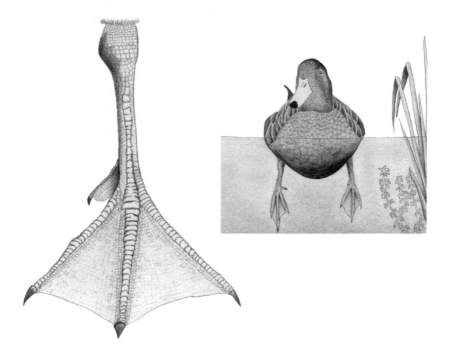

3. Secure the dowel in the ball, and seal against water leakage, by applying a dab of silicon or rubber cement around the dowel where it contacts the ball. If using a cork, bore a hole on one end about ¼ in. (0.6 cm) deep and firmly insert one end of the dowel.

4. Use a thumbtack to affix the duck "foot" itself to the base of the longer end of the dowel ("leg"). (Note: multiple models can be made, each with different material for the foot; the efficacy of the different materials in adhering organisms can then be compared.)

5. The aim is to float the model duck foot in water, with the leg and foot dangling down into the water column. Since the dowel leg itself is buoyant, it will likely be necessary to attach a small fishing weight to the shaft of the carpet tack affixing the foot. Test by placing your duck foot model into water (aquarium or filled sink) sufficiently deep to allow the leg to bob upright, foot down. Attach additional tackle weights as necessary to the base to get the model to float upright, but being careful not to add so

Fishing with model duck feet, in hopes of catching hitch-
hiking aquatic organisms. Photograph by Leslie C. Costa.

much weight that the model sinks altogether. When you have
the weight in balance, your duck foot is ready for deployment!
A fishing line or light string can be attached to the top of the
model using a thumbtack.

Deploying the model:

1. Like a fishing drop line, your duck foot can be deployed into a
   pond or lake by carefully tossing it into the water and tying off
   the line so it doesn't float away.
2. Permit your duck foot model to float in the pond for at least 2 or
   3 days (duration is open-ended). When you recover your model,

have a pan or sealable bag ready to receive the model immediately after lifting it out of the water. Don't collect extra water, just remove the foot from the water directly into the pan or bag. At home, carefully rinse the webbed "foot" over the pan using a squirt bottle filled with fresh water.

3. Use a magnifying glass or dissecting microscope to inspect the foot for any evidence of plant or animal life. Similarly, samples of the pan water can be transferred using a pipet or baster to a petri dish or similar object for observation under the dissecting microscope. It may be difficult to distinguish some small organisms like protists and algae from debris, but careful observation will reveal their structure, and show protozoa continually moving around. Larger organisms—insect larvae, minute snails, etc.—should be obvious. Record the organisms observed, and attempt to count or estimate their numbers.

4. This experiment has a flexible run-time, from days to weeks or longer. With a number of copies the models can be retrieved and analyzed a few at a time—at, say, 1-week intervals over the course of 1–3 months. The number and diversity of organisms adhering to the feet can be tabulated for individual "ducks" and for the group as a whole. Those deploying multiple duck feet can graph the results over time, and/or graph the yield from different foot materials. Do some materials do a better job of attracting colonists? Does the type or abundance of organisms vary among materials, with different materials being richer in different groups?

See also:

J. T. Costa, "Sailing the backyard *Beagle*: Darwin-Inspired Voyages of Discovery in Backyard and Schoolyard," in *Darwin-Inspired Learning*, ed. M. J. Reiss, C. J. Boulter, and D. L Sanders (Rotterdam: Sense Publishers, 2015), 131–146.

"Biogeography" at the Darwin Correspondence Project: www .darwinproject.ac.uk/learning/universities/getting-know-darwins-science /biogeography.

6

The Sex Lives of Plants

In the late 1830s Darwin was preoccupied with sex, a preoccupation with a curious parallel between his work and his personal life. The fall of 1838 saw him discover the mechanism of species change, and a month later propose to his cousin Emma Wedgwood (the culmination of a period of deliberation complete with a list of pros and cons of marriage which happily concluded "Marry–Marry–Marry Q.E.D."). They were married just two months later, in January 1839, and moved to Upper Gower Street in Bloomsbury, London, a little place with such gaudy decor that the couple dubbed it "Macaw Cottage." By the following April Emma realized she was expecting a child—William Erasmus was born later that same year to the Darwins' great joy. All through this period Darwin was also actively exploring the principles of "generation," as reproduction was called, in relation to the variation, crossing, fertility, and vigor of individuals in nature and under domestication, a line of inquiry that began more or less with his conversion to the heretical notion of transmutation in the spring of 1837.

That spring it didn't take Darwin long to realize that agricultural improvement held the key to the process of species change. After all, animal and plant breeders are in the business of creating and improving novel varieties or breeds, and he had a sense that this was related in some way to the formation of species—varieties were, he was convinced, "incipient species" and the difference between variety and species was one of degree and not kind. He immersed himself in the agricultural literature: the writings of William Marshall, John Wilkinson, John Sebright, William Youatt, Robert Bakewell, and others—a "Who's Who" of leading breeders largely responsible for the first great agricultural revolution. Without knowing anything about the basis of heredity, these men built upon the animal and plant breeding techniques that humans had been practicing more or less unconsciously for millennia, elucidating these into principles for agricultural improvement on an unprecedented scale.

Darwin took their lessons to heart, as we saw in Chapter 2. He filled his notebooks with entries bearing on crossing, hybridization, hermaphrodism (having both male and female reproductive organs), interfertility (full breeding capability), and sterility, trying to discern some signal in the noise. The breeders used terms like "picking," "choosing," and even "selecting" in describing the tricks of the trade, including Sir John Saunders Sebright, a giant of agricultural improvement who developed new breeds of cattle, sheep, and poultry. In his 1809 pamphlet *The Art of Improving the Breeds of Domestic Animals*, Sebright declared that "the alteration which may be made in any breed of animals by selection, can hardly be conceived by those who have not paid some attention to this subject; they attribute every improvement to a cross, when it is merely the effect of judicious selection."[1] In response, Darwin jotted in the spring of 1838 how the "Whole art of making varieties may be inferred."[2] When Darwin discovered nature's version of Sebright's "judicious selection" a few months later, he likely thought up the term "natural" selection with Sebright in mind.

Bound up with questions about the origin of species was the origin of the variation that selection acted upon. Environment played a role, he thought, but so too might the reproductive process itself, a line of inquiry that predated his discovery of natural selection. If so, sexually reproducing species should vary more than asexual species, something his investigations seemed to confirm. A few months later, he was convinced of the ultimate benefit, and maybe even the long-term necessity, of sexually reproducing.

He wrote experts for their opinion of this idea. Their replies were not always heartening. The venerable Rev. William Herbert (1778–1847) was a classical scholar, linguist, and clergyman as well as a noted experimenter in plant hybridization. Darwin wrote him wondering about Herbert's declaration that certain plants do not cross: "I ask because I have been led to believe, that . . . every individual occasionally, though perhaps *very rarely*, after long intervals is fecundated by [another individual]."[3] Herbert's reply was equivocal, and he pointed out that if outcrossing was essential, hybridization would be rampant, something we do not see in nature. Darwin second-guessed himself. He wrote in his notebook: "The weakest part of my theory is the absolute necessity, that every organic being should cross with another."[4] Budding and other asexual forms of reproduction meant low variation, so that these species would not be able to change and adapt very quickly. And self-fertilization meant inbreeding, which also led to low variation and a suite of ills ranging from deformity to sterility. Of the two, inbreeding was perhaps the greater problem for species in the long term.

Darwin well knew that the breeders cautioned against close or long-continued inbreeding, which at the time was termed breeding "in-and-in." Plant and animal improvement was (and is) an art as much as a science, finding the right balance between breeding like with like to preserve the desirable characteristics that stem from good combinations of genes and their variants (alleles), and not disrupt these by the inevitable necessity of intercrossing—like a team that plays especially well together but easily loses their mojo when a newcomer joins the group. In the long term, organisms and teams might benefit from at least periodic injections of new blood, but initially that comes at the price of disrupting group dynamics. (Today we understand the genetic basis of the problems stemming from inbreeding: recessive alleles with negative effects are expressed when two copies are inherited, and the incidence of this increases dramatically with inbreeding.) In Darwin's day there was only the empirical observation that inbreeding usually led to reduced fertility and vitality, and that outbreeding led to health and vigor. Darwin was sure that the different outcomes of inbreeding and outbreeding had a bearing on the formation of varieties and species. It's also likely that he felt unease in the knowledge that his own marriage was consanguineous, since he and Emma were first cousins.

Yes, sex weighed heavily on Darwin's mind in more ways than one. But how did this lead him to pry into the secret lives of plants? It started with that nagging question of crossing.

Darwin, Knight, and the Law

Darwin eventually realized that Herbert's objection was not the problem it first appeared because outcrossing was a matter of degree. Hybridization was not rampant because individual animals or plants would only produce healthy offspring if mates were of sufficiently close relationship. At the other end of the spectrum, if individuals are *too* closely related, then the ill effects of inbreeding are felt. Successful intercrossing, or outbreeding, is between these extremes, that golden mean where the maximum benefit is realized.

The very year of Darwin's discovery saw the passing of the breeder who convinced him about the universality of intercrossing. Thomas Andrew Knight (1759–1838) was a renowned Herefordshire horticulturalist. In 1799 the *Philosophical Transactions of the Royal Society* published Knight's "An account of some experiments on the fecundation of vegetables," a work that Darwin seized upon in his quest for general principles of reproduction. Like Mendel later, Knight performed crossing experiments with different varieties of the common garden pea. In one such experiment Knight transferred pollen from large and luxuriant pea plants to diminutive ones, and vice versa. The results on vigor and seed size in the offspring were as unexpected as they were impressive: "I had, in this experiment, a striking instance of the stimulative effects of crossing the breeds; for the smallest variety, whose height rarely exceeded two feet, was increased to six feet."[5] Knight went on to relate mixed success doing similar experiments with wheat, apple, and grape varieties. Nonetheless, he concluded that "improved varieties . . . may be obtained by this process, and that nature intended that a sexual intercourse should take place between neighbouring plants of the same species."[6]

What's more, Knight observed, the correctness of this conclusion is apparent when one considers the many and ingenious ways devised by nature to transfer pollen. In some species pollen is so light that breezes can carry it far and wide. In others the stamens are strategically placed so as to shower visiting insects with pollen, and "the villous coat of the numerous family of bees, is not less well calculated to carry it." So, despite the proximity of male and female parts in the same flower, nature "intended" cross-pollination and had devised a variety of contrivances to ensure that this occurs. But why? Knight thought he had the answer: sexual intercourse "certainly tends to confine within more narrow limits, those variations which accidental richness or poverty of soil usually produces."[7]

In other words, Knight believed, first, that environmental conditions (the quality of the soil) induce variations, something that Darwin also believed and that was not decidedly rejected until the twentieth century with the discovery of Mendel's work. Knight seemed to further believe that crossbreeding acts to spread variation around, a homogenizing process that prevents those environmentally induced variations from getting out of hand. Darwin agreed, but was especially interested in the stimulative effects of crossbreeding. Unlike Knight, Darwin believed that the very *act* of crossing engendered variations. Relying on Knight's work, Darwin developed a working hypothesis: crossbreeding, even if accomplished only occasionally, was essential for the long-term adaptability of species. In 1844, he wrote in his species *Essay*:

> This power of crossing will affect the races [breeds or variants] of all *terrestrial* animals; for all terrestrial animals require for their reproduction the union of two individuals. . . . I cannot avoid suspecting (with Mr Knight) that the reproductive action requires, at *intervals*, the concurrence of distinct individuals. Most breeders

of plants and animals are firmly convinced that benefit is derived
from an occasional cross, not with another race, but with another
family of the same race; and that, on the other hand, injurious consequences follow from long-continued close interbreeding in the
same family.

He called this "Knight's Law," and became so convinced of the critical
importance of crossing that he later dedicated a chapter of over 100
manuscript pages to the subject in his *Natural Selection* manuscript.[8]
While he could observe that "the crossed offspring of two varieties, &
even of two individuals in hermaphrodite plants, have their vigour &
fertility increased," he admitted not knowing why this should be so,
acknowledging "how utterly ignorant we are in regard to Life & its
Reproduction."

Still, Darwin was getting at the very reason behind the phenomena
of sex and sexes. This was the backdrop to many of his experiments and
other investigations through the 1840s and beyond. Darwin's case for
the near-universality of crossing was so compelling that "Knight's Law"
was called the "Knight-Darwin Law" by naturalists in later years, and
he became fascinated by the stunning creativity of nature when it came
to methods of achieving cross-pollination. In the garden, greenhouse,
and meadows of Down House a bevy of experiments and other investigations kept Darwin and his family as busy as the bees he came to
appreciate as the plants' main pollination agents—researches culminating (after literally thousands of experiments and related investigations)
in his remarkable treatises of the late 1870s: *The Effects of Cross and
Self Fertilisation in the Vegetable Kingdom*, published in 1876, and *The
Different Forms of Flowers on Plants of the Same Species* published
the following year. At first glance these books may seem to treat unrelated botanical topics, but to Darwin the significance of crossing was of
a piece with the phenomenon of separate sexes in flowering plants. He
tried to think like a plant, considering reproduction from their point
of view, and came to appreciate the central importance of those six-
legged go-betweens whose abundance and ubiquity make them most

excellent pollen couriers. His investigations proceeded along two partly overlapping paths: he first began a program of documenting insect-mediated cross-pollination, and next documented the beneficial effects of cross-pollination.

Not-So-Secret Agents

Darwin built his case bit by evidentiary bit in characteristic fashion: close observation, meticulous experimentation, extensive inquiries, and, eventually, what we might call crowd-sourcing. One of his early acquaintances and correspondents was Robert Brown (1773–1858), a giant of British botany (and of Brownian motion fame) who, in spring 1841, recommended that Darwin consult Christian Konrad Sprengel's 1793 book *Das entdeckte Geheimniss der Natur im Bau und in der Befruchtung der Blumen* (translated as *Nature's Secret in the Structure and Fertilization of Flowers Unveiled*). In this book Sprengel (1750–1816) argued for something that we take for granted today, namely, that the object of showy flowers is in most cases to attract insects for the purpose of pollination. He meticulously documented a bevy of plant-pollinator relationships but didn't bother with the "why" of these relationships, something that Darwin found disappointing: "Has no notion of advantage of intermarriage," he scribbled in the margin; "Seems to think fact of insects being required at all, does not deserve any explanation."[9] Darwin dismissed as hopelessly naive Sprengel's suggestion that stamens and pistils were found together for the convenience of visiting insects. For all of his criticism, however, he was enthralled with Sprengel's book. He systematically followed up on the botanist's observations and added a great many of his own; there may have been "some little nonsense," but the book was nonetheless "full of truth," he declared to Asa Gray.[10] Years later his son Frank said that Sprengel's book "not only encouraged [his father] in kindred speculation, but guided him in his work . . . It may be doubted whether Robert Brown ever planted a more fruitful seed than in putting such a book into such hands."[11]

Initially, his insect- and flower-watching was done on the side, a relaxing distraction from other pursuits in the early 1840s. At that time Darwin was in the middle of the *"Zoology of the Voyage of the* Beagle" project, which was ultimately published in five taxonomic sections between 1838 and 1843. He effectively acted as editor-in-chief for these volumes, even as he labored over geological projects stemming from the *Beagle* voyage. As we saw in Chapter 1, Darwin had become a rising star in the London geology scene, elected a Fellow of the Geological Society immediately upon his return to England in 1836 and then Secretary just two years later. He gave up the secretaryship in 1841 though—it was all becoming too much, not least owing to the arrival of a new addition to the family: a daughter, Anne Elizabeth, born March of that year. Charles and Emma were exhausted, and he was already showing signs of the mysterious illness that would plague him for the rest of his life. So in late spring 1841 they fled Macaw Cottage for a much-needed respite with family at Maer Hall in Staffordshire, where Emma grew up. They arrived at Maer in late May, 2-year-old William ("Doddy" as they called him then) and infant Annie in tow, to the delight of Emma's parents (and Charles's uncle and aunt) Jos and Bessie Wedgwood.

Charles and Emma both had extremely fond memories of Maer, with its grand Georgian country house set on a low hill, and extensive park and gardens (originally laid out by renowned eighteenth-century landscape designer Lancelot "Capability" Brown). A long, curved, rhododendron-lined drive led to the house, against a knoll overlooking a placid lake ringed by a sandy path the Darwin's loved taking walks on (and would later inspire Darwin's thinking path, the sandwalk, at Down House). There were also acres and acres of surrounding woods and farmland. The Wedgwoods and Darwins were very close—recall that Charles's mother was a Wedgwood, and it was his uncle (and future father-in-law) Jos who had convinced Charles's father to give him a fair hearing on the benefits of voyaging round the world on HMS *Beagle*, an idea that his father initially dismissed as "a wild scheme." Emma was the youngest of eight children, and Charles and his siblings visited often growing up; together the kids had the run of the place.

Emma with her sisters and mother took great pleasure in planting the flower beds, with exuberant, layered clouds of color and texture: reds, yellows, purples, whites, the spreading masses punctuated by gracefully branched beauties and showy spikes. Many years later Charles and Emma's daughter Henrietta (Etty) commented that "Maer Hall . . . was so deeply beloved by the whole group that their children even have inherited a kind of sacred feeling about it. . . . My father used to say that our mother only cared for flowers which had grown at Maer."[12]

And what a palette those flowers made: yellow daylilies, deep purple geraniums, pink azaleas, blue monk's hood, larkspurs, lupines, and campanulas, white fraxinellas, scarlet poppies, multicolored speedwells, heartsease, and many more. On a sunny day the beds were alive with the hum of innumerable busy bees—which meant that Maer's beautiful gardens offered more than peace and quiet for Darwin's frazzled nerves; they were also an outdoor laboratory for testing his latest ideas. Bees and flowers were on his mind, and maybe more: Sprengel's volume was a veritable Kama Sutra of the plant-pollinator world, and Darwin studied the manual assiduously. Every flowering plant had a story, a particular relationship to reveal, and Darwin probed and peered from petal to pistil like a paparazzo to learn the secrets of each. He spent many an hour observing floral structures and the movements of bees as they went from flower to flower, how they probed for nectar, and noting which ones they stopped at and which they passed over, how long their visits lasted, and timing how many blossoms could be visited in a set interval of time. Here is a typical notebook entry:[13]

> Maer. June 15 /41/. Watched plants of Fraxinella, with seven flower stalks for ten minutes, it was visited by 13 Bees—& each examined very many flowers. = 22^d—/during several succeeding days ~~many~~ most numerous bees visited this same bunch & on this day in five minutes eleven Humbles came & each visited many flowers—
> Saw Bees frequent these flowers till late in evening—On rough calc. 280 flowers—allowing each Bee visits 10 flowers in minute each

flower will be visited in 28 minutes—say then each flower is visited 30 times a day is considerably under mark, & this has now gone on 14 days. (except some wet ones/ & w^d go on longer—

His stream-of-consciousness notebooks convey the excitement of discovery. Fraxinella (*Dictamnus albus*), a perennial from the Mediterranean region, is also called "burning bush" or "gas plant" for its highly flammable aromatic oils. Its flamboyant perfumed flowers have white to pink backswept petals and long, protruding stamens—a favorite of bees. Darwin watched a plant with seven flower-laden spikes visited by 13 bees in 10 minutes, but they worked so fast it was difficult to tell just how many individual flowers they visited. Turning that around and estimating how many flowers were visited by a bee in a minute and then extrapolating to the number of available flowers, he figured that each flower was visited about 30 times per day, as a conservative estimate. At that rate transfer of pollen by insect emissaries was virtually guaranteed.

Later that summer, back in London, Darwin weighed in on a discussion in the pages of the *Gardeners' Chronicle*, touched off by a letter signed "Ruricola." The broad-bean crop had been doing poorly that year, and Ruricola laid this at the multiplicitous feet of bumblebees. He spied these bees chewing through the base of broad-bean flowers to extract the nectar, and concluded that this rendered the flowers incapable of fruiting. His advice was both remedy and punishment for the iniquitous bees: "it would . . . be a very desirable object with the bean-grower to avoid such a loss, and no better means could be adopted than to destroy the Humble-bees' nests at the end of summer, and employ children to catch and kill the females in the bean-fields as soon as the first blossoms have expanded."[14] In the course of his bee-watching that summer Darwin had also observed flower-perforating behavior (now called "nectar robbery"), but he had a different take on how it related to failure of the beans.

That August he wrote to the *Gardeners' Chronicle*:[15] "Perhaps some of your readers may like to hear a few more particulars about the humble-bees which bore holes in flowers, and thus extract the nectar," he began. He proceeded to give a litany of examples he had observed: snapdragons and penstemons, salvias and stachys. The rhododendrons he observed at Maer were instructive: while he did witness nectar robbery with these flowers he concluded that most bees nonetheless obtained nectar from them in the usual way, dipping into the corolla, as evidenced by the fact that the specimens were sterile hybrids that could produce no pollen of their own, yet their pistils were covered with many fine grains. He put his finger on the real reason the perforated broad-bean flowers failed to develop fruit. The bees were responsible, not in the way Ruricola imagined, "but by their not extracting the nectar in the manner nature intended them," and therefore failing to transfer pollen. He continued, half-jokingly:

> If all those flowers, even hermaphrodite ones, which are attractive to insects, almost necessarily require their intervention . . . to remove the pollen from the anthers to the stigma, what unworthy members of society are these humble-bees, thus to cheat, by boring a hole into the flower instead of brushing over the stamens and pistils . . . ! Although I can believe that such wicked bees may be injurious to the seedsman, one would lament to see these industrious, happy-looking creatures punished with the severity proposed by your correspondent.

On the contrary, Darwin pointed out with tongue firmly in cheek, florists should praise the bees for their "ingenious method" of obtaining the nectar, since in that way they save wear and tear on the delicate petals, which become ragged when the bees constantly scramble in and out of the flowers: "The little orifice which the bees make, in order to avoid clambering in at the mouth, is hardly visible; whereas all the flowers in some beds of the Mimulus, at the Zoological Gardens, are sadly defaced." (That comment revealed what Darwin was up to when visiting the zoo: rather than viewing the animals, he was in the flower beds making observations.)

Rube Goldberg Flowers

It may seem odd to a modern reader that Ruricola didn't see that the reason for the failure of the beans to develop was want of pollination. But even as late as the mid-nineteenth century the role of insects in pollination was far from clear, or generally accepted. The prevailing assumption for centuries was that plants self-fertilize, and the side-by-side placement of stamens and pistils in flowers of many types reinforced this view and was even cited as evidence of the wisdom of the Creator. Darwin sought to not only establish that insects, especially bees, were the agents of pollination, but even more importantly that the pollen they transferred was by and large from *other* individuals of the same species.

Darwin became more and more intrigued with the quirks and oddities of pollination itself. He noted at Maer in 1841 how in certain flowers the stamens and pistil curved upward so as to place their tips precisely in the path of bumblebees heading for the nectaries (what he called the "fairway" or "gangway"). Rhododendron was a good example, and Fraxinella with stamens that actually moved into position in the gangway as they ripened. Starting out straight as pins, that changed when on a hot day the "anthers all burst . . . & I found their positions all changed & their tips now stood in the direct gangway to the nectary, & were brushed by every Bee which entered."[16] Flowers like those of rhododendron and fraxinella were the simple cases, and no one in Darwin's day would have batted an eye at the seemingly providential placement of the stamens and pistil. That was true of other flower "contrivances" too, but before long Darwin went down the botanical rabbit hole into a world of cross-pollination mechanisms that few in his day could have even *conceived* existed.

There were curious spring-action stamens to fling pollen, lever-action structures where pressure on a petal here results in a stamen poking out over there, flowers where stamens and pistils ripen at different times, and elaborate structures and chambers that force hapless insects down one-way pathways in the flowers or even imprison them for a time like

Great laurel rhododendron (*Rhododendron maximum*) flowers. Darwin studied the up-curved stamens and pollinator "gangway" of rhododendron and a host of other flowers. Photograph by the author.

floral houses of horror. Darwin shed light on odd flower morphs, discovering in some cases as many as three coexisting in the same species, and the fact that many of our favorite botanical beauties supplement their showy blossoms with small hidden flowers that never open, which they seem to use as insurance policies against poor insect pollination years. Rube Goldberg, the master of intricately complicated devices (incidentally born the year after Darwin's death), would have appreciated the odd contrivances for pollination that Darwin elucidated.

Sprengel was Darwin's initial guide, and one of the plants he discussed was the common barberry (*Berberis*). Barberries are the often-disparaged hardy plants associated with uninspired plantings at malls and gas stations (further disparaged by knowing biologists because these are aggressively invasive plants in North America, and host of the parasitic stem rust fungus that attacks wheat). But they have their charms, as Darwin appreciated: barberries provide one of the best examples of spring-action stamens. Small clusters of yellow flowers dangle from slender petioles. As the flower matures, the stamens cock back like catapults, loaded with pollen. At the slightest touch by a bee the stamens spring, showering the visitor. Darwin credited the German botanist

Joseph Gottlieb Kölreuter (1733–1806), of Karlsruhe, with showing this mechanism as early as 1788. (In fact today Kölreuter, more than Sprengel, is considered the father of insect pollination biology.)

Darwin realized that the maturation of male and female parts of the flower were timed to reduce the chances of self-pollination, and also served as a device to control the timing of pollen pickup and delivery by insects. That's what happens in salvia: the stamens ripen first, and dab visiting bees with pollen via a lever mechanism. The immature pistil is held out of the way at this stage, but as the flower ages the stamens wither while the pistil has a growth spurt, elongating dramatically to hang down at the front of the gangway. Pollen-bearing bees visiting flowers at this stage need to squeeze under the pistil, the stigma scraping pollen from the bee's back. Darwin marveled at the contrivance: "I can no more doubt the final cause [purpose] of this structure than I can of a certain mouse-trap."[17] He marveled too at a rather different and bizarre form of pollination found in certain arums. This group has an inflorescence characterized by a *spadix*—a distinctive column or spike crowded with tiny flowers—and a leafy bract called a *spathe*, which often surrounds the spadix like a hood or cape. The common Peace lily (*Spathiphyllum cochlearispathum*) of florists' shops is a good example of the general morphology, with a spadix hooded by a snow-white spathe. Darwin was intrigued by a common British woodland species, *Arum maculatum*, variously called lords-and-ladies, cuckoo-pint, friar's cowl, or Jack-in-the-pulpit.

In these arums, the purple, poker-shaped spadix is partly hooded by a green spathe. Unlike the spadix of many species, the part that is visible does not actually bear the flowers. That's where things got interesting for Darwin. Lower down, the spadix disappears into a bulbous chamber formed by a swelling of the spathe base. The spadix shaft in the chamber bears three types of flowers, each arranged in a ring around the shaft: at the very bottom are female flowers, which produce nectar, above which is a ring of male flowers, and above them at the top of the chamber and practically blocking its narrow neck is a ring of sterile flowers with bristly filaments. These filaments prevent large insects from entering, but it had long been known that small ones, especially certain flies, can get past them and into the secret chamber. Like many arums this plant produces a scent repugnant to our nose—one author described it as "foul and urinous"—but irresistible to flies, its main pollinators. The attraction is all the greater because these remarkable plants can generate heat through a process called thermogenesis, which helps volatilize the odorous compounds and provides a cozy refuge for visiting insects. (The early-blooming North American arum skunk cabbage, *Symplocarpus foetidus*, can generate enough heat to melt snow.)

Leading botanical authorities of Darwin's time assumed the hapless flies flew into the chamber and never saw the light of day again. Darwin assumed otherwise since these plants are not carnivorous. In March of 1842, on a visit to see his father and sisters at Shrewsbury, he

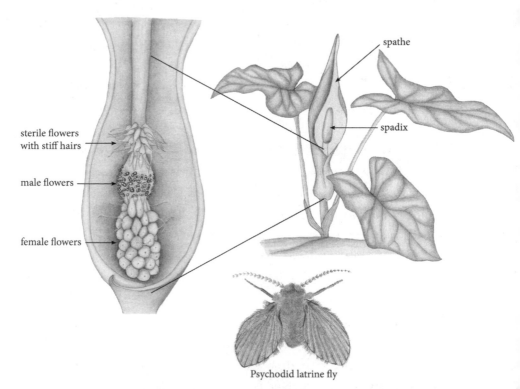

Darwin investigated the pollination biology of wild arum (*Arum maculatum*) in 1842, showing that this plant entices unwary midges into the bulbous chamber below the spadix, releasing them once they are doused in pollen. Drawing by Leslie C. Costa.

dissected the bulbous spathes of a few specimens and found 30 to 60 small midges trapped inside. Many were indeed dead, but others were alive, pollen-dusted and crawling a bit groggily around inside. Would they have been doomed had he not opened the chamber? In *Effects of Cross and Self Fertilisation in the Vegetable Kingdom,* published 34 years later, he explained:

> In order to discover whether the living ones could escape and carry pollen to another plant, I tied in the spring of 1842 a fine muslin bag tightly round a spathe; and on returning in an hour's time several little flies were crawling about on the inner surface of the bag. I then gathered a spathe and breathed hard into it; several flies soon crawled out, and all without exception were dusted with arum pollen. These flies quickly flew away, and I distinctly saw three of them fly to another plant about a yard off; they alighted on the inner or concave surface of the spathe, and suddenly flew down into the flower.[18]

Looking inside, he confirmed that the flies had transported pollen. What's going on with this dipteran chamber of horrors? The ring of filaments of the sterile flowers at the top of the chamber forms an insect trap. Attracted by the odor within, the flies squeeze their way through the barrier, but finding their way out again is not so easy. This arum is *protandrous*, with male flowers maturing first: by the next day the large anthers of the male flowers ripen and shed lots of pollen, covering the duped flies as they walk around the chamber trying to escape. Some die in the process, their bodies littering the floor of the chamber. Within about 24 hours the filaments blocking the entrance wither and the pollen-covered survivors fly free at last . . . only to be quickly duped by another arum. By that time the male flowers have faded and it's the female flowers' turn to mature. Their stigmas become receptive just as the air is filled with the groggy pollen-dusted flies. Chances are small that they'll reenter the same flower that duped them the first time, so when they're conned again they are likely to facilitate cross-pollination.

A different form of protandry that Darwin studied is found in the common foxglove (*Digitalis purpurea*), a perennial garden favorite with its spike of large, deep-throated purple spotted flowers. This species is mainly visited by large bumblebees, and has a neat trick to ensure cross-pollination. A mature plant has from around 10 to dozens of flowers arranged in a spiral up the stem, at the top of which new flower buds are continually produced. Being protandrous, in a given well-developed spike the lowermost flowers are older and so have passed into the female phase, while the middle to upper flowers are still male. Darwin noticed that bumblebees tended to start at the bottom of the spike of one individual (female flowers) and work their way up, picking up pollen from the upper (male) flowers before leaving the plant to find another. At the next plant they go to the lower, female, flowers first again, setting the stage for cross-pollination. (More recent studies found that the female flowers also produce more nectar than do male flowers, a big incentive to ensure the bees start there.)

Ever the experimentalist, Darwin wanted to test if these plants depend on crossing for good seed set and vigorous growth. The opportunity to work with large wild populations presented itself in 1869 while on a family holiday in the scenic coastal town of Barmouth, Wales, on Cardigan Bay. This holiday was meant to be more recuperative than usual for him, since earlier that year he took a serious tumble off of his horse and was lucky not to have been seriously injured or killed. As it was he was pretty sore, but the peace and quiet of field work at the seaside worked wonders. He found a spot thick with foxgloves overlooking the harbor and estuary and set up an experiment: "I covered a plant growing in its native soil in North Wales with a net, and fertilised six flowers each with its own pollen, and six others with pollen from a distinct plant growing within the distance of a few feet. The covered plant was occasionally shaken with violence, so as to imitate the effects of a gale of wind, and thus to facilitate as far as possible self-fertilisation."[19] That would have been a peculiar sight for passing day-trippers, had any spotted a man out in a field seemingly throttling a netted foxglove plant. The little study bore fruit in more ways than one: he found that of the 92 flowers on the netted plant (including the 12 he had hand-pollinated), only 24 seed capsules were produced (and just two of those were full of seeds). All around the plants left uncovered were, in contrast, chock-full of seeds.

He carefully collected the seeds, keeping those from self- and cross-pollinated flowers separate. On returning home he germinated them in his greenhouse and, once they were large enough, transferred the seedlings to the garden. The following summer he measured the flower spikes produced by the surviving plants and found a whopping difference in vigor: the cross-pollinated plants grew to 51.33 inches on average, while the surviving self-fertilized plants were 30 percent smaller, at 35.87 inches. (Mortality was higher with self-pollination too.) This study was yet another striking demonstration of the power of cross-fertilization, another brick in the evidentiary edifice he was building.

To Cross or Not to Cross

Darwin realized that some plants self-fertilize, some even frequently, but he was convinced that most are as indifferent to their own pollen as they are to pollen from distant genera or families. He had little concept of why, but today we understand that self-incompatibility, as the phenomenon is called, is controlled by a simple genetic mechanism involving one or more compatibility genes. These genes consist of many different variants (alleles) in the collective gene pool of the species, but in good Mendelian fashion each individual plant contains just two alleles, one copy of each gene inherited maternally and the other paternally. The self-incompatibility genes have so many variant alleles that individuals are nearly always heterozygous—containing dissimilar variants of the gene obtained from each parent, as opposed to identical variants (in which case they would be homozygous for that gene). In fact, if a pollen grain landing on a stigma possesses the same compatibility allele or alleles as those of the plant bearing the stigma, it will fail to develop a pollen tube and thus cannot fertilize the flower. Since self-produced pollen has by definition the same alleles as those found in ovules of the same individual, self-fertilization is prevented.

This is a basic description of self-incompatibility, but bear in mind that things are more complex than that in the botanical world. While exclusive selfing, as self-fertilization is called, may be rare, in lots of species some degree of selfing is not only tolerated, but ensured. Naturally, Darwin was intrigued, and in May 1862 he wrote Hooker about his investigation of a phenomenon later dubbed *cleistogamy*, production of tiny specialized flowers that self-pollinate. Nearly half of all flowering plant species exhibit "mixed-mating," producing both selfed and outcrossed progeny, and many plant groups even sport specialized flowers that exclusively self-pollinate right in the bud, not even bothering to open. These so-called *cleistogamous* (closed-breeding) flowers are far more common than even the keenest gardeners likely realize: a comprehensive review conducted in 2007 found that nearly 700 species distributed among 228 genera and 50 plant families bear cleistogamous flowers. Among them are a host of garden favorites: many violets, orchids, sorrels, and legumes among others produce cleistogamous flowers along with their showy (*chasmogamous*—open-breeding) flowers. You can be excused if you haven't noticed the cleistogamous ones—not designed for visibility since pollinators are irrelevant, they tend to be small and bud-like, produced on pale stems close to the ground. They either lack petals altogether or the petals are reduced to mere scales, nectaries are dispensed with, and even the stamens and pistils are rudimentary.

At the time, Darwin was cogitating over insect pollination, plant crosses, insectivorous plants, orchids—anything *but* working on his domestication book. "I have been amusing myself by looking at the small [cleistogamous] flowers of *Viola* . . . what queer little flowers they are," he wrote Hooker.[20] He was trying to figure out how the pollen produced in the closed flower gets transferred to the stigma, and had written to the ever-obliging Daniel Oliver at Kew for help. With correspondents like Oliver he investigated seed production by cleistogamy in violets and other flowers, and delved into the literature to see how widespread the phenomenon was. Fifteen years after that letter to Hooker he dedicated an entire chapter in *Forms of Flowers* to this curious phenomenon, including the results of many experiments. He concluded correctly that there is a twofold advantage to cleistogamy. When regular pollination fails due to adverse conditions like bad weather or pollinator paucity, the cleistogamic flowers act as an insurance policy, selfing to ensure seed production. Second, they can produce an abundance of seeds with very little investment since the flowers are small, bud-like, and lack nectar.

The idea of small, closed, selfed flowers as a reproductive insurance policy made sense to Darwin, but selfing by regular flowers strutting their floral stuff to entice pollinators was puzzling. The latter seemed to be especially common in leguminous plants, a group that includes peas, beans, and their relatives—a convenient group for study. In October 1857 he wrote a letter to the *Gardeners' Chronicle* describing the lever-action mechanism of kidney beans: the style is curved in a left-hand loop ("like a French horn") within the modified petal called the keel. The stamens are housed within this looped tube as well. The two lower petals protrude at the bottom and present a landing platform for visiting bees. Their weight depresses these petals, and the stigma at the tip of the style pops out of its looped tube. Darwin noted that just below the stigma the style has a brush of fine hairs that is moved backward and forward each time a bee lands on the flower. Like a botanical pipe cleaner, the brush sweeps shed pollen out of the tube, gradually pushing it onto the stigma. "Hence the movement of the pistil indirectly caused by the bees," Darwin wrote, "would appear to aid in the fertilisation of the flower by its own pollen."[21] He mimicked the effect of the bees by gently pulling the wing-petals, and noticed that, at the same time, a visiting insect would get a dose of pollen too. Some pollination thus likely occurs by crossing in addition to selfing, he surmised. He tested his idea with an enclosure experiment. In each of three replicates he covered two groups of flowers with netting. One group he left undisturbed, but once a day he reached under the netting of the other group and manually tugged on the protruding lower petals of each flower, simulating a landing bee. Then he waited for seed set. "Not one of the undisturbed flowers set a pod," he reported, "whereas the greater number (but not all) of those which I moved . . . set fine pods with good seeds."

In an 1858 follow-up article he gave the results of a scaled-up version of the experiment. The uncovered plants were nearly three times as productive as the covered ones. Darwin found it remarkable that even though the beans self-pollinate, they rely on insect agency to do so—leading him back to Knight's Law and his conviction that they are cross-pollinated at least occasionally by bees. In the long haul of ecological and evolutionary time, even rare outcrossing is all it takes to reinvigorate the population and ensure continued fertility and adaptability. Darwin hypothesized that selfing eventually leads to reduced pollen potency, whereupon pollen from other individuals—nearly always present at low concentration owing to the movement of the bees—will be the only pollen that can fertilize. The reinvigorated progeny can then get away with selfing again for a period of time. We don't quite look at it that way today—it's not that the potency of pollen declines, but rather that too much selfing permits the expression of deleterious recessive genetic traits. But his instinct was basically correct that cross-fertilization is where it's at, evolutionarily speaking. To Darwin this seemingly simple idea had far-reaching implications, from the origin of sexes to coadaptation to what we would call ecological interconnectedness in the web of life.

One of Darwin's most famous examples of interconnectedness found in *On the Origin of Species* stems, in fact, from a pollination experiment. Indeed, this work on bees and red clover may be the earliest discussion of an ecological food chain in the scientific literature. Insofar as the food chain idea underpins the seminal concept of ecological food webs, Darwin's clover experiment is foundational in the history of the science of ecology. He described a food chain based on flower-visiting bees: based on experiments, he said, "I have found that the visits of bees, if not indispensable, are at least highly beneficial to the fertilisation of our clovers." Insofar as bumblebees are the main visitors to common red clover and heartsease (violets), "I have very little doubt, that if the whole genus of humble-bees became extinct or very rare in England, the heartsease and red clover would become very rare, or wholly disappear." But the chain of influence extends further than that:

> The number of humble-bees in any district depends in a great degree on the number of field-mice, which destroy their combs and nests . . . Now the number of mice is largely dependent, as every one knows, on the number of cats; and Mr. Newman says, "Near villages and small towns I have found the nests of humble-bees more numerous than elsewhere, which I attribute to the number of cats that destroy the mice." Hence it is quite credible that the presence of a feline animal in large numbers in a district might determine, through the intervention first of mice and then of bees, the frequency of certain flowers in that district![22]

Darwin's concluding exclamation mark is an indication of the importance that he ascribed to this unexpected conclusion. In pitching his argument in the *Origin* for the reality of evolution by natural selection he needed his readers to appreciate how little we understand about interconnectedness, how individuals and populations and species each have an impact and influence on others in myriad ways we can only dimly envision. This simple food chain is but one tiny thread of connectedness in an extraordinarily complex tapestry of nature, a small exemplar that represents an infinitude of threads, most unseen.

The clover experiments were conducted in mid-August of 1859 in Great House Meadow just behind Down House, maybe as a pleasant diversion as he was working away correcting the proofs for the *Origin*. Bordered by the kitchen garden and the sandwalk, Great House Meadow is a classic English meadow that has probably never been violated by a plow, teeming today as in Darwin's day with grasses and forbs—including lots of red clover, its low patches revealing a scarlet gleam here and there in the grass. In season the meadow swarms with bees, their Doppler hums and buzzes noting their comings and goings. It was another enclosure experiment, as he reported in more detail later in his 1876 *Effects of Crossing* book: "One hundred flower-heads on plants protected by a net did not produce a single seed, whilst 100 heads on plants growing outside, which were visited by bees, yielded 68 grains weight of seeds; and as eighty seeds weighed two grains, the 100 heads must have yielded 2720 seeds"[23]—a potent demonstration of the effects of bee visitation.

As informative as the experiment was for Darwin, by modern standards there are, of course, a number of problems with his design. For one thing the heavy gauze used for his enclosures would have done more than exclude bees; it would have shaded the plants too. Starved of adequate light, perhaps these plants just didn't have energy enough to set seed even by selfing. Another problem is the single covered plot; this was a one-off experiment, informative enough, but today both enclosed treatments and uncovered control plots (both of standard dimensions) would be replicated, and their placement in the clover field might be randomized for added statistical rigor. Such details of experimental design were not appreciated in Darwin's day, and it's not fair to hold him to modern standards. But an even bigger potential problem threatened to break his chain. Darwin maintained that only bumblebees visited red clover since the probosces of honeybees are too short to reach the nectar. If that was not the case, however, then the final link in his causal chain leading from cats to field mice to bumblebees to red clover was in jeopardy. Soon after the *Origin* came out an old acquaintance wrote Darwin to report that he observed bees of different kinds, including honeybees, visiting red clover, but he qualified this by noting that he observed them on mown clover, which reputedly resprouts with smaller blossoms that are perhaps more accessible to shorter-tongued bees like honeybees.

Darwin started looking out for cases of honeybees visiting red clover. While he, Emma, and the younger kids visited William at Southampton in the summer of 1862, he naturally made time for fieldwork. He scoped out a field of red clover with lots of honeybees, and found that some dipped into the clover flowers the usual way while others sucked the nectar through holes left by nectar robbers. A thought suddenly struck him: maybe some of these bees had long probosces while others had short probosces; the long ones could reach the nectar from the mouth of the flower, but the short ones had to resort to nectar robbery instead. This could be a form of division of labor— potentially a very interesting discovery. He collected a number of bees of both "types" (at the flower mouth versus nectar robbing) before the family departed Southampton for Bournemouth, where they planned to continue their holiday. He couldn't find any clover fields there to continue his observations, however, much to his frustration. He was so sure he was onto something that he excitedly wrote to Lubbock, a devotee of ants, bees, and wasps, to observe bees on clover and collect specimens for him: "for Heaven sake catch me some of each & put in spirits & keep them separate—I am almost certain that they belong to two castes, with long & short probosces. This is so curious a point that it seems worth making out."[24] His letter was premature, alas: when he finally got around to examining the bees collected in Southampton he realized there was no difference in proboscis length after all. He hastily fired off another letter to catch the evening post: "I beg a million pardons," he wrote Lubbock. "I do so hope that you have not wasted any time for my stupid blunder.— I hate myself I hate clover & I hate Bees."[25]

Darwin made mistakes too, and that's the nature of research. Long- and short-tongued honeybee castes don't exist, but it was a neat idea. As for the bumblebees, although Darwin was basically correct that these bees play a leading role in the pollination biology of red clover, he seemed to assume that they all had long probosces, whereas some four species are known to occur at Down House, two longs and two shorts. But how is it that sometimes honeybees visit this plant and at other times they ignore it completely? The discrepancy probably lies in *flower constancy*: the phenomenon whereby honeybees tend to focus their efforts almost exclusively on the most profitable flowers available at a given time. They communicate the location and type of flower of interest through their waggle-dance, calling in reinforcements to work with them. Bees finding a good resource patch will dance to recruit their hive-mates to join them there, but the intensity of the dance is related to the quality of the nectar and pollen available. This sets up a dynamic of competing recruiters, where the bees working the best patches recruit bees away from the others, and soon all the foragers of a colony are working on what they perceive as the best available flower patch. So, sometimes red clover is the best thing going and the honey-bees work it like there's no tomorrow (including taking advantage of the holes left by their nectar-robbing bumblebee cousins), and other times there's a better deal out there and they won't give red clover even a passing buzz. Such are the allegiances of honeybees, a bit like a circle of avid bargain hunters who periodically compare notes on the prices they've found; before long they have all flocked to where the best deals are to be had and snub stores with yesterday's bargains. Honeybees, too, shop around. Darwin would have deeply appreciated the marvel-ous honeybee waggle-dance and its competition dynamic—a kind of selection process, after all—but this astonishing phenomenon was dis-covered in the twentieth century.

Insofar as bumblebees are not the sole pollinators of red clover, Darwin's food chain is not, in this instance, quite the linked series of interdependency that he thought. But nature is more complex than we sometimes think. In 2007–2008 the intrepid trio Norman Carreck, Toby Beasley, and Randal Keynes repeated Darwin's bee enclosure experiment at Great House Meadow and found bumblebees visiting the clover, just as Darwin did. When it came to setting seed, the unen-closed red clovers of Great House Meadow didn't fare much better than those in the enclosures. Plants have their good years and their bad years, owing to the vagaries of wind, weather, pollinators, and other exigencies; some years there's a good seed set, other years are a bust, and the reasons why are not always clear. Although Darwin did not replicate his enclosure experiments across seasons or years, he likely would have viewed season-to-season variation as mere noise. Time is of the essence: it's the long haul that matters, he would have argued, sure that insect intermediaries are key. In his *Natural Selection* manuscript Darwin related how Kölreuter was astonished, upon realizing that certain flowers could only be pollinated by insects, that something as important as fertilization would be left to the whims of such creatures. Then Kölreuter thought the better of it, and concluded that with insects the Creator in his wisdom had actually used the surest means possible for transferring pollen. Darwin concurred: "hardly any means, I am convinced, could be surer" than ubiquitous, abundant, ever-active insects.

Modern evolutionary biologists would largely agree, in view of the fact that the dramatic evolutionary diversification of the flowering plants in the Cretaceous period is largely coincident with the rapid diversification of bees and other major insect pollinator groups. Striking specialist insect-floral relationships such as those found in orchids, yuccas, and figs practically scream coevolution, where two organisms become mutually adapted to one another. But the relationship in most cases is more diffuse, with several different kinds of pollinators capable of pollinating a given plant species and several plant species capable of providing nectar and pollen to a given species of pollinator. Even so, features seemingly tailored to entice and (usually) reward insect visitors abound. The most obvious are the flowers themselves, the ones with showy petals, sepals, or bracts, often with nectar guides, those stripes and streaks pointing the way to the reward center like runway lights and painted lines guiding pilots. Tellingly, nectar guides are intended for insects' eyes, being especially prominent (and in some cases *only* visible) at ultraviolet wavelengths, which insects can perceive while vertebrates like us cannot. The insightful Sprengel saw the stripes and other patterns on petals as advertisements for insects. Darwin missed this initially, but later he acknowledged that there may be something to it based on a little experiment. "There can be little doubt," he wrote in *Natural Selection*, "that C. C. Sprengel has pushed his views to quite a fanciful degree; as for instance when he accounts for all the streaks of colour on the petals, as serving to guide insects to the nectary. Nevertheless some facts could be given in favour of such a view: Thus in a

patch of the little blue Lobelia, which was incessantly visited by Hive
Bees, I found that the flowers from which the corolla, or the lower
streaked petal alone had been cut off, were no longer visited."[26] He
wasn't sure if removing those streaked petals made the flowers seem
past-peak to the bees and so not worth visiting, or simply deprived
them of a convenient landing place, but it was possible that the flowers
no longer had a signpost advertising the nectaries.

Legitimate and Illegitimate Marriages

Another of Darwin's adventures in pollination is the subject of perhaps
the most important of the "multiplicity of experiments" that he began
in the early 1860s: flower polymorphism, or the occurrence of two or
more distinct flower forms in the same plant species. That may not
sound momentous, but in fact Darwin succeeded in shedding light on a
phenomenon that had puzzled botanists for centuries.

It all started in the spring of 1860, in the months following the publication of the *Origin* in November 1859. It was a tumultuous period
for Darwin, at least internally. November found him seeking hydropathic treatment in Yorkshire while he waited for the *Origin* to come
out. If his illness was brought on by stress he surely had plenty, judging by his rashes and churning stomach. He thought the *Origin* would
bring down a firestorm of rage and condemnation on his head. That
never materialized, but it did touch off controversy as naturalists and
clergy weighed in for or against his theory. He was still in Yorkshire
when one of his old Cambridge mentors, Rev. Adam Sedgwick, wrote
a scathing, if heartfelt, letter filled with pity, outrage, and disappointment. Henslow on the other hand said that Darwin should be given
a fair hearing, and other clerics, like Charles Kingsley (author of *The
Water Babies*), wrote to register support and declared that Darwin's
ideas were emphatically *not* at odds with religion. Darwin took heart;
already hard at work on corrections for a second edition, he sent thanks
to Kingsley and asked if he could quote him in the new edition. Kingsley was happy to oblige.

Darwin headed home on December 7, 1859. Emma, the girls, and the youngest of the boys welcomed him—21-year-old Willy was at Cambridge by then, occupying his father's old rooms at Christ's College, and 15-year-old Georgy and 12-year old Franky were at Clapham School. (In keeping with the chauvinism of the times, Etty, then 16, and Bessy, just 12, were tutored at home and did not receive the depth or breadth of education that their brothers enjoyed.) He continued to field letter after letter over the newly published *Origin* and labor over corrections; the second edition came out on January 9, 1860, while Darwin continued to edit material for the first American edition.

That spring found him investigating how stamens and pistils bend, picking up on his observations from the 1840s. By the end of April 1860 he had sent a lengthy list of flowers to Hooker for review. He wondered if he was correct in thinking that, as a rule, pistils tend to be bent in the direction of the nectaries. "Why I care about it," he wrote, "is that it shows that visits of insects are so important, that these visits have led to changed structure."[27] But just a few days later he was interrupted in this project when he received a letter from Henry Doubleday, an authority on primroses, the sunshine-yellow tubular flowers of woodland, glade, and garden. There was an ongoing debate at the time over whether the common primrose (*Primula*) and the similar cowslip and oxslip represented three distinct species or if one of these was the product of hybridization of the other two. In the *Origin* Darwin cited the group as examples of strongly marked varieties ranked by some naturalists as distinct species—those "doubtful forms" that so blurred the line between species and variety that even botanists disagreed on their status. In fact the genus *Primula* is complex, with as many as 500 species recognized by some botanists. Many hybridize, and myriad cultivars have been developed. Today the three that were of interest to Darwin are recognized as separate species: common primrose (*Primula vulgaris*), cowslip (*P. veris*), and the rare oxslip (*P. elatior*).

Evidently Doubleday's letter prompted Darwin to take another look at his primroses, but this time he noticed something that he was only dimly aware of before: there appeared to be two distinct flower morphs. Some plants had long stamens and a short pistil while others had short stamens and a long pistil. On May 7th he dashed off an excited letter to Hooker:

> *I have this morning been looking at my experimental Cowslips & I find some plants have all flowers with long stamens & short pistils which I will call "male plants"—others with short stamens & long pistils, which I will call "female plants". This I have somewhere seen noticed, I think by Henslow; but I find (after looking at only two sets of plants) that the stigma of male & female is of slightly different shape & certainly different degree of roughness, & what has astonished me the pollen of the so-called female plant, though very abundant, is more transparent & each granule is exactly only ⅔ of size of pollen of the so-called male plant.— Has this been observed??[28]*

It dawned on him that Henslow had pointed this out years and years ago, but he had somehow forgotten about it. In fact, his own children knew about the phenomenon—long-pistiled flowers are called "pin-eyed" as the stigma is round like the head of a pin, poking just outside the corolla tube, while the long-stamened "thrum-eyed" flowers had elongated anthers somehow reminiscent of those bits of weaving-loom yarn waste. Probably for millennia kids had known that the pin-eyed flower morphs were best for stringing the blossoms into flower chains by threading the corollas over the long pistils.

These flowers made him sit up straight; he dissected several and peered at their pollen under the microscope. "It may turn out all blunder," he said to Hooker, but he planned on tracking the seed production of each morph. He noticed that the female flowers seemed to produce smaller pollen grains than the male flowers, and it struck him that he may have stumbled upon plants caught in the act, as it were, of evolving into separate sexes. "It would be fine case of gradation between an hermaphrodite & unisexual condition," he enthused in his letter to

Hooker. He wrote to Henslow too, hopeful his old mentor could give him other cases to investigate.

This phenomenon, called *heterostyly*, is found in many plant groups but it was a stroke of luck that it occurred in so widely cultivated a group as primroses. Darwin sent the kids (except Etty, who had just come down with typhus) on a mission to collect cowslips. He opened his experiment book entry of May 13th with the results: "My children gathered great bunch of Cowslips 79 stalks were male flowers, & 52 were female flowers."[29] By the next day they had added many more, for a grand total of 281 males and 241 females. His working hypothesis was that these plants were transitioning to dioecy, where male and female flowers are borne on different individual plants of the species. Long pistils and small pollen grains suggested this morph was becoming more female, while short pistils and larger pollen signaled this one was becoming more male. It was a neat and tidy hypothesis, and an exciting one too since, if correct, these primroses would provide a textbook example of an evolutionary dynamic. Unfortunately, however, the primroses ended up presenting a different kind of textbook case: exemplars of that "great tragedy of Science," as Huxley so memorably quipped in his 1870 presidential address to the British Association for the Advancement of Science, namely "the slaying of a beautiful hypothesis by an ugly fact." Darwin expected that the "male" form would exhibit lower seed production, but in fact he found the opposite.

Undaunted, Darwin soon appreciated what was actually going on: the dimorphism of flowers promoted intercrossing *between* the morphs—it was a mechanism to minimize selfing and promote crossbreeding. He explored this systematically over the course of the next year, performing crosses and making observations . . . at least, as systematically as he could, given his simultaneous infatuation with orchids, insect pollination mechanisms, tendrils, and sundews! He noted that both morphs produce nectar and are visited by bees and other insects. However, they are not visited by the same insects, and in any case pollen of the two morphs is deposited in different locations on the probosces of those visiting: around the base for insects coming to the long-stamened form, and nearer the tip for visitors coming to the short-stamened form. Thus, at subsequent floral stops any basal, long-stamened pollen is most likely to be left atop long-styled pistils, and apical short-stamened pollen is most likely to be deposited on short-styled pistils. Did this matter? He hand-pollinated in every pairwise arrangement—long stamen/long style, short stamen/short style, long stamen/short style, and short stamen/long style—and then counted up the number of resulting seed capsules, and weighed the mass of tiny seeds they yielded. The results were clear: nearly twice the weight of seed was produced in long-long and short-short fertilizations than in the long-short and short-long cross-bred fertilizations.

Darwin was excited about his discovery, and wrote his most trusted botanist friends to share the news and pump them for information on other cases of floral dimorphism. Asa Gray wrote back with a list of candidates, and enthused that Darwin's "discovery that the *pollen of one is good for the pistil of the other, but not for the pistil of its own flower, is most important*" (emphasis Gray's).[30] That was October 1861, by which time Darwin had conducted exhaustive crosses and was putting the finishing touches on a paper for the Linnean Society: "On the two forms, or dimorphic condition, in the species of *Primula*, and on their remarkable sexual relations" was read on November 21, 1861, and published in the Society's journal the following year. Remarkably, this was one of some 15 papers, notes, and letters Darwin published that year, on diverse subjects reflective of his expansive curiosity and research program: fertilization of orchids and other species, horse breeding and coloration, causes of variation in flowers, mating behavior of bumblebees, the influence of brain shape on bird behavior . . . all besides revisions to the *Origin* (the third edition came out in April) and endless correspondence over issues raised by the *Origin* and other matters. No wonder progress on his domestication book project proceeded at a snail's pace. As we shall see in the following chapter, it was in fact orchids that took center stage immediately following the *Origin*, but he plugged slowly away on cross-pollination and the significance of flower morphs.

How he would have loved to have understood the light shed on heterostyly by modern genetics. His beloved primroses became the model system, largely owing to his extensive work with them. There is a fairly straightforward genetic mechanism behind pin and thrum plants. Initially the phenomenon was thought to be controlled by a single gene, called S, but more sophisticated study revealed that the S locus itself consists of at least three closely linked genes that essentially act as one: G (controlling pistil length), P (controlling pollen size), and A (controlling anther length). In general, pin morphs can be treated as *ss* homozygotes for the recessive *s* allele, while thrum plants are *Ss* heterozygotes. Crosses between pin (*ss*) and thrum (*Ss*) plants yield a 1:1 ratio of pin to thrum offspring, something that Darwin recognized.

Darwin dubbed the fruitful unions "legitimate" and the less fruitful "illegitimate," and true to form he was determined to test the generality of his *Primula* results. The pioneering German botanist Karl Friedrich von Gärtner (1772–1850) had long since joined Sprengel and Kölreuter as a sort of Holy Trinity of plant pollination and hybridization in Darwin's pantheon. In 1863 he even wrote a letter published as "Vindication of Gärtner" in the *Journal of Horticulture* in response to disparaging remarks made by a commentator. Darwin repeated a multitude of Gärtner's crosses and pursued clues to interesting phenom-

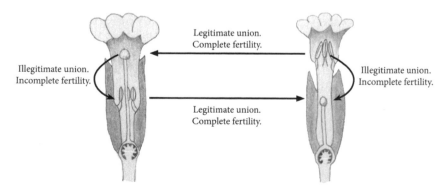

"Legitimate" and "Illegitimate" crosses between long-styled "pin" and short-styled "thrum" *Primula veris* flowers. Redrawn from Darwin (1877), p. 27, by Leslie C. Costa.

ena suggested in his work. He especially admired Gärtner's meticulous crossing experiments and diligent follow-through with the necessary but often neglected step of counting the seeds produced by experimental crosses to quantify fertilization success. He would adopt Gärtner's approach in his heterostyly experiments.

In the meantime, with Gray's help Darwin confirmed heterostyly in Partidgeberry (*Mitchella repens*), a creeping evergreen common in eastern North America and a nineteenth-century favorite for yuletide decoration. This woodland plant bears small paired white tubular flowers in the leaf axils, giving rise in fall to a single bright red berry formed by the fusion of the two carpels. Gray also had his student Joseph Trimble Rothrock make observations and experimental crosses on the common bluet (*Houstonia caerulea*), confirming another case. When Darwin's attention was drawn to common flax, genus *Linum*, his son William assisted by posting over 200 specimens from the Isle of Wight. This interesting case warranted yet another paper for the Linnean Society.

Mad over *Lythrum*

In late 1861 he made another startling discovery: a species with *trimorphic* flowers. It was hiding in plain view: colorful stands of purple loosestrife (*Lythrum salicaria*) and its relative thyme-leaved loosestrife (*L. thymifolia*) are a familiar sight in England along rivers, streams, and other wet places, their tall waving spikes of conspicuous purple flowers giving vivid splashes of color amid the lush green banks. (Purple loosestrife is now a familiar sight along American waterways too, where it is considered a noxious weed.) He came across the account of multiple flower morphs in loosestrifes while making his way through the encyclopedia of botanical geography by Henri Lecoq, director of the natural history museum and botanical garden at Clermont-Ferrand, in central France. The nine volumes were tough going, but well worth the effort to find a prize like this. He had to investigate at once: he thought this might be the most complex plant reproductive system yet known, and what better evidence that heterostyly has nothing to do with evolution of dioecy (separate sexes) than a species with not two but *three* distinct flower morphs!? Frustratingly, winter was bad timing for this revelation. By late March, Hooker supplied him with plants. Darwin's orchid book was now with the printer (it came out on May 15, 1862), so he could happily turn his attention to the secret life of *Lythrum*.

In no time he confirmed that there are indeed three distinct floral morphs: short-, medium-, and long-styled. There are three types of corresponding stamens too, and for each pistil length two of the three stamen types are found in the same flower: the long- and mid-styled flowers both bear long and short stamens, while the short-styled morph bears long and medium stamens. Thus, as Darwin later described it:

> In the three forms, taken together, there are thirty-six stamens or males, and these can be divided into three sets of a dozen each, differing from each other in length, curvature, and colour of the filaments, in the size of the anthers, and especially in the colour and diameter of the pollen-grains. Each of the three forms bears half-a-dozen of one kind of stamens and half-a-dozen of another kind, but not all three kinds. The three kinds correspond in length with the three pistils: the correspondence is always between half the stamens borne by two forms with the pistil of a third form.[31]

Whew! But if you think that's complicated, just imagine keeping track of all the possible crosses between stamens and pistils of different length. There are 18 crosses based on stamen and pistil length, but he also thought he detected differences in the pollen of some of the morphs, giving many additional permutations. Darwin was also determined to repeat each cross 20 times to be absolutely sure of the results. Before long he was "stark staring mad" over *Lythrum*, as he put it to Gray: "it is grand case of Trimorphism with 3 different pollens & 3 stigmas; I have castrated & fertilised above 90 flowers, trying all 18 distinct crosses which are possible within limits of this one species!"[32]

That June, in the middle of these experiments, he and Emma grew increasingly concerned about their son Lenny, 12 at the time. Darwin wrote Willy, in Southampton, about his brother's symptoms: "his kidneys hardly act & his urine is tinged with blood—His liver is much disordered & he vomits. Poor dear little man, he is so patient."[33] He commented on his plans for purple loosestrife, but acknowledged that he didn't have much heart for botany just then. A few days later Lenny was diagnosed with scarlet fever, which could be deadly. In fact their youngest child, 18-month old Charles, had succumbed to the disease almost exactly 4 years earlier. As July wore on Lenny seemed to be out of the woods but was very weak and recov-

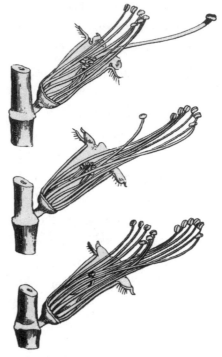

Trimorphic flowers of purple loosestrife, *Lythrum salicaria*, based on a sketch Darwin made for William Darwin in 1862 to illustrate "legitimate" and "illegitimate" crosses. Note (top to bottom) the long, medium, and short pistil lengths, and different size classes of stamens. From Darwin (1864), p. 171.

ering slowly. Gray kindly sent US postage stamps for Lenny's collection to cheer him up; Darwin wrote Gray with gratitude: "He actually raised himself on one elbow to look at them. It was the first animation he showed. He said only 'You must thank Prof. Gray awfully'—In evening after long silence, there came out the oracluar [*sic*] sentence 'He is awfully kind'."[34]

Darwin worked away at his *Lythrum* studies, and wondered if the proportion of the three morphs varied geographically. Hearing that his Wedgwood nieces, Katherine, Margaret, and Lucy, were on holiday in north Wales, he wrote to enlist their help. Margaret replied in early

August: "Dear Uncle Charles, of 256 specimens of Lythrum gathered this morning from different plants, we find 94 with long pistil, 95—middle length pistil, 69—shortest pistil."[35] These were all found in one field, but they would check other locales as well. "My dear Angels!" came Uncle Charles's swift reply; "I can call you nothing else.—I never dreamed of your taking so much trouble; the enumeration will be invaluable."[36]

Willy helped too. His father asked him to check out at least 100 plants in the field, tallying up the three morphs and marking flowers with long styles with three strings, medium-styled ones with two strings, and a single string for short styles. "I much want a few pods, (wrapped up separately when ripe) to count seeds in each, to know natural product, & to compare shape of pods." The idea was to see if some crosses were more productive than others. He had instructions for Georgy as well, a "rare hand at watching insects," his father said. "Beg him to observe carefully what part of body of bee the knobbed stigma of long & mid-styled rubs against."[37] His sister-in-law Sarah Elizabeth Wedgwood managed to procure specimens of the rare Lythrum hyssopifolia from several countries, making for a valuable comparison with specimens Oliver sent from Kew. Darwin was in crowd-sourcing mode, and everyone was eagerly lending a hand.

Charles and Emma decided by early August that with Lenny well enough to travel they would head to Bournemouth, near the coast, stopping en route to see Willy in Southampton. In the meantime they sent Etty, Bessy, and little Horace off with their former nurse, Brodie, now a dear family friend down from Scotland for a visit. They planned to meet up in Southampton. Charles's sister Susan had gone ahead separately with Franky and Georgy, arriving in Southampton on the 2nd of August. Darwin wasted no time enlisting the two of them as field assistants: Willy reported in a letter home later that week that he had "packed off the boys" to observe and collect Lythrum flowers, instructing them to take care to collect from different plants. Charles and Emma traveled with Lenny to Southampton the following week, but en route the boy suffered a relapse of scarlet fever and Emma ended up contracting it too. Charles sent all the other kids on to Bournemouth, tending to Lenny and Emma in Southampton for another 2 weeks before the family was together at last.

Enforced holidays were not endured gladly by Darwin, but he always made the best of them. He went on countryside rambles, and despite complaining to Hooker that Bournemouth was "most barren country," some things caught his eye. Earlier in this chapter I related how honeybees visiting red clover caught his attention. He also came across his old friend the sundew and performed a few experiments (see Chapter 8), inspected local flowers, and pondered *Lythrum.*

The Darwins returned from Bournemouth at the end of September 1862. Everyone was well at last, and Darwin was keen to get back to work. He immediately returned to *Variation*, "dull, steady work"—but his heart lay with loosestrife. "My geese are always at first Swans," he wrote to Gray, "but I cannot help going on marveling at Lythrum."[38] That summer he had managed to complete 94 crosses before Southampton and Bournemouth and the results were spectacular, but now it was getting too late in the season to do much more. He picked up where he left off the following summer, and by August he was able to write to Gray that he had "worked Lythrum like a Trojan," having just completed a heroic 134 crosses. "No slight labour," he pointed out with understatement, "but the case seems to me worth any labour, for I declare I think it about the oddest case of reproduction ever noticed— a triple marriage between three hermaphrodite."[39] And there was more labor to come; in the end his paper was not read at the Linnean Society until June 1864. It was a masterful work of 30-odd pages, with a multitude of tables summarizing the results of hundreds of experimental crosses and meticulous seed counts, and an overview of flower morphs in a host of other species with trimorphic flowers. All told, he had amassed incontrovertible evidence that the evolutionary imperative of intercrossing is the point behind heterostyly. That is, trimorphic flowers exist to maximize the likelihood of mating with a genetically distinct individual.

Darwin's 1864 paper on the three forms of *L. salicaria* was the tip of the iceberg when it came to this botanical hobbyhorse—over the ensuing decade he produced another dozen papers and notes bearing on pollination, cross-fertilization, flower structure, and further studies of heterostyly amid the many other subjects he wrote on, even while managing to finally complete the two-volume *Variation of Animals and Plants Under Domestication* (1868); two volumes on human evolution (*Descent of Man* and *Expression of the Emotions in Man and Animals*, published in 1871 and 1872, respectively); another three *Origin* editions (they culminated in the sixth edition of 1872); and his volume on insectivorous plants (1875; see Chapter 8). So perhaps he can be forgiven for taking until 1876–1877 to come out, at long last, with the magnum opus duo on cross- and self-fertilization and heterostyly.

The *Forms of Flowers* volume, dedicated to Asa Gray, is more than a botanical classic. Most of the conclusions Darwin drew from his observations and experiments are still valid well over a century later. That's impressive, but more to the point of the present book, botanist Herbert G. Baker of UC–Berkeley noted that Darwin's volume is especially valuable "in illustrating the principle that carefully thought-out but simple experiments can substantiate or falsify quite important ideas"—a welcome message for would-be Darwin-inspired experimentisers who might, as Baker put it, "be discouraged by the seeming necessity for high technology in all branches of contemporary science."[40] Baker rightly pointed out that plenty can be done with the simplest of equipment and, above all, a good eye. Darwin's sentiments exactly. The incomparable Henry David Thoreau, who was fascinated with Darwin's ideas and whose untimely death in 1862 meant he never knew of Darwin's exciting investigations into the secret lives of plants, captured the spirit of this philosophy in one lovely journal entry: "Nature will bear the closest inspection. She invites us to lay our eye level with her smallest leaf, and take an insect view of its plain."[41]

Experimentising: Darwinian Encounters of the Floral Kind

Darwin was convinced that the benefits of outcrossing in plants are many—the number, vigor, and fertility of offspring produced by out-crossing are superior as compared to self-fertilization—and he mar-veled at the means plants use to entice, reward, or dupe insects into visiting their flowers and carry their pollen. Here is a Darwin-inspired selection of easily observed pollination "contrivances." Like Darwin, you can learn to see once-familiar plants with new eyes.

I. Pollination Machinations

A. Materials

- This investigation is seasonal in regard to the availability of flowers. With enough lead time you might grow your own, or try to obtain the following plants in flower from a local nursery: kidney bean (*Phaseolus vulgaris*), sage or salvia (*Salvia*), moun-tain laurel (*Kalmia latifolia*), barberry (*Berberis*), foxglove (*Digitalis*)
- Dissecting probe, toothpick, or stiff camel-hair brush
- Forceps or tweezers
- Craft knife or razor blade
- Magnifying lens and/or dissecting microscope

B. Procedure

1. Kidney bean (*Phaseolus vulgaris*)

    Darwin described the lever-action mechanism of kidney bean flowers: the style is curved in a left-hand loop ("like a French horn," he thought) within the modified petal called the keel. The stamens are housed within this looped tube as well. He realized that the two lower petals protruding at the bot-tom present a landing platform for bees. When they land their weight depresses these petals, popping the stigma (tip of the pistil) out of the tube. You can imitate the action of a bee like Darwin did. Carefully tug on the lower petals (you may have to hold the flower steady), and observe how the stigma tip emerges from the coiled tube. Note the brush of fine hairs just below the stigma. Tug and release the lower petals a few times: note the brush moves backward and forward, as it would each time a bee lands on the flower. The bees indirectly help the flower self-pollinate, as the brush sweeps shed pollen out of the tube and gradually pushes it into contact with the stigma.

2. Sage or salvia (*Salvia* spp.)

Stamens can move as levers too, as in *Salvia* species—salvia, sage, and related garden plants in the mint family (Lamiaceae). Each functional stamen is joined near its base with a modified sterile stamen. Shorter and stouter than the fertile elongated member of the duo, the sterile stamen takes the form of

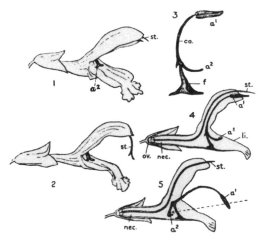

Flower of yellow sage (*Salvia glutinosa*). No. 1 shows a male-stage flower with the immature short stigma (st) and process at the base of the fertile anther (a^2). No. 2 shows a female-stage flower, with an elongated and receptive stigma (st). A fused fertile + sterile stamen is shown in No. 3, showing the ripe anther (a^1) and sterile anther reduced as a spur or process (a^2). The stamen in No. 4 is poised to deposit pollen; when a visiting bee (dotted line) pushes on the spur, the fertile anther is tipped down by lever action. From Ray Lankester's *Science From an Easy Chair: A Second Series* (New York, Henry Holt & Co., 1913), p. 4.

a basal spur or process. These are situated right in the gangway, so bumblebees trying to get at the nectaries behind must push these spurs and in the process tip the overarching fertile anthers down onto the back of the bee, dabbing it with pollen—a simple but effective lever. Using the dissecting probe, toothpick, or camel-hair brush, carefully push at the base of the sterile stamen spur and note how the attached fertile stamen dips down by lever action.

3. Mountain laurel (*Kalmia latifolia*)

The showy flowering shrub mountain laurel (*Kalmia latifolia*), a relative of rhododendron native to eastern North America and now widely planted, presents a nice example of spring-loaded stamens. Mountain laurel is prized for its showy clusters of pinkish-white flowers, each about the size of a nickel. Peer more closely and you'll see that the flowers are beautifully sculpted bowls, the filament of each stamen gracefully arched back, the anthers tucked securely into little puckered compartments or pouches along the bowl's edge.

The flower is a pollen-showering trap waiting to spring on unwary bumblebees as they try to get to the nectaries. Their weight slightly deforms the bowl-shaped corolla, releasing stamens. They spring up, showering their pollen on the bee. You

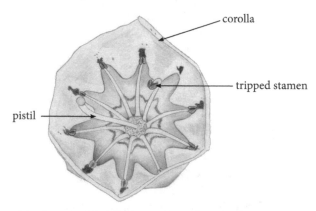

Mountain laurel (*Kalmia latifolia*) flower, with 9 reflexed and 1 sprung stamen. Drawing by Leslie C. Costa.

can simulate this and trip some stamens yourself by carefully deforming the shape of a flower by gently pushing on opposite sides of the inner edge of the corolla bowl.

4. Barberry (*Berberis* or *Mahonia*)

A different form of catapulting stamens is found in the barberry genus *Berberis*, also classified as *Mahonia* by some authorities. Common ornamentals include the Mexican barberry (*B. trifoliata*), Oregon grape (*B. aquifolium*—Oregon's state flower), common (European) barberry (*B. vulgaris*), and Japanese barberry (*B. thunbergii*). These plants produce clusters of small pale white or yellow six-petaled flowers. Be careful handing them: they often have small thorns along the stem, and in some species the leaves are stiff and armed with sharp points like hollies. The flowers can look doubled since the sepals are about as well developed as the petals. Alongside each petal stands a stamen, arrayed around the single central pistil, which often has a mushroom-shaped stigma. Barberry stamens are "irritable" (technically termed *seismonastic*)—sensitive to touch. Nectar organs are found at the base of each stamen. When an insect comes probing, contact causes the stamen to spring toward the pistil. After about 20 minutes the stamens

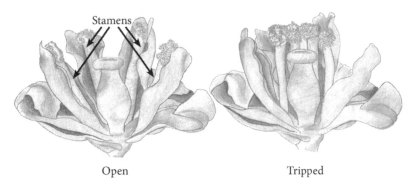

Open   Tripped

Barberry (*Berberis*) flower cross-sections showing four of the six stamens and the central pistil. (Left) Open "armed" position of stamens. (Right) Tripped stamens have moved toward the central stigma. Drawing by Leslie C. Costa.

are slowly retracted, ready to spring anew—an indication that the movement is caused by rapid change in internal (turgor) pressure in certain cells. It's easy to trip the stamens yourself, even one at a time: just gently probe the filament (stalk) of the stamens with a toothpick or camel-hair brush tip and they will spring forward. You can time their refractory period: how long does it take for them to return to the "armed" position?

5. Common foxglove (*Digitalis purpurea*)

The uppermost flowers in a spike will be male-phase, while the lowermost flowers will be female. Select a flower and, with the craft knife or razor blade, carefully slice the flower longitudinally slightly off-center to leave the pistil in place, as pictured. Are stamens or pistils held over the gangway? Does that correspond to the flower's position along the spike? The long style with the bifid (forked) stigma is positioned along the middle of the corolla tube (gangway) in the female stage, but in male-phase flowers it is held out of the way along the roof of the corolla tube. When the anthers mature they move to the gangway, where they can easily shower bees with pollen. Feel the inside lower surface of the flower. Darwin described how the floor the flower is carpeted with coarse hairs. One of Darwin's correspondents suggested that the hairs might provide bees with a foothold as they clamber up the corolla tube.

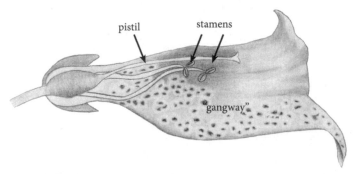

Cross-section of a foxglove flower (*Digitalis*), showing the position of the stamens and pistil along the upper part of the tubular flower and the bee "gangway" along the spotted lower part. Drawing by Leslie C. Costa.

II. Exploring Flower Morphs

One of Darwin's greatest contributions to botany was his discovery of the significance of *heterostyly*—flower morphs bearing stamens and pistils of very different length. These morphs ensure outcrossing. Here we'll observe heterostyly in some familiar plants—once you know what to look for, you will find that this phenomenon is far more common than you might have realized.

A. Materials

- Obtain one or more plants with **dimorphic** flowers (two morphs, long-styled *pin* and short-styled *thrum*): primrose (*Primula* spp.), lungwort (*Pulmonaria* spp.), forsythia (*Forsythia* spp.), and yellow Jessamine (*Gelsemium sempervirens*) are readily available from garden centers and nurseries.
- Obtain one or more plants with **trimorphic** flowers (three morphs: long-, medium-, and short-styled): purple loosestrife (*Lythrum salicaria*) and the aquatic pickerelweed (*Pontederia cordata*).
- Craft knife or razor blade
- Forceps or tweezers
- Ruler
- Magnifying lens and/or dissecting microscope

B. Procedure

1. In each of these genera the flowers have a corolla tube, and since the stamens and pistil typically do not extend beyond the tube a careful dissection of the flowers will be necessary: gently cut along the corolla tube of one or more <u>dimorphic</u> flower with the knife or razor blade and peel back to expose the stamens and pistil within.
2. Is your flower pin or thrum? If you have access to both, measure the length of the stamens and pistils in each morph and compare.

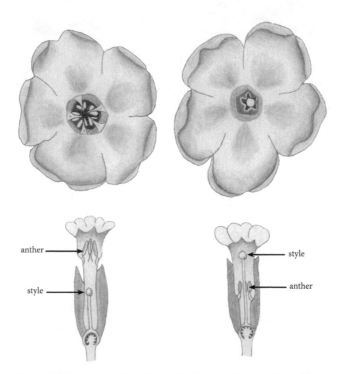

Primrose (*Primula*) flower morphs: thrum (*left*) with short style and long anthers; pin (*right*) with a long style and short anthers. Drawing by Leslie C. Costa.

3. Note that a given individual will have pin or thrum flowers, never both.
4. Repeat, carefully cutting along the corolla tube of one or more <u>trimorphic</u> flower to expose the stamens and pistil within. Is your flower long styled, medium styled, or short styled?
5. Carefully remove each stamen at the base with the forceps or knife and measure the length. Ideally, repeat with several flowers to increase the stamen sample size. What do you notice about the distribution of stamen lengths?

See also:

"Floral Dimorphism" at the Darwin Correspondence Project: www.darwinproject.ac.uk/learning/universities/getting-know-darwins-science/floral-dimorphism.

7

It Bears on Design

I t must have struck many as odd that the epochal work *On the Origin of Species*—a seismic book, quintessentially big picture with its bold expansive theory of species change and provocative philosophical implications—was followed 2½ years later by a slender volume meticulously detailing the pollination of orchids. Had Darwin gone mad? If so there was method to his madness, though it's equally true that it wasn't his original plan. In January 1860 he had been home a month from the hydropathic spa at Ilkley, Yorkshire, where he had taken refuge during the publication of the *Origin*. The battle lines were being drawn for and against his theory, and naturalists and clergy alike (often one and the same) were found on both sides of the question. Darwin's intention now was to write a three-part series of volumes expanding on the key arguments of the *Origin*. He hoped to make good on his promise to "show that I have not been quite so rash as many suppose,"[1] as he put it in a letter to Charles Lyell that spring. He had to get to work right away. And he did, except for the interruptions. Of course, the interruptions were in no small part of his own doing, and as we shall see, at one phase orchids were the main culprit.

In the course of Darwin's investigations into flower structure and pollination, a line of research begun in the 1840s, he studied orchids off and on in the late 1850s and into 1860, performing experiments to

Epipactis helleborine orchid growing at Orchis Bank, Kent, a favorite spot for picnics and botanical explorations by Darwin and family. The most widely distributed of the approximately 25 *Epipactis* species, this one is pollinated almost exclusively by vespid wasps. Photograph by the author.

look at fertilization success and seed set. Or tried to: an experiment book entry dated July 2, 1857, records an ill-fated experiment cross-pollinating bee orchids, ending with a hastily scrawled note in pencil, "All Killed by Cows."[2] He had better luck the following year. It was only a matter of time before he was drawn in more deeply by these beguiling plants, as a number of lovely species grew locally. In fact, one of his and Emma's favorite retreats was a peaceful site teeming with these botanical beauties just a half-mile from his home, high on a hillside of wood and meadow overlooking the fertile Cudham Valley. "Orchis Bank" was the family's private name for Downe Bank. Many years later Etty recalled the site fondly: "Just on the other side of the narrow steep little lane leading to the village of Cudham, perched high above the valley, was 'Orchis Bank' where bee, fly, musk and butterfly orchises grew. This was a grassy terrace under one of the shaws of old beeches, and with a quiet view across the valley, the shingled spire of Cudham church shewing above its old yews."[3]

Now preserved by the Kent Wildlife Trust, Orchis Bank has a timeless beauty evocative of Darwin's day—a verdant landscape dotted with wildflowers where it is easy to imagine Emma and Charles on a family picnic, some of the children exploring and others marauding, their governess in pursuit. In the spring of 1860 five children were home: the girls, Etty (17) and Bessy (13), and the three youngest boys, Franky (12), Lenny (10), and Horace (9). Darwin shared his botanical enthusiasms with them on those excursions, and encouraged a closer look at the weird and wonderful orchids to be found there: compact purple clusters of pyramidal orchid (*Anacamptis pyramidalis*); tall, pale helleborine (*Epipactis helleborine*), mysterious man orchid (*Aceras anthromorpha*) with its rust-colored and odd human-shaped lip; strange insect-mimicking fly and bee orchids (*Ophrys insectifera* and *O. apifera*); and more. Some nine orchid species are found at Downe Bank today, and more grew there in Darwin's day—like the bizarre parasitic bird's-nest orchid (*Neottia nidus-avis*), which Emma noted seeing only twice in the beech wood lining the bank. The orchids of Orchis Bank would soon become study subjects, but it was closer to home that Darwin made the startling discovery that got him thinking about orchids in a different way.

The Rube-Goldbergest Flowers

Orchids, as Darwin well knew from his Cambridge days studying botany with Henslow, were most unusual flowers. The often bizarre floral structures of this largest of plant families are such a departure from the typical, generalized, flower form that they have their own terminology: stigma, rostellum, column, labellum—the product of parts fused and elaborated to create intricate passageways and chambers, often subtended by a protruding or puckered lip. Orchid pollen is unusual too, packaged into paired sacs called pollinia (singular: pollinium) rather than being freely released by the stamens. This is a trait shared by only one other plant group, the milkweeds (Asclepiaceae), and it means that for each flower pollen transfer is all or nothing: the entire sac must be delivered. Such precious cargo is rarely entrusted to generalist couriers; a great many orchids have very specific insect pollinators whose anatomy fits the orchid like a key fits its lock. Many orchids can be pollinated by insects on the wing while others invite the unsuspecting insects inside, enticing them with intoxicating fragrances or oils in special chambers but then forcing the groggy insects to exit through one-way tunnels where the pollen packets are affixed. Others like the bee and fly orchids of Europe (*Ophrys* spp.) or hammer orchids of Australia (*Drakea* spp.) entice male insects with the lure of sex, mimicking sex pheromones or even the appearance of the female insect. Yes, of all the Rube Goldberg contrivances that flowering plants have evolved to achieve cross-fertilization, orchids must be the Rube Goldbergest of all.

The exotic beauty and intricate structure of orchids make them irresistible, and not just to horticulturists. If roses are the flowers of love, orchids are the flowers of sex, and not necessarily of the most straightforward kind; more *Rocky Horror Picture Show* than *Kiss Me Kate*. It's not just that the name "orchid" derives from *orchis*, Greek for testicle (for its paired ellipsoid tubers, a shape inevitably taken as a sign of aphrodisiacal properties). No, it's the flowers that have long excited admiration and passion, with a whiff of the dangerous: what other flower could inspire frenzied collectors, brooding comic book characters, artists, and escort agencies? Only orchids could lie at the center of that Venn diagram.

Long before orchids loomed large in Georgia O'Keefe's art or inspired the dark superheroine Black Orchid of DC Comics, these plants drove Victorian collectors to a form of madness called "orchi-delirium." *Chambers' Journal of Popular Literature, Science, and Art* for April 28, 1894, ran a story recounting the trials, tribulations, murder, and mayhem connected with "The Romance of Orchid-Collecting," perhaps inspiring H. G. Wells to write his short story "The Flowering of the Strange Orchid" a few years later—a macabre tale of a mysterious tropical orchid that proves to possess a deadly combina-tion of irresistible beauty, intoxicating perfume, and a thirst for human blood. This murderous orchid would have been the stuff of nightmares for Oxford artist and critic John Ruskin, who had died 5 years before Wells's story appeared in 1905. In the 1870s and 1880s the perhaps by then slightly deranged Ruskin had already railed against orchids, a "strange" order, as he put it in his scheme of "moral botany" published as *Proserpina: Studies of Wayside Flowers* in 1875. Orchids seem more wayward than wayside to Ruskin. Of both orchids and the whole of Darwin's investigations into the sex lives of plants, Ruskin declared that "with these obscene processes and prurient apparitions the gentle and happy scholar of flowers has nothing whatever to do. I am amazed and saddened . . . by finding how much that is abominable may be discov-ered by an ill-taught curiosity."[4]

Ruskin was railing against modernity, scientific inquiry being little more than "ill-taught curiosity" to him. It was true enough that the discovery of the evolutionary process by Darwin and Wallace led to new insights into the natural world at once exhilarating and frighten-ing to many of their contemporaries. It's difficult, perhaps, for a mod-ern reader to realize that it was not so long ago that wildly different assumptions about the natural world prevailed: moralizing from nature was standard fare from Aesop onward. The lives and behavior of ani-mals played out as so many parables, and plants presented similar les-sons in a perfectly designed world. The beautiful and fragrant blossoms were for *our* benefit, and it was the height of absurdity if not perver-sity to hold that such things have a sex life. By the time of Linnaeus in the eighteenth century plant sexuality was becoming better under-stood by naturalists, even as some felt that botany was an unsuitable subject for young ladies to pursue as a result. But it is also true that even well into the nineteenth century the key role of insects in cross-pollination was not widely accepted, let alone the idea of flowers as sexual advertisements.

In rebelling against the new botany, Ruskin came up with a *newer* botany in *Proserpina*, replacing Linnean binomials with ones of his own device based on traits reflecting the supposed virtues (or not) of each plant. Orchids were renamed order "Ophryds," and most fell into the dubious suborder of "Contorta" (for their twisted petioles, but also for their twisted habit of growing mainly in "torn and irregular ground, under alternations of unwholesome heat and shade, and among swarms of nasty insects"[5]). It may not be coincidental that the orchid Ruskin used to illustrate the group as a whole was the one that Darwin first studied, and the first species illustrated in his 1862 orchid treatise: the early purple orchid, *Orchis mascula* in the Linnean system but renamed by Ruskin *Contorta purpurea*.

Beautiful Contrivances

Darwin was on his customary stroll about the sandwalk one day in May 1860 when he noticed some early purple orchids in bloom. Ever the opportunistic experimentalist, he decided to perform an insect-exclusion experiment, covering some of them and leaving others exposed to visiting insects. In the course of these experiments Darwin poked and probed the purple flowers as an insect might. To his surprise he triggered the flower's pollen release mechanism: retracting his probe, two tiny stalked yellow sacs, the pollinia, were firmly affixed. That's when he first witnessed a decidedly odd thing: the stalked sacs came out of the flower standing straight up, sometimes singly and sometimes both, like diminutive horns, but within seconds they bent forward nearly ninety degrees, like taking a bow. I imagine Darwin sitting bolt upright and peering at his unexpected catch like a serene fisherman suddenly jolted into action when the line jerks. Was it a fluke, or a maybe a defect? He probed another flower and out the pollinia came, again first upright and then bowing forward and holding that position. He was familiar with the "irritability" found in some flowers, with their spring-loaded stamens, but this was another form of irritability altogether—two forms in fact. First the rostellum, a thin, flexible, cup-like membrane in which the pollinia are nestled, is spring-loaded: when ever-so-slightly touched as a probing insect might, the rostellum ruptures, freeing the pollinia with their stalks firmly attached to small sticky discs. The discs immediately contact that probing insect—or pencil—and in the few seconds it takes the insect to back out of the flower they are cemented firmly in place. Seconds more and the next wonderful expression of irritability is on display: certain precisely positioned cells found on the lower front surface of the stalks suddenly deflate, causing the stalks to collapse in that direction, rather like a felled tree tipping in the precise direction of a skillfully placed notch. Darwin immediately realized the significance of this trick: when the pollen-bearing insect visits another orchid flower, pollinia pointing upward would fail to contact the stigma and pollinate the flower. But by tipping forward at 90°, the pollen sac is ready for action: it will be rammed onto the stigma as the insect probes the new flower. The precision of the mechanism enthralled him: it was "a beautiful contrivance" that took plant cross-fertilization to a whole new level.

He excitedly turned to other orchids in the vicinity: the pyramidal orchid (*Anacamptis pyramidalis*), aptly named for its tight pyramid of purple flowers, grew profusely at Orchis Bank and was soon revealed to transfer its pollinia a bit differently. Here the two pollinia were joined at the base by a narrow strap or saddle, somewhat resembling those perky

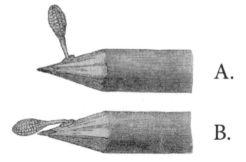

A.

B.

Pollinium of *Orchis mascula* stuck to a probing pencil. Each pollinium is in an upright position upon removal from the flower (A), but within about 30 seconds they tilt forward by 90° (B), well positioned to contact the stigmatic surface of the next flower. From Darwin (1862a), fig. II.

antennae of any respectable Martian costume: the pollinia are like the upright antennae affixed to the curved headband. Darwin soon realized why this shape proved useful: once the structure latches itself onto the proboscis of an unsuspecting insect, the curved band pinches tight, making it all but impossible to dislodge. Perhaps the connecting band is an adaptation to ensure that both pollinia are transferred simultaneously, since in species where they are not attached to one another it's very common for just one to adhere to an insect. In any case, as with its relative the early purple orchid, the stalks of pyramidal orchid pollinia take the same 90° bow, positioning the pollen sacs on either side of the proboscis, ready to contact a receptive stigma from another orchid. Darwin exulted to Joseph Hooker: "You speak of adaptation being rarely *visible* though present in plants: I have just recently been looking at common Orchis, & I declare I think its adaptations in every part of flower quite as beautiful & plain, or even more beautiful, than in Woodpecker."[6]

This comment, from a letter written in June 1860, is revealing of Darwin's growing dual interest in orchids: their pollination was part and parcel of his broader investigations into the mechanisms and beneficial effects of cross-fertilization in plants, but now he began to marvel at just how elaborate their pollination tricks are, as beautiful adapta-

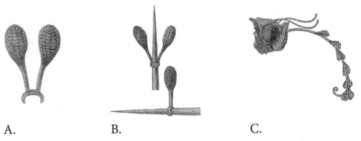

A.   B.   C.

(A) Paired pollinia of the pyradmidal orchid, *Anacamptis pyramidalis*, with sticky strap-like base. (B) Dorsal and lateral views of pollinia attached to a needle probe. (C) Head and proboscis of the four-spotted moth (*Tyta luctuosa*, Noctuidae) caught by Darwin bearing seven pairs of pyradmidal orchid pollinia attached to its proboscis. Note that the pollinia are aimed in the forward direction. In his orchid book Darwin listed 23 species of butterflies and moths with *A. pyramidalis* pollinia attached to their probosces. From Darwin (1862a), figs. III and IV.

tions. It was fortuitous that Darwin happened to work with *O. mascula* and *A. pyramidalis* first, as the tilting of the pollinia of these species is especially easy to observe owing to their long stalks. He eventually confirmed that this occurs in a great many orchid species, but the movement is not as obvious in those with short-stalked pollinia.

Darwin published a long open letter on orchid pollination in early June. It sheds a light on Darwin's orchid-inspired activities and musings of the previous month since his surprising discovery of pollinia movement. We see the experimentalist in Darwin come to the fore: spying wild orchids at the sandwalk and in the orchard he used a bell glass to prevent insects from visiting some while leaving others accessible. He meticulously checked the pollinia of new flowers as they expanded, showing that the pollinia were always removed from the flowers left exposed but never from those under the bell glass. Approaching this a different way, on plants bearing flowers in a long spike or raceme, Darwin took advantage of the "age gradient"—flowers lower down are oldest, while the uppermost ones are youngest. The older ones tended to be missing pollinia, while the younger ones still had them intact. Finally, he recorded cases where he found pollinia stuck to the stigma of orchid flowers that still had their own pollinia intact: clear evidence that pollen was being transferred *between* flowers. (Remember that in Darwin's day many people assumed that flowers self-pollinate as a rule.)

This was all of a piece to make the case that these plants must cross to reproduce, and that crossing is officiated by insects. The mystery in many cases was *which* insects, and when did they visit the flowers? At least, that was a mystery for certain species like the bee orchis (*Ophrys apifera*), a species no one could recall ever seeing an insect visit. The bee orchis does vaguely resemble large bees, and Robert Brown once speculated that this was to deter insects from visiting, as opposed to the usual floral ploy of drawing insects in. Darwin came to disagree, unsurprisingly, but he never did learn the secrets of its pollination. Unlike other orchids this one seemed to be always self-fertilized, even though the pollinia sport those sticky ends that in other orchids function to glue the pollinia to insects. If this was a useless feature, selection might be expected to have eliminated it from the population. "What are we to say with respect to the sticky glands of the Bee Orchis," Darwin asked in his article, "the use and efficiency of which . . . in all other British Orchids are so manifest? Are we to conclude that this one species is provided with these organs for no use?" He did not think so: "I . . . would rather infer that, during some years or in some other districts, insects do visit the Bee Orchis and occasionally transport pollen from one flower to another, and thus give it the advantage of an occasional cross."[7]

Darwin supposed that insects visit these flowers at least occasionally, and he asked the readers of the *Gardeners' Chronicle* for help in his typical crowd-sourcing fashion. Fellow orchid enthusiast Alexander Goodman More tried and failed to document insect pollination. "All facts clearly point to eternal self-fertilisation in this species," Darwin acknowledged to More. "Yet I cannot swallow the bitter pill."[8] Darwin ultimately had to admit that the bee orchis seems to largely self-pollinate. His close observations revealed to him how. The sticky discs are not so useless after all: the pollinia are attached to them by a long, thin, flexible stalk. They develop in an upright position, but when they mature they fall out and hang limply by their stalks, still firmly attached by their sticky feet. In that position the pollinia are directly in front of the orchid's stigma, so that even slight breezes can cause them to sway and sooner or later come into contact with it, where they stick fast, thus self-pollinating.

The bee orchis is indeed the only member of the genus *Ophrys* that relies mainly on self-pollination, but Darwin's hunch that they must outcross at least occasionally (maybe regularly in some parts of its range) is correct, though he never learned how. This plant is distributed from central and southern Europe, and around the Mediterranean in North Africa and the Middle East. In the northernmost portions of its range—including England—they are almost exclusively self-pollinated, but solitary bees are known to visit them in the Mediterranean populations. This is a fairly common bet-hedging strategy among flowering plants, combining the certainty of selfing with chancy but beneficial outcrossing, if only occasionally. But in any case there is more to the pollination story of these orchids: although Darwin figured that the bee-like flowers somehow attracted insects, it was not until the early twentieth century that the phenomenon of floral insect mimicry was described by the French Algerian botanist Maurice-Alexandre Pouyanne and his Swiss colleague Henry Correvon. Pouyanne, in particular, discovered that bees not only visit the flowers, they attempt to copulate with them. This duping of amorous male hymenopterans is called *pseudocopulation*, and has been found in a wide variety of orchids worldwide, including perhaps most spectacularly the rare hammer orchid (*Drakea* spp.) of Australia, which is pollinated solely by male thynnid wasps. The Swedish biologist Bertil Kullenberg in the mid-twentieth century built upon and greatly extended the work of Correvon, Pouyanne, and others. He discovered that the mimicry employed by hammer orchids extends beyond the visual and tactile to the chemical: they produce a fragrance mimicking the very sex pheromones of the bees and wasps they dupe, which explains the frantic repeated attempts at copulation by the hapless love-struck males. Luckily for them the orchids are merely duplicitous and not Circean—they are not drawn in and trapped; after all, the now-pollinia-studded males must live to visit other flowers.

All *Ophrys* orchids seem to have some degree of self-pollination (the bee orchis perhaps most of all), maybe as an adaptation to the ups and downs of local bee and wasp populations over evolutionary time. But Darwin would be fascinated to learn of another orchid that takes self-pollination even further. In 2006 a team of Chinese scientists found that pollinia of the high-elevation orchid *Holcoglossum amesianum* can rotate a full 360° in a remarkable feat of contortionist self-pollination. They go from their initial upright position to a hanging position like those of the bee orchis and then back up again to firmly press against the stigma. Where the bee orchis relies on gravity and wind to effect pollination, *H. amesianum* can rotate its pollinia against gravity. This unusual form of "assured selfing" may have evolved in this species owing to the virtual absence of pollinating insects in its harsh, dry mountainous habitat. The team studying *H. amesianum* inspected nearly 2,000 flowers and always found this same manner of pollination. But are they perpetually inbred? Darwin would surely hold out for the discovery that sometimes, somewhere tucked away in a remote corner of its range, an insect performs pollination services for this orchid; after all, even a little bit of outbreeding goes a long way.

Orchidelirium

As spring of 1860 turned to summer Darwin continued multitasking as usual, juggling sundews, primroses, and orchids. Several correspondents sent him insect specimens with protruding pollinia, while others made field observations for him on Bee Orchis and other species or sent him fresh specimens for study. Etty apparently suffered typhoid fever, and as her recovery was very slow the family decided to pack up and head off to Hartfield, Sussex, to stay with Emma's sister Sarah Wedgwood in the hopes that a change of environs would be beneficial to their sick child. He took a week-long detour first to take the "water cure" at Sudbrook Park—and it may not have been entirely coincidental that the Oxford meeting of the British Association for the Advancement of Science was held at just this time. This was the meeting that saw fireworks over the Darwin-Wallace theory, not least the famous exchange between Bishop Samuel "Soapy Sam" Wilberforce and Thomas Henry Huxley. Darwin preferred to keep a low profile, letting his friends rally to his defense; his chronically churning stomach was perhaps a casualty of the heated debate over the theory, or provided a good excuse to stay out of the fray.

In the meantime he kept busy. Although he bemoaned having done nothing researchwise in a letter to Hooker at the end of July, in the same letter he mentioned making "some good observations" of a species of *Malaxis*, adder's mouth orchid, kindly provided by a local doctor named William Wallis. These curious orchids secrete a sticky droplet on the rostellum that serves the purpose of, first, catching the pollinia as they ripen, and then adhering them to probing insects, similar to the *Listera* orchids (twayblades) that Hooker had studied back in the mid-1850s. It was there, too, that he inspected the pollinia of *Orchis pyramidalis*, the ones united at the base, and enthused to Hooker how they outdid *Listera* as a marvel of adaptive contrivance: "it almost equals, perhaps even beats, your Listera case . . . I never saw anything so beautiful."[9]

And so it went, with Darwin increasingly excited about orchid after orchid, the rich variety of structures and pollination mechanisms, each odder than the next. It was almost *too* rich: orchid flowers vary in such extreme ways in different orchid groups it's very difficult to see commonalities between them. Yet to be able to relate them in some kind of series that reflects their evolutionary history, underscoring how even the most bizarre orchid reflects modification of the same basic parts, was of paramount importance to Darwin. And they could be bizarre indeed, like the tropical *Catasetum* orchid with its "dull and coppery and orange-spotted tints—the yawning chasm in the great fringed labellum,—the one antenna stuck out with the other hanging down," giving to these flowers, Darwin wrote, "a strange, lurid, and reptilian appearance."[10] (More on this orchid to come.) It was Hooker who pointed out to him that he could use the spiral-shaped veins or vessels to trace homologies of different orchid parts, and he at once tried to teach himself the finer points of orchid anatomy.

Darwin returned home on August 2nd with the boys in tow, Emma and the girls following a few days later. Etty was showing signs of improvement, to her parents' relief, but she remained poorly and so in mid-September they decided to head to Eastbourne for a month-long holiday for sunshine and balmy sea breezes. He made the best of it but local orchids were scarce, so he poked around for some sundews, those diminutive carnivorous plants that had ensnared his interest ever since coming across them at Hartfield that summer [see Chapter 8]. He really should have been at work on his domestication book, but was enjoying the orchids and sundews too much. He guiltily fessed up in a letter to Lyell: "I have been of late shamefully idle; ie observing instead of writing & how much better fun observing is than writing."[11] Many a scientist can relate to the fun of doing research and discovering things, as opposed to the sometimes tedious process of writing it all up. But Darwin being Darwin there was even more vying for his attention, such as his recent discovery of primrose flower polymorphism, continuing pollination and crossing observations, and keeping up with a steady stream of correspondence over the *Origin* and the battle over his and Wallace's theory.

What to do . . . so many interests, so little time! Research-wise the orchids won out for now. Down House soon became Orchid Central, with specimens from the dainty to the dazzling arriving from around the British Isles and (via Kew and various nurseries) the world. From the Scottish Highlands his correspondent George Gordon sent ladies' tresses (*Goodyera*) with their spiral wands of delicate white flowers, while lurid orange-spotted "lizard orchids" (*Catasetum saccatum*) with their grotesque antenna-like processes arrived courtesy of James Veitch's Royal Exotic Nurseries in Chelsea. George Chicester Oxenden of Canterbury and his friend Bingham Malden obliged him with consignments of rare *Orchis* species among others, and Alexander More, orchid maven of the Isle of Wight, continued to provide specimens and field observations. Darwin made his way steadily through an array of species worthy of a splashy botanical garden orchid show, except rather than ogling his prizes, he got to work peering, poking, and dissecting to work out how each is pollinated and trace the relationships of their parts.

There was order to his orchidelirium: he aimed to examine representatives of each of the seven great orchid tribes described by John Lindley, noted botanist and orchid specialist, in his book *The Vegetable Kingdom*. Published in 1846 and reissued in several expanded editions, Lindley's book was the first attempt at a comprehensive classification of orchids and provided Darwin with the best available guide to their relationships, a useful framework for testing his ideas about the diversification of their complex structures and pollination. Nearly all the British orchids were confined to just two of Lindley's tribes, which led to his great need for orchids from abroad. His helpful neighbor George

Henry Turnbull even lent him the use of his greenhouses, as Darwin wasn't to build his own for a few years yet.

In the meantime Etty had continued to do poorly, and so during the following summer the family once again found themselves heading to the seaside. They arrived at a rented cottage at Torquay on July 2nd, and William, who was finishing his first year at Cambridge, joined them a week later. Darwin, keeping up his correspondence as usual, learned that Hooker and his family were planning a seaside respite too. Hooker's wife, Francis Harriet Hooker, née Henslow, was deeply grieving the recent loss of her father, John Stevens Henslow, Darwin's great mentor from Cambridge. Darwin memorialized the much-loved Henslow, recalling his "simple, cordial, and unpretending . . . encouragement which he afforded to all young naturalists."[12] Henslow had certainly encouraged Darwin, and was responsible for Darwin's invitation to join the voyage that changed the course of the young naturalist's life—and the course of science.

Hooker, too, was encouraging of young naturalists, especially where his friends were concerned. Darwin's son William was set to become a partner in a Southampton bank, but he had a love of botany that his father was keen to encourage. Darwin sought Hooker's advice on nurturing it. Among other things, Hooker invited William to join him on a botanical excursion. The timing didn't work out, however, and so that summer Willy amused himself with dissecting and drawing plants during his holiday with the family at Torquay. Hooker sent a selection of orchids for the Darwins, father and son, to work on, as the Torquay countryside proved to have a limited selection in the orchid department. Darwin was particular about which ones he sought: "I much want a Cattleyea [sic] or some one of the Epidendreae [an orchid tribe], as I have examined a Humble-Bee with the pollinia of [Cattleya] attached to its back. Really the contrivances in Orchids beat, I think, any animal."[13] Not 2 weeks later, a box arrived—being close friends with the director of the greatest living plant conservatory in the world clearly had its perks!

Darwin was thrilled: "My dear Hooker, you cannot conceive how the Orchids have delighted me," he enthused. He oohed and aahed about each orchid Hooker sent, and then: "I most specially want to know what [the] little globular brown Orchid is . . . surely have you not unintentionally sent me what I wanted most . . . viz one of the Epidendreae?!"[14] It was indeed a *Cattleya* orchid, and in view of Darwin's veritable pleading for one its inclusion was surely intentional. This genus was named for Englishman William Cattley, businessman and horticulturist who first coaxed a blossom from this tropical orchid in 1824. Darwin's rapture was complete. Though the "globular brown" description may sound unpromising, this orchid represents a sizable genus from the American tropics celebrated for large, vivid flowers. He longed to see a *Cattleya* since the previous month when entomologist Frederick Smith at the British Museum sent him a bumblebee bearing several puzzling pollinia stuck to its back with a smear of dried glue-like material. Darwin realized that they came from no British orchid, and asked Smith if the bee was caught near a greenhouse. Indeed, it came from *within* a greenhouse, housing *Cattleya* orchids.

Now that he had a *Cattleya* to study, representing a whole new tribe, he proceeded to work out the peculiarities of its anatomy and pollination. The reason he wanted this orchid so badly is revealing: the structure of its pollinia was unlike any he had seen before, and so he knew the orchid's mode of pollination would be unlike any other too. More importantly he thought that *Cattleya* pollinia were intermediate in structure between the two general types he had already studied, giving insight into stepwise pollinium evolution. This was an approach he followed to good effect in the *Origin*, showing how intermediate forms could be seen as evolutionary stepping stones of sorts, or precursors illuminating alternative pathways that natural selection might have taken. In either scenario intermediate forms help to link even the most highly divergent structures. So too with the highly modified pollination mechanisms of orchids, one aspect of which is their curious pollen packets. Darwin knew that one great group of orchids has pollen in a waxy ball without thread-like stalks, while another has a mass of pollen joined by elastic threads that form the stalks, or caudicles.

The two are quite different, but *Cattleya* pollinia linked them. And so as an intermediate form he believed that *Cattleya* held the key to the very origin of pollinia: by eliminating this or that structure of *Cattleya*-type pollinia, virtually any other type could be made. "Hence," Darwin declared, "I am inclined to look at this as the prototype."[15]

Pollen packets are one thing; another important matter is how they get transferred. He got to work mapping out the structure and release trigger of the flower. This is far easier said than done. Consider, for example, pollinia delivery in this species. First he needed to minutely map out where the pollinia are lodged—the dryness of his description a good indication of how he pored over the minutiae of structure and function: "The rostellum . . . is a broad, tongue-shaped projection, which arches slightly over the stigma: the upper surface is formed of smooth membrane; the lower surface and the central portions . . . consist of a very thick layer of viscid matter. This viscid mass is hardly separated from the viscid matter thickly coating the stigmatic surface which lies close beneath the rostellum. The generally projecting upper lip of the anther rests on, and opens close over, the base of the upper membranous surface of the tongue-shaped rostellum."[16] Ugh. But then, cleverly using a dead bumblebee as a sort of model, he mimicked how bees entered the flower, working out just how the pollinia end up on the bee's back and how they are delivered to the next flower. The detailed description he eventually gave of the complex moving parts of these orchids would be eyeball-glazing to any but the most devoted orchid (or pollination) aficionado; suffice it to say that painstaking work went into ferreting out the secrets of each and every one of these complex flowers.

At the start of his stay at Torquay, Darwin thought he would write a paper for the Linnean Society on the weird and wonderful pollination mechanisms of these plants. Long before the end of his stay he realized he had too much material: "My paper, though touching on only subordinate points will run, *I fear*, to 100 M.S. folio pages!!!" he lamented to Hooker; "The beauty of the adaptations of parts seems to me unparalleled."[17] It was about this time that he decided to present his findings in book form, which meant there'd be even *more* orchid studies. Having by then studied all available British orchid groups and a handful of foreign ones, he was keen to expand his study of orchid exotica and complete his examination of the rest of Lindley's orchid tribes. But just as importantly, it gave him space to expand upon what he took to be the philosophical significance of orchids—an important subtext of his book as we shall see later in the chapter.

Shaking the Foundation

The family returned home near the end of August, an enjoyable holiday despite the weather being a bit damp and chilly. While on the coast, Uncle 'Ras, Charles's brother Erasmus and a great favorite of the kids, had come down from London for a spell, and his cousin Fox's son visited them too. Emma took Etty and one of her Wedgwood nieces off on an excursion to Dartmoor, the strikingly beautiful and barren moorland of Devon, in southwest England (now a national park). Charles managed country walks, collecting, writing, orchid observing, experiments (he could not resist the sundews), and even a social call or two, visiting his friend Thomas Wollaston, the entomologist, who lived not far from Torquay. For another entomological friend, John Lubbock, he and the boys hunted for bristletails—the following year Lubbock published a study of these curious wingless insects.

Etty had steadily improved, probably not helped along by the heavy doses of cod-liver oil prescribed by Hooker. If you're wondering about a botanist prescribing medication, recall from Chapter 5 that Hooker was an M.D. who had served, in his early 20s, as assistant surgeon aboard HMS *Erebus* on its famous Ross expedition to Antarctica. The expedition enabled Hooker to establish his botanical bona fides, and with his medical bona fides he dispensed advice and prescriptions for his family and friends—hence Etty's cod-liver oil regimen, in her case rubbed on the skin as she wasn't able to stomach it.

As summer gave way to fall, things continued to look up on the home front: September saw Frank and George back at Clapham School, and things were falling into place for their older brother William and his banking partnership. Etty was well enough to play croquet on the back lawn with Bessy and Horace, the healthiest she had been in over a year. In early October they enjoyed a visit from Miss Pugh, the children's former governess, and later that month Emma took Etty to London to see a concert featuring the acclaimed Swedish opera star Jenny Lind and the pianist Felix Mendelssohn-Bartholdy. (A lover of music, Emma was a talented pianist herself, even taking a few lessons from Frédéric Chopin in her youth.) That fall Charles and Emma no doubt enjoyed a sense of relief they hadn't felt in quite a while, especially with their daughter out of the woods. On the scientific front, however, there were ups and downs.

Soon after returning home he approached his publisher, John Murray, about publishing a small book on orchids. Characteristically modest, he ventured that although he was "very apt to think that my Geese are Swans," yet "the subject seems to me curious & interesting."[18] He even offered to pay for the illustrations himself, fearing such a book might be a financial loss to Murray. Murray had no hesitation and immediately offered generous terms, illustrations included. Darwin no sooner had a contract for the book than an illustrator was engaged: artist George Sowerby arrived at Down House on October 7th and stayed for the next 10 days preparing orchid drawings, in some cases working from live specimens and in others adapting illustrations from an orchid treatise borrowed from Kew. All was well as he forged on ahead with his orchid studies, though in the 2 weeks between settling terms with Murray and Sowerby's arrival there is one revealing comment in a letter indicating he had down days too: on October 1st he closed a letter to Lyell with a rather startling declaration (perhaps half-joking, but only half): "I am very poorly today & very stupid & hate everybody & everything. . . . One lives only to make blunders—I am going to write a little book for Murray orchids & today I hate them worse than everything."[19]

What happened? The word "blunder" may be a clue: the letter was in reply to one from Lyell relating the latest work of the energetic young geologist Thomas Francis Jamieson, who was working on the glacial landscape of Scotland. Jamieson examined the so-called "parallel roads" of the Glen Roy valley, curiously parallel terraces of cobble seemingly etched into the green heath and circling the entire valley like a bathtub ring—a simile that is especially apt in this case as they are fossil beaches marking ancient water lines at different levels. Encouraged by Lyell and Darwin, and building on the work of Agassiz and others, Jamieson provided the nail in the coffin of Darwin's explanation for the parallel roads, which Darwin had published over twenty years before (see Chapter 1). Jamieson's painstaking reconstruction of the action of the former glaciers that dammed the valley and formed the parallel roads won him election to the Geological Society of London the following year. Darwin always called his misinterpretation of the site his "greatest blunder." In that letter to Lyell he magnanimously called Jamieson "a capital observer & reasoner" but also gave himself a kick in the pants: "A nice mess I made of Glen Roy!"

Alternatively, the source of his frustration may have related more to the subject at hand: orchids. He could have been confounded trying to figure out his latest orchidaceous Rube Goldberg contrivance, or maybe he was kicking himself because he neglected to preserve any exotic orchid flowers he had studied and now he desperately needed them for illustrations. With Sowerby soon arriving he applied to Hooker for help, explaining that he hadn't saved the flowers because at the time he had no intention of producing a book on the subject. But now, he beseeched Hooker, he needed them for illustrations. "If you can send me any, send them by Post in tin-cannister on middle of day of Saturday Oct. 5th; for Sowerby will be here."[20] Hooker came through, as usual. A large package of orchids soon arrived, followed a few days later with a single spectacular blossom of a rarity much coveted by Darwin: *Catasetum saccatum*, the bizarre lizard orchid of South America. Darwin called it "the most wonderful orchid I have seen,"[21] and feared touching the flower and tripping its pollinia delivery before he could study it. This is an orchid modified to such a degree that new structures needed naming, in this case the long, slender processes of the rostellum that Darwin dubbed "antennae." He showed that one of these was an ultrasensitive trigger that forcefully fires the pollinia. John Lubbock recounted what happened when "Mr Darwin [was] so good as to irritate one of these flowers" in his presence: "the pollinium was thrown nearly three feet, when it struck and adhered to the pane of a window."[22] Presumably people keeping these orchids in their greenhouses were careful not to accidentally set one off and get a pollinium in the eye—a botanical BB-gun that a horticultural Ralphie might have been warned about in a different kind of Christmas story.

But there was more to this orchid genus than projectile pollinia. Darwin had also acquired specimens of *Catasetum tridentatum*, a relative of the lizard orchid that abounds along the lush banks of the wide Essequibo River in what is now Guyana, and the source of much wonder. Back in 1836 the German-British naturalist and explorer Sir Robert Hermann Schomburgk (1804–1865) presented to the Linnean Society a botanical curiosity: an orchid specimen bearing the flowers of not merely different species, but species from two different genera.

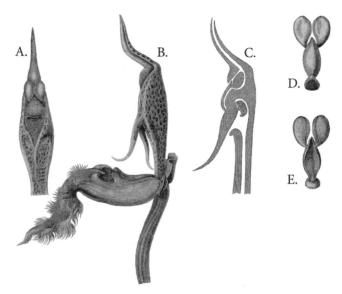

Catasetum saccatum's bizarre flower. (A) Front view of column. (B) Side view of flower, with all the sepals and petals removed except the labellum. (C) Section through the column, with all the parts a little separated. (D) Pollinium, upper surface. (E) Pollinium, lower surface, which lies in contact with the rostellum. From Darwin (1862a), figs. XXV and XXVI.

The bulb produced several stems, some of which bore *Monachanthus viridis* flowers and, astonishingly, others with those of *Myanthus barbatus*. These were no Piltdown-style doctored specimens, that was clear. The plant was meticulously drawn and preserved in spirits so that botanists could inspect it. Schomburgk documented a second case too, suggesting the first wasn't just a freak of nature, and then, writing from the former Dutch colony of Demarara along the river of the same name in the Guianas he alerted the Linnean Society to a further discovery: in the orchid collection of a Mr. Wortman was found a "vigorous plant" bearing flowers of *Monachanthus viridis* that then grew a stem producing those of *Catasetum tridentatum*. "This I saw myself," Schomburgk wrote; "the bulb was young, but the flowers in every respect quite perfect."[23] At a time when the nature of species and varieties was still being puzzled and the idea of transmutation—evolutionary change—was anathema to most naturalists, Schomburgk's revelations were astounding: could three species, in altogether different genera, somehow develop from one plant? Was this a rare case of transmutation in action? Could new species spring forth wholesale in some sort of monstrous budding process? In *The Vegetable Kingdom* John Lindley declared that "such cases shake to the foundation all our ideas of the stability of genera and species, and prepare the mind for more startling discoveries than could have been anticipated."[24]

Schomburgk's keen eye noticed something that would prove to be the key Darwin needed to solve this botanical mystery. In his letter to the Linnean Society Schomburgk commented that though he had seen hundreds of *C. tridentatum* orchids along the Essequibo River, not one bore seeds, while the *M. viridis* orchids "astonished me by their gigantic seed-vessels." These sexual differences were significant, Darwin realized—the putative three-species-in-one plant was not some chimera but represented the sexual forms of the same, single, species. Studying the flowers he determined that what they called *Catasetum tridentatum* was a male plant, *Monachanthus viridis* female, and *Myanthus barbatus* both—an hermaphroditic flower with male and female parts. It's easy to see how botanists presented with such different flower morphs separately and at different times would have considered them different species and even genera—the same sort of thing has happened with sexually dimorphic insects where one naturalist named a species based on one sex while another conferred an altogether different name based on the other sex. Only later, by rearing or finding the two in copula, was it realized that the two "species" were really one and the same.

Darwin was so excited about his discovery—and his botanical friends at the Linnean equally delighted at the light he shed on Schomburgk's orchids—that he was invited to read a paper there on the subject, which he did on April 3rd.[25] George Bentham, then president of the society, declared to Darwin that the paper "put us on a new and unexpected track to guide us in the explanation of phenomena which had before appeared so irreconciliable with the ordinary prevision and method shown in the organised world."[26] In the broader scheme of things his discovery might have been considered little more than an interesting case of observation triumphing over assumptions based on incomplete information, but Darwin realized it was more than this. He documented how *Monachanthus*, the female plant, nevertheless bore nonfunctional, rudimentary pollinia. He dissected and described these defunct pollinia, recognizing them as yet another object lesson against the still widespread misconception of "design" in nature. "We thus see that every single detail of structure which characterises the male pollen-masses is represented," he wrote, "with some parts exaggerated and some parts slightly modified, by the mere rudiments in the female plant. Such cases are familiar to every observer, but can never be examined without renewed interest." And then the kicker:

At a period not far distant, naturalists will hear with surprise, perhaps with derision, that grave and learned men formerly maintained that such useless organs were not remnants retained by the principle of inheritance at corresponding periods of early growth,

but were specially created and arranged in their proper places like
dishes on a table (this is the comparison of a distinguished naturalist) by an Omnipotent hand "to complete the scheme of nature."[27]

"People, get real," he was effectively saying. Here Darwin was repeating a key point made in the *Origin*: to him it was plain that rudimentary structures were clues to evolutionary change, but those opposed to
his theory of evolutionary modification over time must invent all sorts
of excuses for such apparently useless traits. Perhaps the most laugh-
able of these was the assertion that rudimentary traits were specially
created placeholders, facsimiles of structures which although not functional nonetheless have a role to play in preserving pattern or symmetry in nature—like setting every place at a table even though there are
fewer diners, for the sake of completeness or symmetry. In this line of
reasoning the Creator presumably felt that an absence of these traits, a
lack of symmetry, would be somehow jarring to humans. Darwin had
little patience for this kind of logic, and in chapter 13 of the *Origin* his
disdain is clear. Reflecting on the rudimentary structures of orchids, he
said, we can only be astonished, "for the same reasoning power which
tells us plainly that most parts and organs are exquisitely adapted for
certain purposes, tells us with equal plainness that these rudimentary
or atrophied organs, are imperfect and useless."[28] He pointed out that
it was silly to maintain that rudimentary structures were created "for
the sake of symmetry" or "to complete the scheme of nature." Such so-
called explanations were worse than no explanation at all, he fumed,
and merely a restatement of fact. With a hint of sarcasm he asked if
it would be "thought sufficient to say that because planets revolve in
elliptic courses round the sun, satellites follow the same course round
the planets, for the sake of symmetry, and to complete the scheme of
nature?"[29] This is in fact the subtext of Darwin's orchid book, the *Origin*'s sequel of sorts: as with his cherished barnacles (see Chapter 2),
orchids were a case study in evolutionary modification, with the same
parts varied in seemingly endless ways from group to group, some
even reduced or aborted, yet each well adapted for cross-pollination by

insects. By the same token an argument *for* descent with modification is an argument *against* special creation and design, and he saw this as a key lesson taught by orchids. Or was it? Not everyone otherwise sympathetic to Darwin's evolutionary ideas dispensed with the idea of design in nature.

A Flank Movement on the Enemy

Darwin's friend Asa Gray might be the best spokesman for those who saw no incompatibility between evolution and the doctrine of natural theology, with divine design at its heart. Gray and Darwin had sparred over this issue ever since the *Origin* came out. In today's terms Gray would be labeled a theistic evolutionist—one who accepts the fact of gradual evolutionary change over deep time and an ever-ramifying tree of life, but sees this as, at some level, a divinely directed process. How the Creator might direct evolutionary change is usually left unsaid; perhaps by divine guidance of the right mutations at the right time and place, or maybe by influencing the action of natural selection. As with every aspect of the theistic view this is necessarily speculative. Gray was nearly as much of a bulldog in defending the Darwin and Wallace theory as Huxley, but the devout botanist saw the hand of the Creator in the process. This still put him at odds with more conventional natural theology adherents, like Gray's Harvard colleague Louis Agassiz, who denied any kind of evolutionary change. Gray debated with Agassiz in public, and through a series of spirited and well-argued articles in *Atlantic Monthly* magazine.[30] Darwin was so impressed with these pieces that he helped bankroll their republication in pamphlet form in the United States and Britain, appearing in 1861 under the title *Natural Selection Not Inconsistent with Natural Theology*. That was a calculated move on Darwin's part—he may not have agreed fully with Gray's take on things, but at least his friend was arguing eloquently for natural selection and species change—a step in the right direction as far as Darwin was concerned.

It is against this backdrop that Darwin's orchid book, *On The Var-*

ious Contrivances By Which British and Foreign Orchids Are Fertilised By Insects, and On The Good Effects of Intercrossing, came out in May 1862. Beyond meticulously documented studies of the pollination of orchid after fantastic orchid, the book was an argument against design. He declared his intention in the introduction: "This treatise affords me also an opportunity of attempting to show that the study of organic beings may be as interesting to an observer who is fully convinced that the structure of each is due to secondary laws, as to one who views every trifling detail of structure as the result of the direct interposition of the Creator."[31] By "secondary laws" Darwin meant laws of nature.

Darwin sent complimentary copies of the book to Gray and numerous others, from fellow naturalists to the nurserymen and amateur botanists who assisted him in various ways. His expectations were modest, but warm and enthusiastic accolades soon poured in. Most commended Darwin for his methodical and ingenious techniques in documenting the oddities of orchid pollination, and some grasped the broader evolutionary lessons taught by these plants. Wallace, for example, lavished praise on the book, declaring himself eager to get out and observe orchids himself in the new evolutionary light Darwin put them in, and, ever perceptive, further commented that he was "quite as much *staggered* by the wonderful adaptations [Darwin showed] to exist in them as by the *Eye* in animals or any other complicated organs"[32] (emphases Wallace's).

That was exactly it—those Rube Goldberg orchids were like botanical versions of the eye, complicated but clearly jury-rigged machines, parts modified this way and that to yield beautiful, intricate, often bizarre pollen delivery systems, all traceable along clear pathways of gradual modification. Is this the way an intelligent designer would operate? Unlikely, he thought. Why create a thousand mix-and-match contrivances when a single perfect one would suffice? Darwin was curious to see what his most theistically minded friend, Gray, thought of his book. In one letter Gray commented that he had read George Bentham's recent presidential address at the Linnean, in which Bentham's initial opposition to the Darwin-Wallace theory seemed to be weakening as a result of Darwin's orchid work. Gray was amused, he said, "to see how your beautiful *flank*-movement with the Orchid-book has nearly overcome [Bentham's] opposition to the Origin."[33] But he also said that if Darwin grants that an intelligent designer has a hand anywhere in nature, he could be sure it has much to do with his wonderful orchids. Darwin was delighted with Gray's perceptiveness: "Of all the carpenters for knocking the right nail on the head, you are the very best: no one else has perceived that my chief interest in my orchid book, has been that it was a 'flank movement' on the enemy." But as for an intelligent designer having a hand in orchid contrivances, he went on to ask Gray what he thought "about what I say in last Ch[apter] of Orchid Book on the meaning & cause of the endless diversity of means for same general purpose—It bears on design—that endless question."[34] ("Endless" indeed. It stubbornly persists today, more than 150 years after both *Origin* and orchid book appeared.) Gray wasn't taking the bait: "it opens up a knotty sort of question about *accident or design* (his emphasis), which one does not care to meddle with much until one can feel his way further than I can."[35] The trans-Atlantic friends agreed to disagree.

Pattern, Process, and Prediction

The final chapter of Darwin's orchid book can be read as a manifesto for the manifestly clear (to him) lessons of orchids: remarkably intricate adaptive mechanisms of pollen transfer, with flower structures that indicate evolutionary modification of the same parts in different ways in different groups. He opened with recounting homologies (a concept introduced by the antievolutionist Richard Owen but which took on new, evolutionary, significance thanks to Darwin and Wallace), then discussed gradation of orchid structures, genealogical affinities, and the importance of rudimentary traits among other topics. Amplifying key points of the *Origin*, he pointed out how "old" (pre-existing) structures can be repurposed by natural selection for new functions.

An example would be making a machine for some special purpose, but reusing old springs, pulleys, wheels, etc. originally devised for a different purpose to do so.

In this chapter Darwin came full circle twice over, underscoring the real motivation of his orchid hobbyhorse but also concluding with the key topic that led him to examine orchids to begin with: the vital importance of cross-fertilization. Those initial investigations gave him an appreciation for how these plants promote crossing to a high art: few flowers come close to the breathtaking intricacy and diversity of structure seen in orchids, with their "almost endless diversity of beautiful adaptations," and he marveled at "the endless diversity of structure—the prodigality of resources—for gaining the very same end, namely, the fertilisation of one flower by the pollen of another."

But beyond documenting their elaborate pollination mechanisms Darwin was extending the new field of "evolutionary botany," the field he inaugurated when he began to probe the secret lives of plants (Chapter 6). With orchids Darwin's evolutionary botany was less experimental than observational, as the study of form and function often is by necessity, but no less significant for that. Seen through the lens of natural theology, exquisite adaptations are just that, while seen through Darwin's evolutionary lens they took on a new significance: exquisite, yes, but their jury-rigged mix-and-match qualities and their anomalies such as rudimentary structures speak of history.

There is a predictive aspect to the study of evolutionary form and function too, like in any area of science where an understanding of pattern and process yields new insights. The nineteenth-century Russian chemist Dmitri Mendeleev discerned a pattern in the properties of known chemical elements and, formulating the first periodic table, used this pattern to predict the existence of elements yet to be discovered. Similarly, a generation earlier the Frenchman Urbain LeVerrier (and independently Englishman John Couch Adams) used knowledge of orbital mechanics to correctly predict the existence of a yet-unknown planet based on the orbital perturbations of Uranus. The planet subsequently named Neptune was triumphantly discovered on September

23, 1846, within a single degree of where LeVerrier predicted it would be. Darwin's orchid work, too, led to a celebrated case of predictive insight: the mystery of the pollination of a remarkable orchid from Madagascar. Wallace in fact used the Neptune example in praise of Darwin and the case of a beautiful orchid with an appendage of puzzling proportions.

The comet orchids of the genus *Angraecum*, prized for their large, brilliant white star-like flowers and prominent nectary spurs, number about 220 species in Africa and the Indian Ocean islands. A great many are found on Madagascar, where, like their (very) distant relatives the lemurs, they have diversified spectacularly. One in particular took naturalists' breath away: *Angraecum sesquipedale*, variously known as the star-of-Bethlehem orchid, Christmas orchid, or, now, Darwin's orchid. The botanical name of this orchid, which grows high overhead on trees, is clear to those conversant in Latin: *sesquipedale* means "foot and a half"—referring to the stunning length of the nectar spur that dangles down from the shining white flower like a power cord. Many orchids and other flowers house their sweet nectar in a pouch or spur well positioned for pollen transfer when insects come calling. Typically the nectar pools deep within the nectary, so the pollinator must get up close and personal to get the treat. Darwin received specimens of this orchid in full bloom in January 1862. He excitedly wrote to Hooker: "I have just received such a box full [of orchids] from Mr Bateman with the astounding *Angraecum sesquipedalia* [sic] with a nectary a foot long—Good Heavens what insect can suck it."[36] Although short of the 18 inches promised by the species' name, a foot in length is impressive nonetheless.

Darwin knew that so exaggerated a trait could hardly be functionless, and of course he also knew that long spurs must mean a pollinator with a long tongue. But there are spurs and there are SPURS . . . what insect or other animal could possibly have a tongue long enough to get at the nectar of this plant? Again, let form and function be the guide. "What a proboscis the moth that sucks it must have!" he enthused to Hooker. He knew that in all likelihood it was not only a moth, but a sphinx or hawk moth, family Sphingidae. These moths are known for their large, robust bodies, long tongues, and pollination services. Darwin held that some as-yet-unknown sphinx moth must possess a nearly foot-long tongue. Such a pollinator would be a spectacular example of coevolution in Darwin's view, with natural selection favoring, over the eons, successive genera-tions of orchid flowers with ever-slightly-longer nectar spurs and their moth partners' ever-slightly-longer tongues to match.

Five years later George Campbell, the eighth Duke of Argyll, long hostile to any notion of evolution, criticized Darwin's explanation for *A. sesquipedale*'s long nectar spur in his book *The Reign of Law*. Dar-win had maintained that through natural selection "we can thus par-tially understand how the astonishing length of the nectary may have been acquired by successive modifications." The Duke seized on the word "partially," sneering that "it is indeed but a partial understand-ing." On the contrary, "Purpose and intention . . . are what meet us at every turn . . . We know, too, that these purposes and ideas are not our own, but the ideas and purposes of Another—of One whose man-ifestations are indeed superhuman and supermaterial."[37] Darwin's use of orchids as an argument against design was not lost on Campbell any more than on Gray.

Wallace rallied to the defense. In a long critical review of the Duke's book he marshaled a spirited counter-argument asserting that "the laws of multiplication, variation, and survival of the fittest . . . would under certain conditions *necessarily* lead to the production of this extraor-dinary nectary"[38] (emphasis Wallace's). To Wallace the coevolutionary scenario Darwin posited was clear as day, the case of *A. sesquiped-ale*'s pollinator so compelling that he commissioned an illustration of a hypothetical sphinx moth visiting comet orchids for the piece. In the illustration the moth hovers to one side, proboscis extended toward the flower like a long slender straw—aptly, as the prosces of moths and butterflies are indeed hollow drinking-straw-like tubes.

Wallace even narrowed down the best candidates for the pollina-tor. In the insect collection at the British Museum he scrutinized the sphinx moths with measuring stick in hand, searching for champion prosces. He found a *Macrosila* species from South America with a 9¼-inch proboscis, and another species in this same genus from tropical Africa measuring 7½ inches—not far from *A. sesquipedale* proportions. He wrote in his review, "That such a moth [with a long enough proboscis] exists in Madagascar may be safely predicted; and naturalists who visit that island should search for it with as much confidence as astronomers searched for the planet Neptune—and they will be equally successful!"[39] He was more correct than he realized: at last, in 1903, a population of this very moth species was found in Madagascar boasting those crucial extra inches. Wallace knew it as *Macrosila morganii*, now changed to *Xanthopan morganii*. It was a new subspecies of *X. morganii* soon dubbed "*praedicta*" by the eccentric English naturalist and financier Walter Rothschild and his colleague Karl Jordan in their monumental treatment of the Sphingidae published in 1903. It is commonly assumed that Rothschild and Jordan were honoring Darwin's prediction with this name, but in fact they were honoring Wallace's more precise prediction based on his moth proboscis measurements. So, although this moth is often referred to as "Darwin's sphinx moth" it should be called "Wallace's sphinx moth"—a name more resonant as well as accurate: it is fitting that "Darwin's comet orchid" should be pollinated by Wallace's sphinx moth.

It's remarkable to think that Darwin's book on orchids was published just 2 years to the month after coming across an orchid on the sandwalk and deciding to look more closely at it. And his book was far from the last of his work on those endlessly fascinating flowers. Even as the book was published he continued to make observations and gather data, often aided by sons William, Frank, and George. The boys mainly helped record the kinds of insects that visited different orchids. He continued to encourage William's botanical interests, and complimented George, then 16, for his "splendid work in watching orchids." He also made corrections to his book: Gray pointed out that he was mistaken about lady's slipper orchid pollination, at least for North American species.

The pouched labellum ("slipper") of these celebrated orchids has a slit-like opening along the top. Gray suggested that insects enter through the slit but cannot exit that way and are forced to escape through small holes at the rear of the structure, where they pick up and deposit pollen. Darwin confined some flies in the pouch of some lady's slipper orchids, to no avail: "they were either too large or too stupid, and did not crawl out properly."[40] He then tried small solitary bees and, lo and behold, out they came through the small orifices, covered in pollen. He repeated the trick again and again with a number of hapless bees; this little exercise showed the value of letting the insects do the probing rather than the experimenter. These observations and more were duly reported, some in foreign-language translations of the orchid book, others in a follow-up article on orchid fertilization in the *Journal of Horticulture*, and most comprehensively in the next English edition of the book in 1877. Significantly, orchids took center stage in later editions of the *Origin*, ammunition for an argument on evidence for gradual modifications of structure in a new chapter inserted to address "miscellaneous objections to the theory of natural selection."

As usual he forged ahead on the orchid front even as he was advancing a bevy of other botanical studies. Darwin's "orchidelirium" may have been a hobbyhorse, but he certainly rode that horse for all it was worth. As he realized the deeper significance of their astounding pollination contrivances, orchids became object lessons in evolution by natural selection, illustrating diversification, stepwise gradualism, trait reduction or loss, co-optation, and more. All bearing on design, that endless question.

Experimentising: Orchidelirium

Prized for their extraordinary beauty, diversity, and intricacy of structure, orchids also present us with extraordinary lessons in the evolution of form and pollination contrivances, as Darwin saw. Following Darwin's lead you can investigate these Rube Goldberg flowers yourself.

Let's take a look at representative orchids with an eye to the "beautiful adaptations" that so captivated Darwin.

A. Materials

- Forceps or tweezers
- Toothpicks
- Hand lens or dissecting microscope, if available
- Orchids

It is best to work with commercially available orchids that are propagated sustainably. Many temperate and tropical orchids are available commercially, and some, like *Phalaenopsis* and *Dendrobium*, are even common supermarket or garden center plants. Any of the following genera will do, all common in the orchid horticultural trade although availability can vary: *Cattleya* (tribe Epidendreae), *Dendrobium* (tribe Podochileae), *Cymbidium* (tribe Cymbidieae), or *Phalaenopsis* (tribe Vandeae). Others may require going to a specialist (e.g., Clown Alley Orchids: clownalleyorchids.com; Eldon Tropicals: eldontropicals. com; Flora Exotica: floraexotica.ca; Logee's Greenhouses: logees.com). Dave's Garden (davesgarden.com) lists many other sources.

B. Procedure

1. Orchid flower structures generally come in groups of three, the three petals and three sepals being the most obvious, arranged like two superimposed triangles inverted with respect to one another. Look closely at a *Cymbidium* or *Cattleya* flower. You can easily see the triangular petal and sepal groupings: the outermost three are *sepals*, at roughly 12, 4, and 8 o'clock on the flower dial, and the innermost three are *petals* at roughly 2, 6, and 10 o'clock. Notice that the lowermost petal looks decidedly different from the other two—sufficiently so to be given another name: the *labellum*, or *lip*. Does the lower lip seem inviting? The flower is indeed inviting company, of the insectan kind— the lip is a landing strip.

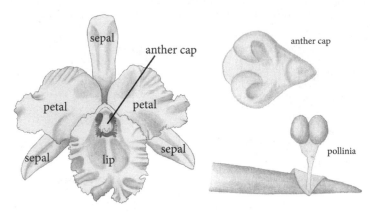

Cattleya orchid, showing anther cap (upper right) and extracted pollinia stuck to a probe (lower right). Drawing by Leslie C. Costa.

2. Just above the lip is the *column*. Representing the fusion of ancestrally separate stamens and pistil(s), the column contains the sexual organs of the orchid. At the top of the column look closely to see the *anther cap*, which acts as a protective shield concealing the *pollinia*.

3. You can expose the anther cap at the end (top) of the column by removing or pushing petals and sepals back and out of the way. Remove the anther cap with a toothpick or forceps, or if you have a steady hand attempt to release the small hook-like structure at the tip of the column. This is a spring-latch of sorts that keeps the anther cap closed. The anther cap comes off very easily, exposing the pollinia beneath: look for the twin sacs of pollen housed at the tip of the column and covered by the anther cap.

4. Try probing the pollinium with forceps or a toothpick. It will readily adhere to your probe and come out intact, same as it would if your probe were an insect. The clear sticky mass stuck to the pollinium is called the *viscidium*. The mucilaginous sticky stuff is viscin, a substance found in several plant groups (for example, this is the sticky pulp that allows mistletoe berries to stick to birds' beaks and tree bark).

5. *Phalaenopsis* flowers have the same basic structure but the

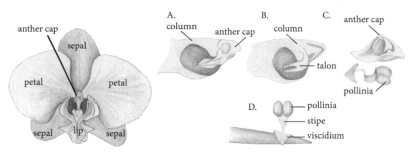

Phalaenopsis orchid. (A) Exposed tip of the column, with anther cap (rounded pollinia visible beneath) and rostellum extending over the hollow stigma chamber. (B) Column with anther cap and pollinia removed, leaving forked rostellum (talon) behind. The talon serves to pry pollinia off a subsequent insect visitor, pollinating the flower by securing the pollinia in the hollow stigma chamber. (C) Anther cap (above) and paired pollinia (below) attached to curved stipe. The other end of the stipe bears the sticky viscidium, which adheres the pollinia to an insect pollinator. (D) Pollinia stuck to a toothpick or similar probe. Drawing by Leslie C. Costa.

labellum, or lip, is often reduced compared with *Cymbidium* and *Cattleya*. Expose the column, which may be straight or slightly curved, by removing the petals and sepals with the forceps or tweezers.

6. Note its domed anther cap with the slender rostellum pointing downward. The rostellum has a sticky strip down its middle. This is the viscidium, working like adhesive tape. When a pollinator touches it, the tape sticks to the insect's body or head. Pulling away, the insect pops the anther cap off and the pollinia, which are attached to the sticky strip by the stipe, remain stuck to the insect. The rostellum remaining behind is a forked structure ("talon") extending in front of a hollowed chamber (the stigmatic cavity). This will help scoop the pollinia off of a later-visiting insect, and pollinate the flower by implanting it into the stigmatic chamber.

See also:

"Orchids" at the Darwin Correspondence Project: www.darwin project.ac.uk/learning/universities/getting-know-darwins-science /orchids.

8

Plants with Volition

It all started with the sundews that he found while the family was visiting Emma's sisters in Hartfield, Sussex, in the summer of 1860. A nice patch of sundews growing in a boggy spot caught Darwin's eye when he was off hunting orchids one day. It might have been the pinkish hue of the dainty leaf rosettes he noticed against the green mosses, or maybe an occasional flash and sparkle as sunlight played off the clear liquid droplets the plants are named for: *Drosera*, from the Greek word for dew. He stopped for a closer look. What appear as reddish indistinct patches from a distance resolve into orderly sets of small spatula-shaped leaves moist with tiny dewdrops, each borne on a crimson stalk; the whole looks like a botanical pincushion with rather stubby pins bearing clear acrylic heads. A few plants may be easily overlooked but not likely a whole glistening reddish bank of them, each with a tall slender wand of a stem bearing white-petaled flowers stirring in the breeze.

The delicate beauty of these plants belied their sinister aspects. The flypaper quality of the sticky leaves of the plant had long been known, but was that accidental? Was their object to capture insects? If so, why, and how did they sense their prey as they bend and move their dew-tipped filaments to ensnare hapless insects that venture too close? After all, this is a stunning thing for a plant to do: the life of a carnivore, yet seemingly lacking a stomach let alone the speed usually associated with "sit-and-wait" predators—and what can a plant do but sit and wait, rooted to the spot? There is something disquieting in the idea of a flesh-eating plant, especially magnified to human scale in our imagination. The horror inherent in the image of a mobile human-eating plant is doubtless what inspired the ambulatory botanical monsters of John Wyndham's *Day of the Triffids*. These little sundews, *Drosera rotundifolia*, are terrors to the insect world, but at least they don't *pursue* prey. Somehow their prey come to them, so, clearly, the plants have some means of enticement. Darwin collected about a dozen plants that day, and found that around half of the fully developed leaves bore dead insects or their remains. He was intrigued; as Janet Browne remarked in her Darwin biography, "by the end of the afternoon he was caught as surely as any fly."[1]

On the surface of it Darwin's fascination with these plants and the series of experiments that ensued might seem like a pleasant distraction, coming as it did in the wake of the *Origin*'s publication and the sometimes bitter early round of reviews and controversies surrounding the book. Certainly Darwin was curious about the question of sense perceptions of these plants, but it was Emma who, perhaps without realizing it, put her finger on the deeper reason for his interest. Back home at Down House later that summer, she mentioned her husband's *Drosera* fascination in a letter to Lady Lyell, Charles Lyell's wife, commenting that she supposed "he hopes to end in proving it to be an animal."[2] That was just it. In Darwin's evolutionary vision, plants and animals likely shared common ancestry in the deep past. As he put it in the *Origin*:

> Analogy would lead me one step further, namely, to the belief that all animals and plants have descended from some one prototype . . . all living things have much in common, in their chemical composition, their germinal vesicles, their cellular structure, and their laws of growth and reproduction. . . . Therefore I should infer from analogy that probably all the organic beings which have

ever lived on this earth have descended from some one primordial
form, into which life was first breathed.[3]

The notion of common ancestry of plants and animals, let alone
other forms of life such as protists and bacteria, was one unwelcome
implication of Darwin's already rather unwelcome theory. But he persevered, and had a hunch that sundews, with all of their apparent "sensibilities," just might be seen as a linking form of sorts. If plants and
animals shared common ancestry, at a fundamental level they must
share some common physiology despite outward appearances. Just as
there might be animals out there with plant-like features, there should
be plants with animal-like features. The sundew might be just such
a plant. He also thought of sundews as another kind of linking form,
remarking to his American friend Asa Gray that he "began this work on
Drosera in relation to gradation as throwing light on *Dionaea*"[4]—the
Venus flytrap, *Dionaea muscipula*.

The two plants are relatives, sharing the family Droseraceae. While
there are a great many sundew species, there is only one Venus flytrap species, and its native range is barely a hundred or so miles across,
in boggy swales dotting a small section of the coastal Carolinas in the
United States. This plant is singular in other ways: it is the only carnivorous plant in the world with a "snap trap" mechanism, modified leaves
so disturbingly like animal jaws that Europeans disbelieved the earliest reports of this plant. The renowned Philadelphia naturalist William
Bartram swooned over this botanical marvel in his famous *Travels* of
1791: "But admirable are the properties of the extraordinary *Dionea
muscipula*! . . . Astonishing production! . . . Can we after viewing this
object, hesitate a moment to confess, that vegetable beings are endued
with some sensible faculties or attributes, similar to those that dignify
animal nature; they are organical, living and self-moving bodies, for we
see here, in this plant, motion and volition."[5]

Such seemingly extreme development, such exquisite adaptation,
was often cited in the natural theology tradition as evidence of special
creation. As we saw in Chapter 4, examples of organic perfection in the

eighteenth and nineteenth centuries—the eye and honeybee's cells—
were tackled by Darwin in the *Origin*. In the "difficulties" chapter,
under the heading "organs of extreme perfection and complication," his
solution to the problem of the eye was to argue for a transitional series.
Look to related species to see what gradations might be possible, he
urged. In the case of the eye, the invertebrate world presents a rich
diversity of eye anatomies. "We can commence a series,"[6] he argued in
the *Origin*, from mere spots of light-sensitive pigment to fairly complex
structures and a great many variants in between. The many and varied
eye structures of invertebrates showed that a continuous gradation of
forms is possible, just what would be expected on his theory. Far from
isolated islands of perfection, such forms were connected to all others
by stepping stones, and this was his interest in the flytraps.

Dietary Supplements

There are many kinds of carnivorous plant, belonging to several plant
families. They thrive in marginal, low-nutrient habitats like bogs and
seeps by supplementing their photosynthetic diet with fresh meat: trapping and digesting small animals. The Botanical Society of America
notes five ways that carnivorous plants trap their prey:

- Pitfall traps. The leaves of pitcher plants (family Sarracenia-
  ceae) form tubular structures that fill with rainwater. The plant
  then secretes digestive enzymes into these leaves, which dissolve insects and other small animals for food.
- Snap traps. Venus flytrap (*D. muscipula*) and the related waterwheel plant (Droseraceae; *Aldrovanda vesiculosa*) have hinged
  leaves that snap shut at the touch of trigger hairs on the leave's
  inner surface.
- Suction traps. The aquatic bladderworts (Lentibulariaceae;
  Utricularia spp.) have small globular bladder-shaped structures on their leaves, hollow with a hinged door lined with trigger hairs. Minute aquatic animals brushing against the hairs

trigger the trapdoor to spring inward, sucking the prey into the hollow bladder where they are digested.

- Lobster-pot traps. The leaves of corkscrew plants (Lentibular-iaceae; *Genlisea* spp.) are labyrinthine tubular channels lined with hairs and enzyme-secreting glands. It is easy to get in, but all but impossible to get out again.
- Flypaper traps. The leaves of sundews (Droseraceae; *Drosera* spp.), Portuguese sundew (Drosophyllaceae; one species: *Drosophyllum filiformis*), and butterworts (Lentibulariaceae; *Pinguicula* spp.) bristle with stalked glands that exude a sticky substance. Once prey is stuck, digestive enzymes are secreted to break down the body.

Not all of these groups have the power of movement to catch prey, and it was the combination of movement and digestive power that most intrigued Darwin. His sundew studies proceeded along two paths. The most immediate, and experimental, was in pursuit of mechanistic questions: how did the dew-tipped "tentacles" move and how did they sense prey? Did the plant produce digestive enzymes? What sorts of matter could it digest? The other path was more comparative: how did the movement and digestion by this plant compare with related plants? This path resonated with his argument about the evolution of eyes in the *Origin*: with enough versions perhaps gradations in the prey-catching and -digesting mechanism could be traced, gradations along a general evolutionary pathway.

An evolutionary link between sundews and Venus flytraps was not farfetched: in Darwin's day it was recognized that the two are in the same taxonomic family. Just as the flytrap's lobes close to form a stomach-like digestive chamber, the leaves of some sundew species inflect to form a similar cavity, sealed by the dewy stalked glands. A plant with a stomach: the very idea of it fires the imagination. A sundew, Darwin would write in *Insectivorous Plants*, "with the edges of its leaves curled inwards, so as to form a temporary stomach, with the glands of the closely inflected tentacles pouring forth their acid secretion, which dissolves animal matter, afterwards to be absorbed, may be said to feed like an animal." Asa Gray agreed that the sundew's mode of leaf movement is akin to that of the flytrap. But are there sundews out there with leaves that fully close upon their prey? Darwin found what he was looking for in John Lindley's 1846 book *The Vegetable Kingdom*. The Asian species *Drosera lunata* did just that, prompting him to swing into crowd-sourcing mode, posting a query to the readers of the *Gardeners' Chronicle*. "In Lindley's Vegetable Kingdom (p. 433) it is stated that the leaves of Drosera lunata 'close upon flies and other insects that happen to alight upon them.' Can you refer me to any published account of the movement of the viscid hairs or leaves of this Indian Drosera?"[7] Lindley himself responded to Darwin's request, giving his source for the observation and even noting that he was so struck by its similarity to the Venus flytrap that he initially gave it the same specific name, *Drosera muscipula*, until he realized that the name "*lunata*" had already been given. It is now known that this species is one of many sundews with somewhat bowl-shaped leaves that can close upon insects, most dramatically in the Australian *Drosera huegelii*, a ground-hugging species with dangling bell-shaped leaves fringed with glandular hairs. How Darwin would have marveled at this species, had he known of it.

Now, this is not to say that the snap trap of the Venus flytrap is strictly homologous to the disk concavity of sundews, but they have some important physiological features in common that reflect their close relationship. For example, the structure of the stalked glandular hairs of *Drosera* and the trigger hairs of *Dionaea* and *Aldrovanda* are very similar and likely have a common origin. And then there's the genetic evidence: DNA sequence data confirm that the two are indeed sister groups; in fact, all *Drosera* sundews (about 150 species) form a single evolutionary group (clade) which is sister to the single-species lineage of *Dionaea*, the flytrap, with *Aldrovanda* falling intermediate between the two.

A Most Sagacious Animal

The Darwin family returned home from Hartfield in early September 1860, sundews in tow, but 17-year-old Etty's continued ill health prompted them to return to seaside Sussex by the end of the month. They stayed 7 weeks at a house on Marine Parade, Eastbourne, where Darwin continued his sundew studies—as much to take his mind off of his daughter's dangerous illness, just a decade after her elder sister's death, as to satisfy his curiosity. He wrote in mid-November to his neighbor John Lubbock: "[Etty] was desperately ill at Eastbourne & we gave up all hope; but she has rallied considerably. God knows what the ultimate result will be. I never passed a more miserable time than at Eastbourne."[8]

Late summer of 1862 found the family back at the seaside, in Bournemouth where they rented a little holiday place called Cliff Cottage. Twelve-year-old Lenny had contracted scarlet fever at boarding school the previous June and was sent home; making matters worse his mother soon came down with it too, en route to the coast. By September both were recovering well, but Darwin was having a difficult time with his enforced holiday between fretting over the family and itching to get back to work: "This is a nice, but most barren country & I can find nothing to look at," he wrote Hooker. "Even the brooks & ponds produce nothing—The country is like Patagonia.—My wife is almost well, thank God, & Leonard is wonderfully improved."[9] He took rambles with the children, and was tickled when Horace, his 11-year-old, showed a good grasp of natural selection. Relating the story to Gray, Darwin recounted how "Horace said to me, 'there are a terrible number of adders here; but if everyone killed as many as they could, they would sting [bite] less.'—I answered 'of course they would be fewer.' Horace [said] 'Of course, but I did not mean that; what I meant was, that the more timid adders, which run away & do not sting would be saved, & after a time none of the adders would sting"—Natural selection!!"[10] Darwin was amused and proud in equal measures, and described Horace's scenario as "natural selection of cowards" to Lubbock.

About this time he realized he could easily procure his beloved *Drosera* in that "most barren country"—just the thing to calm his nerves. He picked up on his digestion experiments, first making a fly-sized bundle of his own hair and watching over the next couple of days how the sundew's filaments first covered the bundle as if to consume it but then, as if the plant realized its error, uncurled and rejected the hair. What other substance to try? One can imagine him looking around, wondering what else was at hand . . . how about a bit of toe-nail? On the morning of Tuesday, September 16th, he wrote: "Put a bit of old nail of my toe on another leaf of same plant."[11] Once again the filaments closed, only to reject the offering. That must have been a curious sight—the esteemed naturalist sitting in a seaside cottage feeding bits of his hair and toenails to sundews transplanted in soup dishes.

Darwin had commented two years earlier, during his initial explorations, that sundews are "first-rate chemists"—the plants seemed to be able to distinguish between organic and inorganic substances. Now he had found out that, like animals generally, the sundews can't digest tough organic material and could distinguish between edible and inedible organic matter. It makes sense to us now that the plants would be well adapted to distinguish edible objects from inedible and so avoid wasting time and energy on substances that aren't digestible or nutritional. But we know this with the benefit of hindsight. No one thought of these little plants as "eating" to gain nutrition—a most unbotanical thing for any self-respecting photosynthetic plant to do—and as no one had thought of the question to ask, naturally no one had thought of experiments to test such a question until Darwin did. And these were only the beginning; over the next few years he launched himself into experiment after experiment with sundews, and fired off letter after letter to friends and associates asking for information on other sundew species, suggesting experiments, and reporting on his own experimental successes and failures.

Experiment builds upon observation, and so Darwin began with a close look: "The whole upper surface is covered with gland-bearing filaments, or tentacles, as I shall call them, from their manner of acting," he later wrote in *Insectivorous Plants*.[12] Note his word choice: "tentacles" are evocative of marine invertebrates—animals, not plants. Consciously or not, that was how Darwin viewed his little sundews, as his playful comment about them in 1863 to Gray indicates: "a wonderful plant, or rather a most sagacious animal."[13] Darwin set about observing his sundews closely, describing in minute detail the production of the viscid droplets that give the group their name, and documenting the structure and movement of the tentacles. He watched in fascination as they bent in unison in response to food, and set about testing just what would tempt them most.

Darwin went about his experiments on *Drosera* systematically, testing the plants' response first to nonnitrogenous and then nitrogenous fluids based on an informed guess that nitrogenous compounds were key. He hypothesized that these plants became carnivores as an adaptation to their nutrient-limited habitat. Given that bogs are especially poor in nitrogen relative to other nutrients, it stands to reason that it is nitrogen-bearing substances the plants are after. If the substances selected for his nonnitrogenous trials seem a bit haphazard—gum arabic, sugar, starch, diluted alcohol, olive oil, even an "infusion and decoction" of tea—that's because they were: Darwin used whatever

Drosera rotundifolia leaf, viewed from above and enlarged 4x. (Left) Resting position. (Center) Filaments inflected after exposure to phosphate of ammonia. (Right) Leaf with filaments on one side inflected over a bit of meat placed on the disk. From Darwin (1875a), p. 3, fig. 1, and p. 10, figs. 4 and 5.

was at hand, pilfering from the kitchen and washroom medicine cabinet. His methods seem modern in some ways, but not in others: he first tried a control by applying drops of distilled water (eliciting no response), varied the concentration of the substances he was testing, and tested the water control and treatments on more than one leaf. In spirit these reflect modern ideas of experimental design, with controls, treatments, and multiple replicates. On the other hand, Darwin could be imprecise and inconsistent by other modern standards. The distilled water was applied to "30 to 40 leaves" rather than a standard number for all trials, for example, and in the gum arabic tests he reported trying four solution concentrations but only quantified one of them. Darwin next tested the drops on 14 leaves, but was casual about the duration of each test, leaving them on anywhere from a day to almost 2 days, but generally about 30 hours. The imprecision was common in that era, but it wouldn't be long before more rigorous standards for experimental method, in the light of statistical rigor, would become the norm.

In all Darwin tested over 60 leaves with nonnitrogenous substances, and in no case did the tentacles respond. He next turned to nitrogenous substances, and the response could not have been more different: drops of milk, egg white, raw meat infusion, saliva, mucus, nitrate of ammonia, even his own urine induced movement of the tentacles of nearly every one of the 64 leaves tested. The short-stalked ones in the center don't tend to move, but the long marginal tentacles bent as much as 180° to position their glistening globe of "dew" onto the object, taking from 1½ to 6 or so hours to inflect fully. Writing to Hooker at Kew, Darwin described these "first-rate chemists" as capable of distinguishing even minute quantities of nitrogenous from nonnitrogenous substances. In another letter, he enthused that the tiniest grain of nitrate of ammonia was sufficient to elicit an effect from the sundew leaves.

Having confirmed his hunch that nitrogenous substances whet the plants' appetite, he came up with more questions: are solids as effective as fluids in eliciting a response? Do the leaves actually have the power of digestion? Can they can break down and absorb organic matter? Those experiments waited a decade, however, as Darwin got sidetracked with a number of other projects: his orchid book was published in 1862, a succession of new *Origin* editions (second through fifth) appeared throughout the decade, and the two-volume work on domestication with which he was struggling, all the while conducting experiments in hybridization, flower structure, and climbing plants. He had given a paper on *Drosera* at the Philosophical Club of the Royal Society on February 21, 1861, and then all was silent on the sundew front for a while. His experiments largely halted until *Variation in Animals and Plants Under Domestication* appeared in 1868, but sundews and their relatives were far from ignored. He continued to receive information and specimens of various sundew species from friends and other correspondents, housing a growing collection in his greenhouse.

Sundew Stomachs

Almost exactly a decade after the family convalescence in Bournemouth, Darwin had a false start getting back to these plants-cum-animals: "Began working at Drosera" his journal read for August 23, 1872. But he had the usual interruptions, so yet a year later, June 14, 1873, he wrote: "Began Drosera again."[14] In the meantime naturalists on the other side of the Atlantic had become intrigued by these plants. Mary Treat of New Jersey was a talented naturalist who published numerous papers on birds, insects, and carnivorous plants. She likely learned of Darwin's interest in sundews in the course of her frequent correspondence with Asa Gray at Harvard, and wrote to Darwin in the early 1870s. Darwin had high praise for Treat's work, encouraging her to publish and enlisting her help in observing the thread-leaved sundew (*Drosera filiformis*) and other species in the eastern United States. She was happy to assist, and many of Treat's observations on sundews and other carnivorous plants are cited in *Insectivorous Plants* (and were also published separately).

Two of the most important lines of investigation Darwin took up once he restarted his sundew studies in earnest included a return to the question of digestive ability, and the motor stimulus that leads to the inflection of the tentacles and leaves. In these he had the assistance of one John Burdon-Sanderson, a professor of physiology at University College, London. Darwin invited him to visit Down House in June of 1873 to discuss *Drosera*, and soon had him hooked. A key question for the digestion experiments was whether the plants could render animal matter soluble: can they digest matter, or simply absorb what is already broken down? In *Insectivorous Plants* he summarized his experiments systematically.[15] First, he did what he called the "acid test"—testing the droplets from 30 leaves in an unstimulated state with litmus paper, he found little to no discoloration. He then tested another set of leaves fed, so to speak, with a range of materials from glass particles to hard-boiled egg white to bits of raw meat. After 24 hours the droplets of the tentacles inflecting toward the material were shown by the litmus paper to be decidedly acidic. Darwin's conclusion was that the viscid secretion from the leaves becomes acidic once the tentacles contact edible objects. Darwin also made an incidental discovery: the dewdrop secretions of the tentacles has antiseptic properties, preventing the growth of mold and other microorganisms and in this way also acts like the gastric juice of animals.

Following the suggestions of Burdon-Sanderson, Darwin next carefully tested for digestion using small cubes of hard-boiled egg. Some cubes were set up on wet moss as a control, while the others were placed on sundew leaves and observed at intervals ranging from 21 to 50 hours. The egg cubes on the leaves turned into liquid globules and disappeared (presumably absorbed). Those left on the moss simply rotted. This was all very suggestive, but the *real* acid test then followed: if he could neutralize the acid and shut down the digestion process, Darwin felt this would be "the best and almost sole test of the presence of some ferment analogous to pepsin in the secretion." Pepsin is an animal digestive enzyme released by cells lining the stomach. It was in fact the first enzyme discovered, isolated by German physiologist Theodor Schwann in 1836. Schwann derived the name "pepsin" from the Greek for digestion: πέψς, *pepsis*. It was later joined by trypsin and chymotrypsin as a trio of proteolytic (protein-degrading) enzymes—

proteases—playing the leading role in animal digestion. The enzymes depend upon the acidic environment of the digestive system for both their own activity and to assist in the breakdown of digestible matter, which is why Darwin's litmus paper tests showing acidity of the sundew droplets were so encouraging. He now wanted to test whether that acidity played a role in sundew digestion.

Darwin used a control and two treatments. With the control plants, cubes of hard-boiled egg were permitted to break down as usual, with Darwin adding only minute water droplets (on the order of 5 microliters) a few times daily. The plant absorbed the egg in 48 hours. He then divided treatments using similarly sized cubes of egg into two groups: he "watered" one with equally minute droplets of a weak hydrochloric acid solution (1 HCl:437 H_2O), and one with a weak sodium carbonate solution (1 Na_2CO_3:437 H_2O). *Voila*: the hydrochloric acid accelerated the absorption of the egg-white cubes to little more than 24 hours, while the sodium carbonate neutralized the droplets altogether. Darwin was then able to neutralize the neutralizer, countering its effects with weak hydrochloric acid until breakdown and absorption proceeded normally. "From these experiments," he concluded, "we clearly see that the secretion has the power of dissolving albumen, and we further see that if an alkali is added, the process of digestion is stopped, but immediately recommences as soon as the alkali is neutralised by weak hydrochloric acid." His experiments "almost sufficed to prove that the glands of *Drosera* secrete some ferment analogous to pepsin, which in the presence of an acid gives to the secretion its power of dissolving albuminous compounds."[16] Darwin again proceeded to test a range of substances both digestible and indigestible to the plants. Protein or protein-rich substances like roast beef, fibrin, cartilage, gelatin, and even pollen were readily digested, while substances like fats, starches, hair balls, mucus, and bits of nail were not, as he had found in Bournemouth years before. (Keratin, the chief constituent of hair and nails, is a protein but its highly ordered structure renders it immune to digestion under the conditions of animal—or plant—stomachs. This is why cats cough up their hairballs instead of simply recycling the protein by digesting them.)

Darwin concluded the "digestion" chapter of *Insectivorous Plants* arguing that the so-called ferment of *Drosera* is very similar if not identical to the digestive juices of animals, specifically a combination of pepsin and some sort of acid. "That a plant and an animal should pour forth the same, or nearly the same, complex secretion, adapted for the same purpose of digestion, is a new and wonderful fact in physiology," he declared. There was something missing, however; something that bothered him as well as his son Frank, by now (in his late 20s) a regular research collaborator with his father. It was very suggestive, sundews producing a pepsin-like "ferment" and dissolving nitrogenous treats. But were they *digesting* per se—gaining nutritional benefit? After all, these plants have chlorophyll and can survive even if deprived of treats. The crucial experiment was attempted—feeding trials to see if "fed" plants grew better, flowered more, and set more seeds than unfed plants. But many of the plants fed and unfed alike unaccountably died, so the experiment was left undone by the time *Insectivorous Plants* came out in 1875. It was soon taken up again by Frank, who devised a new approach.

Planting round-leaved sundews (*D. rotundifolia*) in a set of six Wedgwood soup dishes in the summer of 1877, Frank arrayed these on a tray fitted with a removable wood frame covered with gauze to exclude insects, and so that all the plants received about the same amount of light. He also kept them moist. He fed 86 sundews pieces of roasted meat weighing about 1.3 mg each every 5 days or so. He did not feed the control group, similar in number. Within a few weeks Frank saw results: "The first difference noticed between fed and starved halves of the plates was on July 17th, when the fed side, viewed as a whole, was clearly greener than the starved half. The difference was quite distinct in all six plates, as both my father and I observed."[17] Toward the end of August, he noted the number of flowering stems (a total of 116 in the unfed plants versus 173 in fed), number of stems with at least one flower (19 unfed, 34 fed), and the number of healthy leaves, reckoned by the presence of glandular secretions (187 unfed, 256 fed).

Frank also measured the diameter of a sample of leaves from fed and unfed plants, and obtained their respective dry weights, finding longer leaves and higher weights among the fed sundews. Finally, he turned to the key measures of vigor and reproductive success in a plant: numbers of seed capsules and the weight of seeds produced. In all cases the fed sundews outperformed the unfed controls, and the net weight of seeds produced by fed plants exceeded that of the unfed plants by a factor of nearly 4:1. The stumbling block to realizing this benefit before, he pointed out, was the fact that these plants seem to do okay without such nutritional supplements. Darwin was proud of his son, but didn't live long enough to see the results of Frank's experiment included in *Insectivorous Plants*. The second edition appeared in 1888, 6 years after Darwin's death, and Frank duly appended the results to the first chapter.

Other aspects of the nutritional biology of sundews were beyond the ability of the Darwins (or anyone of the time) to explore. Stable nuclear isotope technology, for example, developed in the following century, found a multitude of applications from medicine and agriculture to geology and ecology. This technology was put to good use in further exploring sundew nutrition by the British and Scottish team of Jonathan Millett, Roger Jones, and Susan Waldron, who approached Darwin's question from another angle. These researchers sought to distinguish between animal- and non-animal-derived sources of nitrogen in two sundew species, *D. rotundifolia* (Darwin's favorite, the laboratory rat of the sundew world) and the related species *D. intermedia*, capitalizing on the fact that the "heavier" of the two stable forms of nitrogen, ^{15}N, is found in a higher concentration in animals than in (most) plants, owing to differences in the way animals and plants accumulate body mass. Animals accumulate this isotope by eating other organisms, while most plants derive theirs more slowly from the soil.

This study found that on average some 50 percent of the sundew's nitrogen was derived from insect prey.

Characterizing the digestive "ferment" of sundews similarly had to await the advent of new technology. A series of studies in the twentieth century showed it to be a species of protease very similar to one called nepenthesin, described from pitcherplants of the genus *Nepenthes* in the 1960s. The proteases of *Nepenthes* pitcherplants and those of sundews and Venus flytraps have since been found to be very closely related—a fact that should not surprise us given that these carnivorous plants are cousins. Darwin would have been delighted to know that his hunch had proved correct: the "ferment" of *Drosera* is indeed chemically similar to animal pepsin, and functions in exactly the same way.

Irritable Plants

Darwin was also keenly interested in the "irritability" of his sundews. Not that his fascination with their animal-like characteristics led him to believe they could become irate; rather, his fascination was irritability in the sense used by physiologists of the eighteenth and nineteenth centuries. The *Oxford English Dictionary* defines it as "the capacity of being excited to vital action (e.g., motion, contraction, nervous impulse, etc.) by the application of an external stimulus: a property of living matter or protoplasm in general, and characteristic in a special degree of certain organs or tissues of animals and plants, esp. muscles and nerves." *Plants* with muscles and nerves? Precisely why he was interested. He tried to figure out the mechanism behind the sense perception and movement of sundew tentacles, tracking their mobilization across leaves in response to point stimuli like carefully placed bits of meat. He experimentally determined that the "motor impulse," as he put it, is transmitted more readily in a longitudinal than transverse direction across the leaves, a pattern that maps onto the main vascular architecture of the leaves.

Placing a morsel on a single tentacle, he watched as the impulse spread to the surrounding tentacles, which promptly bent toward the point of excitement. "Nothing could be more striking than the appearance of [the leaves] each with their tentacles pointing truly [toward the morsel]," he declared in *Insectivorous Plants*. "We might imagine that we were looking at a lowly organised animal seizing prey with its arms." Images of sea anemones come to mind. But he had to conclude that the comparison with animals was only superficial in the case of these movements. He described the motor impulse of tentacular movement as a "reflex action" (and "the only known case of reflex action in the vegetable kingdom" to boot), where the food morsel stimulated the tentacles it was in immediate contact with, the excitation of which in turn stimulated other nearby tentacles to slowly arc toward the food. He had to concede, however, that the mechanism of this "reflex action" was very different from that of the animal nervous system. Rather than electrical conducting tissue (nerves) triggering contractile fibers (muscle), he concluded that the movement was related to the mobilization of protoplasm causing the walls of particular cells to contract and relax.

The prominent plant physiologist Julius Sachs of the University of Würzburg had proposed in 1874 that movement of plant structures result from the osmotic flow of fluid in and out of specialized cells, creating rapid changes in tension that translate into contraction and relaxation (more on this in Chapter 9). This was later confirmed to be an important mode of movement in many plant groups, but Darwin ruled it out in the case of his sundew tentacles, which tend to move rather slowly. It likely plays a role, however, in the movement of the "snap tentacles" of the Australian pimpernel sundew, *Drosera glanduligera*. These specialized tentacles fringe the leaves of this ground-hugging sundew, arrayed horizontally awaiting the footfall of inquisitive insects. They lack the dewy droplets of normal tentacles, instead tapering to a fine upcurved point which functions not so much to pierce as to trap. In milliseconds they snap toward the leaf, flinging their insect prey onto the nearby glandular tentacles where they become entrapped—a phenomenon that would have captivated Darwin too, no doubt.

The Most Wonderful Plant in the World

Darwin's darling of a snap-action plant was the Venus flytrap (*Dionaea muscipula*), the small plant endemic to a narrow band along the coastal Carolinas in the United States. Its bilobed leaves with their fringe of stout spike-like bristles resemble the iron-toothed snap traps that do their job with such devastating effectiveness. Early observers had correctly interpreted them as traps—the bristles function more to imprison than pierce hapless insects. But traps for what? Before the virtues of pollination were appreciated the traps were imagined to protect the flowers from insects. Darwin's grandfather Erasmus, perhaps even more famous for his scientifically inspired poetry than his celebrated medical skills, expressed this interpretation in his botanical poem *Loves of the Plants* (1788). He described the "wonderful contrivance to prevent the depredations of insects" as follows: "The leaves are armed with long teeth, like the antennae of insects, and lie spread upon the ground round the stem; and are so irritable, that when an insect creeps upon them, they fold up, and crush or pierce it to death." Two hundred years later, it likely inspired the monster plant of *Little Shop of Horrors*.

"Astonishing production!" exulted colonial-era American naturalist William Bartram in 1791, contemporary of Erasmus though a generation younger. This was the plant, recall, that Bartram declared was endowed with "sensible faculties or attributes, similar to those that dignify animal nature; they are organical, living and self-moving bodies, for we see here, in this plant,

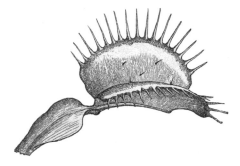

Venus flytrap, *Dionaea muscipula*. In this drawing of an open trap by George Darwin, the triangle of small trigger hairs is visible on the inner surface of the upper leaflet. The long, stiff spikes lining the edges of the lobes look lethal and some early observers thought they functioned to pierce prey. Their lethality stems, however, from their vise-like grip once the trap snaps shut. From Darwin (1875a), p. 287.

motion and volition."[18] Darwin concurred, opening his discussion in *Insectivorous Plants* with the declaration that "This plant . . . is one of the most wonderful in the world."[19]

Remember that Darwin's *Drosera* studies were initially a means to an end, and that end was to understand the Venus flytrap: "I began this work on *Drosera* in relation to gradation as throwing light on *Dionaea*," he wrote in 1860 to Daniel Oliver at Kew. Oliver was able to oblige him with a few flytrap leaves for study, and eventually whole living plants, but not before Darwin had to beg, cajole, plead, implore . . . everything short of grovel:

> *Is it [Dionaea] a very precious plant? Could a living plant be packed so as to come here by Railway & could I keep it alive for week or two in sitting room? If so, would you be so kind as to give me address of any nurseryman where I could purchase a plant. Or if there are several plants at Kew, would you read this note to Sir William [director of Kew at the time] & ask him whether he could lend me a plant, which should be returned (carriage free) to Kew; but I should require to gather & dissect some leaves. I want to compare structure of hairs of Dionæa & Drosera . . . I would just as soon purchase as borrow: I only mention Kew in case the plant is not to be easily purchased & in case of there being several plants at Kew. How I should like to see a fly caught by it![20]*

It *was* a precious plant, a rarity that the gardeners of Kew were not very willing to share (hard to imagine today, when every garden center large and small offers them for sale, each neatly packaged in its own plastic hothouse). Darwin's desperation is almost palpable—he was ecstatic when he finally got a few to study, so imagine how he would have reveled over them in their native environment of the coastal Carolina marshlands. Like the sundews, at a distance natural populations of this diminutive plant form an impressionistic red-tinged green wash, resolving on closer inspection into whorls of bilobed green leaves fringed with those evil-looking (but harmless) spikes, each leaf gaping wide to show its tempting scarlet throat—a lascivious trap set with a botanical come-hither look outdone only by the orchids.

What made Venus flytrap especially wonderful to Darwin was not so much how its snap-trap leaves become a temporary stomach, fascinating and instructive though that was to him. (He tested their reaction to the same diversity of substances as with *Drosera*.) Rather, he was convinced it had a nervous system analogous to that of animals, something he investigated but could not quite convince himself of with the "reflex action" of the sundews. Darwin's initial work done with the flytraps was put on hold along with his sundew experiments back in the early 1860s, but later that decade Asa Gray first drew Darwin's attention to the work of another American naturalist, William Marriott Canby of Wilmington, Delaware. Canby, a businessman and philanthropist, was also an avid botanist and a regular correspondent with Gray on such matters. In 1867 Canby wrote to Gray: "The little *Dionaea* has become exceedingly interesting to me, & I have been making some experiments with a view to test its carniverous [*sic*] propensities." He then went on to describe a host of observations and experiments on prey trapping and digestion, even testing their taste for various substances à la Darwin: "In one leaf I placed a piece of cheese. This was partly dissolved but it was evidently not the food nature intended, as the leaf after going through a considerable part of the operation of digestion, became sick and finally died." Death by indigestion—a horrible end, even in a plant. Gray sent the letter to Darwin. The observations were scarcely new to him, however, and he had already made the same observations. He wrote back to Gray, "This letter fires me up to complete and publish on *Drosera, Dionaea*, etc., but when I shall get time I know not. I am working like a slave to complete my book,"[21] he lamented. Darwin was still bogged down finishing *Variation*, which came out at last the following year. Turning his attention back to his "sagacious" plants in the early 1870s, he recalled Canby's letter to Gray, and Canby's declaration in that letter that he wished "someone would take hold of this who was fully able to elaborate it" and that he would "gladly aid all I can in furnishing specimens, &c." In January 1873, 6 years on, Darwin again wrote Gray about this:

> *You sent me a letter dated Wilmington July 9. '67 about Dionaea. It has no signature but you refer to it as written by Mr. Canlay or Canbay or Cawley. Will you be so kind as to write the name for me distinctly, for some people are so foolish as to say that your handwriting, like mine, is not very legible. The letter has interested me much in some respects & the gentleman seems very kind & willing to oblige.*[22]

(It was an understatement for Darwin to acknowledge that his handwriting was "not very legible"—even today few individuals are versed in the art of deciphering his notoriously illegible scrawl.) Gray supplied him with Canby's address and Darwin soon wrote, but mistakenly thought that Canby lived close to the native habitat of the flytraps—an understandable error, with the plants naturally found around Wilmington, North Carolina, while by coincidence Canby lived in the more northerly Wilmington, Delaware. Canby set Darwin straight, and let him know that in some months' time he hoped to travel to coastal North Carolina and would make observations for him.

In the meantime Darwin subjected his flytraps to a battery of experiments, quickly noticing that the inner surface of each leaf lobe bears three slender filaments arrayed in a triangle. These proved to be the trigger hairs that produce the snap-action of the trap. Unbeknownst to Darwin, the clergyman and naturalist Moses Ashley Curtis (who unlike Canby *did* live in Wilmington, North Carolina) had correctly noted the function of these hairs decades earlier, in an 1834 paper. Curtis had also noted the digestive properties of the leaf, commenting in his paper on the plants' mucilaginous fluid that seemed to dissolve insects.

Darwin explored the sensitivity of the filaments (technically termed "trichomes") by devising a probe: a piece of "very delicate human hair" (probably his own) was affixed to a handle and cut to 1 inch in length, a length "sufficiently rigid to support itself in a nearly

horizontal line." Carefully reaching in and touching the side of a filament lightly with his probe, he induced the lobes to abruptly snap
shut. He could have done this with a needle or some such, but using
the fine hair enabled him to show that the trigger hairs were sensitive
to the lightest touch. Of course, the price of such sensitivity of the leaf
might be needless closing—cases of mistaken identity, when blown
debris happens to touch a filament, or wasted effort when insect prey
minute enough to escape between the fringing bristles touch it off.
Darwin tried to trip the trap by sprinkling water and other substances
like flour on the lobes. He even tried blowing hard, simulating a gale.
The traps remained unsprung. He correctly concluded that in the
plant's natural environment downpours and gales are common, and
the plants adapted so as not to be fooled. Occasionally blown debris
may set one off, but more often than not it is a prospective meal that
springs the trap, and a noteworthy property of the trigger reduces the
error rate: the filaments must be contacted by something solid, and a
given filament must be contacted not once but twice within about 15
seconds to induce the trap to close. Reducing the error rate is important
for the plant because the traps have the ability to snap shut only so many
times. After all the leaves lose their ability to close, the plant can only
rely on photosynthesis to live.

The reason for this gets at the heart of an important difference
between animal movement by muscle action and the movement of these
plants: think of the pull of muscle attached to bone as a force generated
from within the body. In contrast, movement of the flytrap leaf lobes
stems from a force generated outside, on the leaf surface. More specifically, the pull results from changes in the curvature of the leaf surface,
which in turn results from rapid shrinkage of the inner side of the leaf
relative to the outer. But what is behind the spurt of growth of the
leaf cells? Plant cell walls are usually fairly rigid, but Venus flytrap has
a mechanism called "acid growth" that permits rapid cell expansion.
Acidic chemical compounds are produced that relax the usually rigid
cell walls, allowing the cells to grow or expand. Expansion on the outer
side of the leaf makes that side larger in area than the other, so the leaf

flexes in the opposite direction—the lobes close inward. The problem is that this is not reversible. Turgor pressure changes in the cells on the outer leaf surface result in the lobes slowly opening, and resetting the trap, but each time the trap snaps shut the acid growth process results in the cells getting slightly larger; after a while they reach their limit, and that leaf is no longer able to function.

Darwin did not know these details, but he realized that each trap can be sprung a limited number of times. He looked closely at the mode of capture and was immediately struck by the speed and force with which the leaf lobes closed on prey. The lightning-fast closure is helped by the fact that the lobes are at right angles to one another, so each needs to move about 45° as opposed to the 90° arc needed if they were splayed out horizontally. The fringing bristles of each lobe interlace with one another, like the fingers of clasped hands, but he was careful to note that the bristles themselves did not move. Leaves that had been fooled into closing by probing the trigger hairs or feeding them indigestible substances took a day or more to reopen fully. Even when partially open (Darwin observed that of 10 leaves observed most were about two-thirds open after about 7 hours) they are capable, however, of closing again quickly. Another thing that a more casual eye might not notice is how the lobes initially form a concave space around trapped prey, like a stomach, and then slowly but surely become firmly appressed all around the prey until the trapped insect appears as a distinct bulge on either side of the leaf, like a meal within opaque green shrink-wrapping. At this point the lobes are so tightly shut that Darwin could only peer inside by forcefully driving a wedge between them. He had to be careful; with too much force they "are generally ruptured rather than yield."

The quick snap action of the leaves immediately reminded Darwin of reflexive motor impulses of animals. Here, perhaps, was a better candidate for a botanical nerve-pulse "reflex action" than sundews. His first breakthrough in pursuing the idea that the plant had a nervous system came in fall of 1872; as he described in an excited letter to Gray:

The point which has interested me most is tracing the nerves!!! which follow the vascular bundles. By a prick with a sharp lancet at a certain point, I can paralyse ½ the leaf, so that a stimulus to the other half causes no movement. It is just like dividing the spinal marrow of a Frog:—no stimulus can be sent from the Brain or anterior part of spine to the hind legs; but if these latter are stimulated, they move by reflex actions. I find my old results about the astonishing sensitiveness of the nervous system (!?) of Drosera *to various stimulants fully confirmed & extended.*[23]

Ever supportive, Gray replied: "it is wonderful—your finding the nervous system of *Dionaea*!!!" Pursuing this line of inquiry further required the assistance of a specialist. He collaborated again with John Burdon-Sanderson, his physiologist friend. In the summer of 1873 Burdon-Sanderson suggested experiments using a galvanometer to test for electrical activity in the leaves. Working with plants at Kew Gardens, it was not long before Burdon-Sanderson had flytraps wired up like a patient getting an ECG. He first found differences in electrical potential using parts of the leaf and stem, and then, most excitingly, discovered that when a fly stimulates a leaf, a striking change in the leaf's electrical current results: the "leaf-current," as he called it, showed an abrupt spike in electrical potential immediately before the lobes snapped to a close. He telegraphed Darwin with the news, and

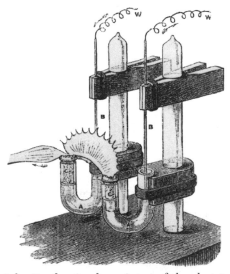

John Burdon-Sanderson's test of the electrical properties of a Venus flytrap leaf using a modified Thomson galvanometer. From Burdon-Sanderson (1874), p. 128.

soon gave lectures on the phenomenon at the British Association and the Royal Society. "These currents are subject," he declared, "to the same laws as those of muscle and nerve."[24] Darwin was so excited that he made a rare effort to get up to London to attend Burdon-Sanderson's lecture at the Royal Institution that June.

Darwin reported Burdon-Sanderson's results along with his own flytrap experiments in *Insectivorous Plants*. He was able to show that the lobes can act independently by "paralyzing" one of the two lobes, and with the help of his son Frank he made detailed study of the cellular structure of the leaves. More than any other plant he had studied to date, Venus flytrap really did seem animal-like not only in its "behavior" and digestive abilities, but also in having a sort of nervous system. He shared Bartram's wonder at these plants, endowed as Bartram said with "sensible faculties or attributes, similar to those that dignify animal nature"—plants with volition, even. Little wonder that Darwin declared these plants "one of the most wonderful in the world."[25]

An Aquatic Flytrap?

A good many other carnivorous plant groups were experimented upon and discussed in *Insectivorous Plants*: other sundews, aquatic and terrestrial members of the bladderwort family, Lentibulariaceae: *Utricularia*, with its tiny underwater "suction traps," and *Pinguicula*, the butterworts common on heathlands where they compete with sundews by trapping minute insects on their sticky flypaper-like leaves. The most striking one, though, turned out to be an aquatic relative of the Venus flytrap: the waterwheel plant, *Aldrovanda vesiculosa*, widely distributed from Europe to Africa and Australia. The botanist Giuseppe Monti named this plant *Aldrovandia* in 1747, in honor of the sixteenth-century Italian naturalist Ulisse Aldrovandi. Evidently Linnaeus made a typographical error when he recorded the name in his *Species Plantarum* of 1753, inadvertently dropping the "i." The error stuck.

Aldrovanda is a curious, rootless plant, with hinged leaves arranged in a whorl around a central, free-floating stem. The leaves are lobed some-

what like those of the Venus flytrap and also snap shut at the touch of trigger hairs, but do so far faster than Venus flytrap, in as little as a tenth of a
second. Hooker sent some over from Kew in September of 1874.

Darwin immediately noticed that the plant seemed to have snap-
action leaves like *Dionaea*, but floated in water and caught minute
free-swimming prey like *Utricularia*, the aquatic bladderwort. On
closer investigation he found that *Aldrovanda* was even more similar
to bladderwort than he thought. The inner surface of the leaf lobes has
lots of minute filaments, each with four tiny arms. Darwin called them
"quadrifid processes," and soon determined that they played a digestive function, like the glands in *Drosera* and *Dionaea*, which secrete
digestive enzymes and absorb the digested matter. Then he noticed
that the lobe edges are lined with similar filaments, which apparently
only absorb and do not secrete enzymes. Now, *that* was a novelty for
a carnivorous plant. It suggested a rudimentary form of the kind of
specialized function found in other carnivorous plants. As he put it
in the conclusion of his chapter on *Aldrovanda:* "if this view is correct, we have the remarkable case of different parts of the same leaf
serving for very different purposes—one part for true digestion, and
another for the absorption of decayed animal matter." If so, this would
have broader evolutionary implications, he realized: "We can thus also
understand how, by the gradual loss of either power, a plant might be
gradually adapted for the one function to the exclusion of the other;
and it will hereafter be shown that two genera, namely *Pinguicula* [but-
terworts] and *Utricularia* [bladderworts], belonging to the same family,
have been adapted to two different functions."[26] In *Aldrovanda* Darwin thus saw the grand themes of gradual adaptation, divergence, and
common descent that his theory of evolution by natural selection so
nicely explained.

The sensitivity to touch, movement, and digestive abilities of carnivorous sundews and flytraps may echo the animal world, but the
mechanisms behind them are not just simpler versions of mechanisms

better developed in animals. They are analogous rather than homologous; generally speaking, they are evolutionary convergences of similar solutions to the same problem. Darwin may have been hoping for
a closer correspondence, but he conceded that phenomena like the
"reflex action" of these plants were probably very different physiolog-
ically from those of the animal world. At another level, Darwin was
right on the plant-animal relationship: the physiology and anatomical
machinery behind movement and digesting food in these plants may
differ from those of animals, yet there are enough similarities to point
to commonalities in physiology, commonalities that point to common
ancestry.

Therein lies an important lesson: the concepts of homology versus
analogy in understanding the "animal-like" qualities of carnivorous
plants, and its bearing on the broader concept of Darwin's Tree of Life.
Darwin's overarching interest in these plants lay in what they might
teach us about universal common ancestry. For Darwin, commonalities in underlying physiology, or cellular processes, was evidence for
such an ancestor. Plants and animals (let alone fungi, protists, and the
various prokaryotic groups) outwardly might seem to have little in common, making it difficult to see them as relatives, but Darwin sought
similarities at the less obvious, physiological, level. He would have reveled in the modern understanding of the ultimate evidence for shared
ancestry: our very DNA.

Experimentising: Feed Me, Seymour!

In pursuing his vision of universal common descent—the idea that all
species are descended from common ancestors—Darwin was inevitably intrigued by plant-like animals and animal-like plants. And there
may be no plants more animal-like than certain botanical carnivores.
Darwin was interested in the sense perception of these plants with
regard to trapping prey: what they respond to, and how they move, and
what they digest.

Obtaining Sundews and Flytraps

For these investigations we want *Drosera rotundifolia* or other suitable sundew species, and Venus flytrap (*Dionaea muscipula*), preferably with fresh leaves. While available in garden centers, these plants are often not well cared for. Carolina Biological (www.carolina.com) is a trusted source. If you want to grow your own sundews or flytraps, an excellent source of information (and seeds) is the International Carnivorous Plant Society (ICPS, carnivorousplants.org/howto) with growing guides by species.

I. Sensitive, Are We?

A. Materials

- *Drosera rotundifolia* or other sundew
- Dissecting probe or stiff paint brush
- Forceps or tweezers
- Hand lens and/or dissecting microscope
- Milk (small quantity—no more than 2 tsp [10 ml])
- Pipet or eye dropper
- Hard-boiled egg white, cubed to $^1/_{16}$ in. (2 mm) on a side
- Toothpick (wood), cut to about $^1/_{16}$ in. (2 mm) pieces
- Hard cheese and gummy bear candy, cut into about $^1/_{16}$ in. (2 mm) cubes
- Litmus paper

B. Procedure

1. Be careful not to touch the filaments on the inner surface of the leaf lobes before you are ready to begin. Using the hand lens, note the difference in length of the hairs (filaments) fringing the leaf versus those found in the leaf center.
2. The sensitivity of the filaments is best demonstrated with the long outer ones, which can be individually excited to inflect. Firmly poke a single long filament at its tip (droplet) three or

four times with the probe or paint brush. (Note the viscous consistency of the droplet.) Did the filament inflect toward the center of the leaf? If so, how long after stimulation? Repeat this step with several filaments on several leaves to observe a good response. How many responded? How many didn't? Darwin found that when he poked a number of filaments four, five, or even six times with enough force to bend the filament, only a portion of the filaments became inflected. "The result was so uncertain," he wrote, "as to seem capricious." He suggested that it is adaptive for the plant not to be overly sensitive to touch, since the leaves are probably often contacted by blown debris, waving grass blades, etc. When a filament did respond to touch Darwin found that it did so quickly (he reported that he had "distinctly seen, through a lens, a [filament] beginning to bend in ten seconds, and . . . strongly pronounced inflection in under one minute" after stimulation. How quickly did stimulated filaments bend in your study?

3. Selecting several fresh filaments, not touched before, apply a minute droplet of milk to each by dipping the tip of the probe or brush into milk and carefully transferring the tiny droplet to the tip of a single filament in turn (a hand lens or dissecting microscope will help). Did inflection of these filaments occur any faster?

4. Select another leaf. Manually stimulate the short filaments at the center of the leaf disk with the probe or brush. Observe the "behavior" of the longer marginal filaments over the course of 1 to 2 hours. How many become inflected, and how long did it take them? An experiment like this was Darwin's first indication of transmission of a "motor impulse" in sundew leaves.

5. Observe the direction of the "motor impulse." With the forceps carefully place a small cube of egg white on filaments near the leaf edge on one side of the medial line of a leaf (X at the lower left of the diagram). Repeat by placing a cube directly on the

medial line, near the upper margin or lower margin (X at the lower right of the diagram). Inspect the leaves after 24 hours and count how many filaments on the side of the leaf *opposite* from where you placed the egg have responded by inflecting. In what direction do they inflect? Do any bend in a different direction? Darwin found that the "motor impulse" is transmitted more readily in a longitudinal than a transverse direction on the leaf; accordingly, we would expect more filaments longitudinally opposite the egg cube to have responded than those located transversely across the medial line.

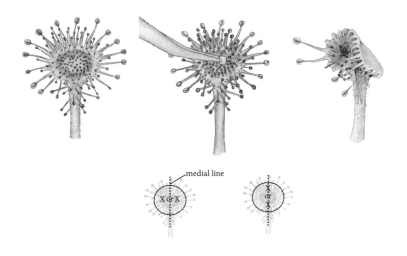

II. Picky Eaters

A. Materials
Same as for Part I.

B. Procedure
1. Can sundews tell the difference between edible and inedible substances? Select two fresh leaves on the same plant, preferably side by side. Place an egg-white cube in the center of one leaf, and place a similar-sized wood matchstick cube in the center of

the other. Make observations on the "behavior" of each leaf at regular intervals—perhaps at 2, 4, and 6 hours—comparing the reaction of the leaf to the egg versus the wood.

2. Select three other fresh leaves. "Feed" one an egg-white cube, the second a bit of cheese about the same size, and the third a piece of the gummy bear (cut into a like-sized piece). Observe the reaction of the leaves to each. After 7 days, gently extract any food or food remains (it may be necessary to carefully pry open the leaf or filaments). Is the sundew a picky eater? Which food was digested and which not, and why? (Hint: *Drosera* needs nitrogen-containing protein, and excessive sugars are not to its taste.)

3. Stimulate two filaments using milk or egg. Carefully touch a piece of litmus paper to a droplet or two from unexcited (uninflected) filaments. Repeat with another piece of litmus paper, this time touching a droplet or two from filaments stimulated by the presence of digestible food. Take care not to touch the paper to the food item itself. What is the color difference of the two pieces of litmus paper? Which one shows more presence of acid? Can you think of why? The droplet from the excited filament should be decidedly more acidic, indicating digestive enzymes.

III. Trigger-Happy Flytraps

A. Materials

- One or more fresh *Dionaea muscipula* plants
- Dissecting probe or stiff paint brush
- $^3/_{16}$ in. (0.5 cm) diameter wood dowel, toothpicks, or wood kitchen matches
- Glue or tape
- Forceps or tweezers
- Short (~1" [2.5 cm]) piece of thread, monofiliment, or hair
- Hand lens and/or dissecting microscope

NOTE: Each trap can close and reopen only about six or seven times before withering. Darwin and others found, however, that traps with insects sometimes never reopened, or if they did were sluggish or unresponsive for a time.

B. Procedure

1. It was the rapidity of the snap-trap action of this plant as much as its "stomach" that fascinated Darwin. Look closely at the inner surface of one of the leaf traps with a hand lens or dissecting microscope, and find the two to five small trigger hairs (trichomes) arrayed in the center of each lobe. Look at two or three other leaves; can you tell visually if the trigger hairs are always in approximately the same position? Compare the length and width of the trigger hairs with the far stouter spike-like projections fringing the lobes.

2. Touching two or more of the trigger hairs in succession will trip the trap. You can repeat Darwin's test of the sensitivity of the hairs by making a probe with a short length of hair. (Someone with longish straight hair might wish to donate one from his or her head; otherwise, try a substitute like a length of moderately

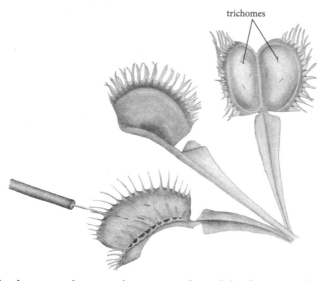

The slender trigger-hairs on the inner surface of the flytrap can be tripped with a probe. Drawing by Leslie C. Costa.

thick thread, dental floss, or fishing monofilament.) Darwin's "hair probe" made it possible to touch the filaments very lightly. To make your hair probe, affix a single hair with tape or glue to the end of a 1 in. (5 cm) length of ³⁄₁₆ in. (0.5 cm) diameter wood dowel, toothpick, or matchstick with the flammable tip removed. This will be the probe handle. Cut the hair to extend about 1" (2.5 cm) beyond the end. (The hair can be glued to the tip of a dissecting probe, alternatively.) Now you are ready to test the sensitivity of the trigger hairs: ever-so-gently brush the hair of your probe against two of the trichomes in fairly rapid succession and observe the result; watch your fingers! (Just kidding.) Allow up to 24 hours for a closed trap to fully reopen.

See also:

A. M. Ellison, "They Really Do Eat Insects: Learning from Charles Darwin's Experiments with Carnivorous Plants," in *Darwin-Inspired Learning*, ed. M. J. Reiss, C. J. Boulter, and D. L. Sanders (Rotterdam: Sense Publishers, 2015), 243–256.

"Insectivorous Plants" at the Darwin Correspondence Project: www .darwinproject.ac.uk/learning/universities/getting-know-darwins-science /insectivorous-plants.

9

Crafty and Sagacious Climbers

V ines and other climbers have always had the power to fascinate. They evoke associations positive and negative: in myth, metaphor, and lore such vines as grapes and hops are ancient symbols of growth, prosperity, fertility, and a good party, and vines play important supporting roles, so to speak, in the adventures of heroes like Tarzan or Indiana Jones. But more often than not we find something vaguely sinister about climbing plants, and our fascination comes with a tinge of fear or wariness. I think the reason is at least twofold: first, vines can easily bring entanglement to mind, if not strangulation. Fictional ones are almost always sinister, from J. K. Rowling's devil's snare to the lethal assassin vine of Dungeons and Dragons. Some are stranglers in lore, like the European swallow-wort (*Cynanchum louiseae*), a milkweed relative commonly known as "dog-strangling vine." Others are stranglers in fact (albeit of other vegetation, not animals), like the notorious kudzu vine of Japan (*Pueraria* spp.). This viney legume was introduced to the southeastern United States in the 1940s to stabilize soil during road construction. However, it proved to be an extremely invasive plant and is now popularly known as the Vine that Ate the South. The second reason that climbing plants can seem sinister is their movements: uncannily animal-like, they inspire an uneasy sense that they are purposeful in their probing; watchful as they stealthily grow and entwine. That movement, a slither in slow motion, may underlie yet a third reason for our uneasiness: altogether too snake-like, they evoke images of perhaps the most (unfairly) feared and reviled of animals.

Darwin had no such negative associations with climbing plants. On the contrary that same uncanny animal-like ability to move, reach, probe, and feel that underlies our unease piqued his curiosity. This was another group of plants with volition, with sense perception that underscored their fundamental relationship with animals, just like his beloved sundews and flytraps. Certainly John Horwood, the neighbor's gardener who helped Darwin with many of his botanical experiments and oversaw the design of his heated greenhouse, thought so. Asa Gray had sent Darwin seeds of *Echinocystis lobata*, the North American wild cucumber plant, an excellent climber with tendrils that always seem to find something to grab in their vicinity. In a letter to Hooker, Darwin marveled at the sensitivity of those tendrils, and, referring to Horwood, commented that "A clever gardener, my neighbour, who saw the plant on my table last night, said 'I believe, Sir, the tendrils can see, for wherever I put the plant, it finds out any stick near enough.'"[1]

Darwin was in the midst of his early flush of excitement about *Drosera*, back in early 1861, when Gray brought tendrils to his attention. They must have been discussing the irritability of *Drosera* leaves. In a now-lost letter Gray evidently reminded Darwin that he had written a paper in 1858 entitled "Note on the Coiling of Tendrils." This paper described the exquisite sensitivity of tendrils of the bur cucumber, *Sicyos angulatus*, which coil at the slightest touch and uncoil again in the space of an hour.

Darwin filed this information away in his mind, for at that moment he was in thrall to sundews as well as flower polymorphisms and crossbreeding, while also hard at work on his orchid book. But a year later he turned to tendrils and wrote Gray to remind him about that paper. "I should like to try a few experiments on your Tendrils; I wonder what would be good & easy plant to raise in pot," he asked. Ever obliging, Gray sent a reply to Darwin later that month, joking about how North American "weeds" are routed by invading species (another subject of interest to Darwin), but also providing seeds and growing instructions (with a jibe about the lack of enough warmth and sunshine in England for the plants):

> *Oh—as to the weeds,—Mrs. Gray says she allows that our weeds give up to yours. Ours are modest, woodland, retiring things, and no match for the intrusive, pretentious, self-asserting foreigners. But I send you seeds of one native weed which corrupted by bad company—is as nasty and troublesome as any I know, viz. Sycios [sic] angulatus,—also of a more genteel Cucurbitacea, Echinocystis lobata (the larger seeds). Upon these, especially upon the first, I made my observations of tendrils coiling to the touch. Put the seeds directly into the ground; they will come up in spring, in moist garden soil. My observations were made on a warm sunny day. I doubt if you have warmth and sunshine enough in England to get up a sensible movement.*[2]

That spring Darwin tried to germinate Gray's seeds in his newly completed greenhouse, designed to John Horwood's specifications and built against the kitchen garden wall. Horwood's boss John Turnbull had kindly lent Darwin the use of his own greenhouses for Darwin's orchid work. Now, with a greenhouse of his own, he could maintain his plants all year long and a lot more of them to boot. Whether it was the limited English sunshine or not, just one *Echinocystis* seed that Gray sent germinated, and none of the *Sicyos* seeds. Darwin made the most of the one specimen he had: "I have been looking at its tendrils & seen with great interest their irritability; it is a very pretty little discovery of yours," he wrote to Gray that June. He also revealed a discovery of his own: "I am observing the plant in another respect, namely the incessant rotary movement of the leading shoots, which bring the tendrils into contact with any body within a circle of a foot or 20 inches in diameter.—If I can make out anything clear about this movement, & do not find that it is known, I will perhaps write a letter to you for the chance of its being worth inserting in Silliman or elsewhere."[3] "Silliman" was a reference to the *American Journal of Science and Arts*, informally known as "Silliman's journal" for geologist Benjamin Silliman Sr. of Yale, the journal's founder.

The regular circular motion of the shoot immediately captured Darwin's attention, and that same month he wrote to Hooker to ask if he or Daniel Oliver at Kew were aware of the phenomenon. Observing *Echinocystis lobata*, he suddenly noticed that the stem between the uppermost leaves slowly rotated, tracing a full circle some 12 to 20 inches across in 1½ to 2 hours. Sometimes it stopped and rotated in the opposite direction. Curiously, the stem didn't appear to become twisted, yet the rotation went on day and night. When a solid object was placed within the circle traced by the probing plant, the tendril immediately grabbed hold. No wonder the plant seemed to be able to see. He asked Hooker if the slow rotary searching behavior was already well known.

Hooker hadn't taken notice of this phenomenon and immediately jumped on the bandwagon and set some Kew staff to observing climbing plants. Darwin then wrote to ask Hooker if he had any climbing plants to spare for his greenhouse, and described in more detail some of his observations. Hooker was encouraging: "Your observations on Tendrils &c are most curious & novel," he said, "& I am delighted that you are going on with them—you are 'facile princeps' [easily the best] of observers."[4] Darwin's new greenhouse was soon home to a host of climbing plants from around the world—he was ideally situated to benefit from proximity to the largest botanical collection anywhere, courtesy of his good friend who happened to be assistant director.

It turned out that although Hooker did not know of the circular motion of the tendrils, Gray in the United States did. Darwin excitedly wrote to Gray that July with his findings, but was deflated when Gray replied that the phenomenon was well known. When he told Gray that he found nothing published on the subject, and nor did Hooker and his colleagues at Kew know anything about it, Gray came down a bit hard on the Kew botanists, telling Darwin that "As to tendrils, What are Hooker & Oliver (the latter a Professor too) about, and where have they lived not to know anything of them?" A slightly exasperated Gray told Darwin that "everybody" must have seen the way the tendrils sweep, and coil up if they fail to contact something after a certain time. Then he directed Darwin to reread the opening sentences of Gray's 1858 paper, the one that initially piqued Darwin's interest, where he would find reference to the work of Hugo von Mohl, the botanist at the University of Tübingen in Germany who first described the rotary movement of tendrils back in the 1820s. Gray long before had encouraged a colleague to translate von Mohl into English, and he reprimanded Darwin tongue-in-cheek for overlooking it: "and be thankful to me for having instigated (in 1851) Van Voorst to get Henfrey to translate this little book.—tho you English pay no attention to it, when you have got it." Ouch. But he encouraged Darwin not to give up his experiments: "do not abandon this subject, for it will be fruitful in your hands."[5]

It turned out that others, too, had described the rotating and twisting of climbing plants: Ludwig Palm, another botanist at Tübingen, and the Frenchman Henri Dutrochet, perhaps most extensively. Chastened but sure that he could make a contribution, Darwin forged ahead; there was more to his interest in these climbers than their apparent sense perception. Besides, observing them in the quiet of his study and greenhouse was the perfect occupation for his increasingly jangled nerves. The years 1863 and 1864 were difficult for Darwin. The *Origin*, out in its third edition in April 1861, was still a lightning rod drawing strong criticism from some quarters, and he agonized over the harsh reviews. He received a constant stream of letters with all manner of questions, comments, criticisms, attacks, and the all-too-occasional congratulations. He was relieved to let friends like Huxley, Lyell, and Gray enter the fray on his behalf—a fray that he literally could not stomach. Darwin's old illness, which seemed to ebb and flow, came on like a fast-flowing spring tide in those years, with near-daily retching and chills. By September 1863 he was in constant pain; in a letter to his friend the Rev. John Innes (formerly the local vicar, who had since moved to Scotland) Darwin commented that he had taken to botany and all he could manage was an hour or two of work each day. Emma insisted that they take a family holiday in Malvern, Worcestershire, where Darwin could undergo treatment at the hydropathic establishment of James Smith Ayerst—their youngest son Horace, aged 12, even took the "water cure" with his father, since he was displaying similar dyspepsia. They stayed several weeks but to little effect; the locale was perhaps not the best for soothing his nerves, considering it was the scene of the devastating death of their daughter Annie back in 1851. He tried different cold-water treatments when he returned home, but they failed, too, and he finally gave up the water cure for good. This period of ill health was the worst that Darwin had ever experienced, confining him to bed for some weeks by early 1864.

Under the circumstances, watching the slow drift of shoots and tendrils in the near-dark of his study was the perfect occupation. "The pace of his life," as Darwin biographer Janet Browne nicely put it, "slowed to that of a plant."[6] His children (most now young adults) helped as they could. Darwin informed Willy, his eldest at 24 years old and working in banking in Southampton, that his current hobbyhorse was tendrils, and asked him to make some observations on dimorphic plants. George, Frank, and Lenny, all teenagers, were boarding at Clapham School, not far away in south London. When home on holidays they once again became faithful research assistants; Lenny recalled how his father, joking about his son being "infernally healthy," sent him off to fetch plants for the sickroom-study, and George proved to be a talented botanical artist. George illustrated his father's Linnean Society paper with drawings of plants in the gardens and greenhouse at Down. In 1864 George graduated Clapham School and entered Trinity College, Cambridge, where the mathematical instruction of Clapham headmaster Charles Pritchard, an astronomer, stood him in good stead. While his start at Cambridge was inauspicious, to the family's surprise he ultimately graduated as "second wrangler" (second-highest scores in the university) on the grueling Mathematical Tripos examination. George went on to become an able astronomer himself, eventually Plumian Professor of Astronomy and Experimental Philosophy at Cambridge.

The Many Ways to Be a Climber

The sensory aspects of Darwin's "crafty & sagacious" climbers were fascinating, but he was just as keen on another great evolutionary principle that they demonstrated: adaptive variation in climbing. Different species accomplished the same thing in different ways; that is, different structures of the plants had been modified to serve similar uses in different but related groups, illustrating principles such as homology, analogy, and convergence. In August of 1863 Darwin made a telling statement to Gray that tendril "irritability is beautiful, as beautiful in all its modifications as anything in orchids." It was the modifications—structural variations on a theme in accomplishing climbing—that became the most important evolutionary story told by vines. Recall Darwin's admission (also to Gray) that his "chief interest" in his orchid book was how "it bears on design, that endless question." There he showed how orchids exemplified the great principle of different parts modified for similar ends, and similar parts modified for different ends. The underlying homologies and analogies of structure in different species pointed to the vagaries of evolution by natural selection. In response to similar selection pressure, different species may evolve the same "solution" but do so in different ways. It was as if anatomical structure were a toolbox, and different parts could be modified to serve similar functions. "Homologous" organs have a common underlying structure and origin (embryological and evolutionary), even if they outwardly look very different once fully developed. A textbook example is the limb structure of mammals: bone-for-bone the wings of bats, forelegs of moles, flukes of whales, running legs of horses, and reaching arms of humans are the same structure, though the bones have different sizes and shapes. Homologous organs may be used in similar ways, or not—most of these limbs are used for locomotion, but very different forms of locomotion, from burrowing to flying to swimming. "Analogous" organs, in contrast, are outwardly similar in structure and function, but they are anatomically different and lack a common origin. Whale flukes and fish fins are a typical example—these have the same

function and superficially look alike, but are very different anatomically. That difference indicates that they evolved independently, natural selection having honed them from different ancestral starting points
into the shape and maneuverability necessary for efficient swimming.
They are convergent, with penguin wings representing another such
convergence: these birds "fly" through the water.

In the last chapter we saw Darwin effectively asking why a Creator
would bother to modify different floral structures in different orchid
groups to function in essentially the same way. Vines, too, present oddities and convergences in regard to the different ways that climbing is
accomplished. This variation, Darwin argued, is more consistent with
descent with modification than with special creation and supernatu-
ral design. The design argument implies that, insofar as there is a best
way to do something, an omnipotent designer would use that best
solution—that is the essence of good design. So, whether we're talking
about orchid pollination or climbing vines, a designer would use the
same excellent design solution throughout the respective groups, not
mixing and matching, or jury-rigging, to accomplish the same ends in
different lineages. That messiness is more consistent with an evolutionary process where each step must build upon what came before.

Climbing adaptations certainly are variations on a theme. Some
ascend by means of grappling-hook thorns; others have modified roots;
still others modified leaves or flower peduncles. Darwin divided them
accordingly in his long 1865 paper "On the Movements and Habits of
Climbing Plants" (taking care to point out that "these subdivisions . . .
nearly all graduate into each other"). First there are the *twiners*. Familiar examples include brewer's hops (*Humulus lupulus*), morning glory
(*Convolvulus*), and honeysuckle (*Lonicera*). In these plants the stem
itself is the climbing and "grasping" organ as it grows. The stem's tip,
or apical meristem, steadily revolves in a counterclockwise direction,
thereby spirally encircling supports like the stems and branches (below
a certain diameter) of other plants. This is the rotary motion that first
captured Darwin's attention. It is now thought that this *circumnutation*
(Charles and Frank Darwin came up with the term) results from dif-

ferential growth and elongation of cells of the growing meristem, which is a zone of cell division. Division and elongation of cells evenly across the meristem would result in straight (outward and upward growth), but if cell division and elongation was happening a little faster on one side, that side would in a sense "pull ahead" of the neighboring cells, forcing the apical portion of the meristem to bend or curve toward the slower-growing cells. What seems to happen is that cell division and elongation move back and forth across the meristem face like a "fan wave" rolling back and forth across the stands at a football game, with the result that the meristem curves continuously in a circle, taking anywhere from 1 to 24 hours to complete a circuit.

The *leaf-climbers* include vines such as virgin's bower (*Clematis*), nasturtium (*Tropaeolum*), and gloriosa or climbing lilies (*Gloriosa*). These plants climb with sensitive petioles that respond to contact by bending and hooking. Once they clasp a support these specialized petioles often undergo a metamorphosis, thickening and often becoming woody. Also included in this category are vines that climb with modified flower stalks, or peduncles. One well studied peduncle-climber is *Strychnos*, a genus of mostly tropical trees and vines known for their potent toxins (one gives us strychnine, the alkaloid used in rodent control). Leaf-climbers also exhibit circumnutation, and often rotate a good deal faster than the twiners.

Then there are the *tendril-bearers*. Darwin studied a great many of these, including grapes (*Vitis*), peas (*Pisum*), creepers (*Parthenocissus*), crossvine (*Bignonia*), wild cucumber (*Echinocystis*), bryony (*Bryonia*), catbriers (*Smilax*), and beans (*Vicia*). Tendrils are specialized climbing organs derived from different plant structures depending on the plant, including leaves, stems, or peduncles. While their circumnutation is similar to twiners, they tend to move faster and in more of an elliptical pattern, and a single individual is capable of circumnutating clockwise and counterclockwise. These plants often grasp in a two-phase coiling movement. Contact coiling occurs when the tendril first touches a potential supporting object. The tendril then has a growth and coiling spurt, leading into the free coiling phase, which tightens its hold on the support. The tendril does this by coiling in the opposing direction, resulting in two helical spring-like structures coiled in opposite directions relative to one another (one a right-hand coil, twisting counterclockwise; the other a left-hand coil, twisting clockwise).

The term for the left-hand-coiling process is *perversion*. (Most of these climbers, like humans, are right-handed, and so left-handedness is generally seen as an oddity; consider the word sinister, from Latin for left-handedness.) The opposite-helix phenomenon found in many tendril-bearers has been studied by physicists, who have shown that as tension in a cord or stem (or tendril) is decreased by helical coiling, at a certain point an instability (termed "curvature-to-writhe" instability) spontaneously sets in, resulting in two helices of opposite handedness. This is the first step toward the formation of super-coils (coils of coils) and can be demonstrated by grasping a length of rubber band and twisting both ends simultaneously. First, at a critical juncture of tension, twists of opposite handedness will form. Then with increased twisting a super-coil will form.

Some tendril-bearers will produce adhesive disks at the end of their tendrils, which attach to surfaces with glue-like secretions. The properties of the adhesive disks, which allow the plant to climb smooth surfaces without the need to coil around supporting branches to ascend, was of great interest to Darwin. The Ivy League schools have those

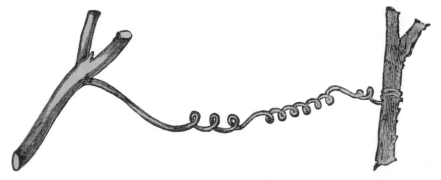

Tendril of bryony (*Bryonia dioica*), showing coil "perversion." Note the counter-clockwise coiling in the proximate portion of the tendril, and reversal to clockwise coiling in the more distal portion. From Darwin (1865), p. 96, fig. 13.

disks to thank for their nickname: they are the reason that Boston ivy (*Parthenocissus tricuspidata*—a species Darwin studied) can adorn those stately limestone and brick walls.

Hook- and root-climbers, finally, consist of a grab bag of climbers, including ramblers and scramblers that ascend with the aid of grappling-hook-like thorns (multiflora rose, *Rosa rugosa*) or specialized rootlets that produce sticky secretions (English ivy, *Hedera helix*). Darwin didn't find their behavior very interesting, so he shoehorned them into one category to acknowledge the group, but discussed them only lightly. I will do the same.

It was the leaf-climbers and tendril-bearers that he found most illuminating, groups that provide a good analogy with some of the swimmers mentioned earlier: cetaceans and penguins are related vertebrates that have separately evolved organs similar in structure and function, from a similar ancestral pentadactyl (five-toed) limb starting point. Tendrils, flower stalks, and leaves and leaf petioles, Darwin realized, have a point of common ancestral origin, too. In most plant groups, they serve different functions and so have different structural appearances. But in climbers, they have been modified into similar structures, for similar functions.

Darwin worked slowly but steadily on his climbing plants during 1863 and 1864, and by the end of 1865 he had studied scores of climbing plants. This work is breathtaking in its scope, a first in this line of research. He proceeded with observations and experiments on the motions and touch sensitivity of the plants; and he also observed their structures, with an eye toward comparative anatomy. At first, mostly he watched.

Do the Twist

As a student Darwin had been taught that climbing plants twist as they grow owing to the natural tendency for plants to grow spirally, by twisting the stem on its axis. As he now realized, that was something of a nonexplanation, and on close inspection he found that twiners and other climbers do *not* wholly twist on their axis as they grow. In the first section of his 1865 paper he took on the nature of the rotary movement of twiners, including the rate of rotation and the way spiral growth is achieved. He explained it with an analogy: "If we take hold of a growing sapling, we can of course bend it so as to make its tip describe a circle, like that performed by the tip of a spontaneously revolving plant. By this movement the sapling is not in the least twisted round its own axis." That is, rotating the sapling does not twist it up like a wound rubber band. Rather, the sapling is flexible, so it just bends in each direction it's pulled.

He devised a characteristically simple experiment to test for twisting in twiners. He put a dab of paint on the outer curved or bowed part of a twining shoot, and watched. As the twiner grew and slowly rotated around, the dot of paint seemed to change position. First it inched over to one side of the bowed tendril, then in the concavity beneath the tendril, opposite the outer bow. Later still it appeared on the other side of the bowed tendril, and finally ended up back where it began, on the outermost bowed surface. "This clearly proves," Darwin wrote, that tendrils "during the revolving movement, become bowed in every direction. The movement is, in fact, a continuous self-bowing of the whole shoot, successively directed to all points of the compass."[7] He explained it again as a thought experiment:

> As this movement is rather difficult to understand, it will be well to give an illustration. Let us take the tip of a sapling and bend it to the south, and paint a black line on the convex surface; then let the sapling spring up and bend it to the east, the black line will then be seen on the lateral face (fronting the north) of the shoot; bend it to the north, the black line will be on the concave surface; bend it to the west, the line will be on the southern lateral face; and when again bent to the south, the line will again be on the original convex surface.[8]

This process is counterintuitive; most people likely assume that the tendril gets twisted up when it rotates. Not so. But what is happening at the cellular level? Darwin explained this by imagining a wholesale contraction of cell growth on each side of the stem in turn; this would have the effect of bowing the sapling in the direction of the contracting cells and forcing it to point in different directions. If the "zone" of cell contraction rolls around the sapling stem, the whole stem rotates without twisting. "In fact," he concluded, "we should then have the exact kind of movement seen in the revolving shoots of twining plants."

Darwin was largely correct. Changes in turgor pressure (the pressure within plant cells created by an inflow of water, like filling a balloon) leading to rapid expansion and contraction are indeed thought to lie behind circumnutation and many other plant movements. It took many years, however for plant physiologists to appreciate the significance of osmosis (movement of water across a semipermeable membrane like a cell's membranous wall) for circumnutation and other plant movements. A considerable step in that direction was made by Hugo de Vries in Holland and Wilhelm Pfeffer in Germany, working on cell turgor and what we call today osmotic pressure. The work of these experimentalists was foundational for the modern understanding of how the elasticity of young plant cells and the active movement of substances like sugars into and out of them permit rapid and repeated changes in turgor pressure, and how this in turn can lead to rapid movement when cells act in concert. Until about the mid-nineteenth century botanists thought that bending movements in plants were due to an increase in cell growth on the convex side, opposite the bending. That can happen, but the physiologists showed that increased turgor pressure on the concave side is more important. As a set of cells in one area puff up like balloons or deflate in others they collectively exert pressure on neighboring cells, enough to bend a tendril.

Inspiring Revolutions

The slow rotation of stems and tendrils in wild cucumber, *Echinocystis lobata,* that first caught Darwin's attention proved to be nearly universal in climbing plants, though wild cucumber seemed to move especially fast. He recorded 35 revolutions of the upper internodes (sections of stem between leaves) and tendrils, and found that they could complete a circuit in an average of an hour and 40 minutes. He cut off the tendrils, yet the shoot continued to rotate. He then turned to tendril-free brewer's hops, grown in abundance in beer-loving Britain: "When the shoot of a Hop (*Humulus lupulus*) rises from the ground, the two or three first-formed internodes are straight and remain stationary; but the next-formed, whilst very young, may be seen to bend to one side and to travel slowly round towards all points of the compass, moving, like the hands of a watch, with the sun."[9] The average rate of rotation of hops shoots was just over 2 hours.

Darwin summarized the direction and rate of movement of over 40 twiners in a long, five-page table; some rotated with the sun, some opposite the sun, some fast, some slow. The champion of this botanical track-race was *Scyphanthus,* an attractive vine from Chile, lapping in just an hour and 17 minutes. He later found two climbers that rotated even faster: tendrils of the purple-flowered Mexican vine *Cobaea* completed a circuit in about an hour and 15 minutes, and a passionflower, *Passiflora gracilis,* had tendril speeds averaging 1 hour and 1 minute. Now, that's a vine you could set your watch by!

Darwin needed a way to graphically represent the movement of shoots and tendrils. Working with Frank, his first approach was to put a plant inside a large bell jar, a sort of terrarium of domed

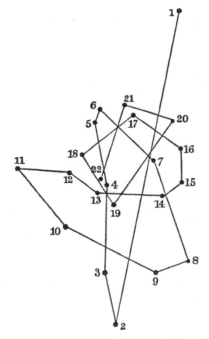

Elliptical movement of the upper internodes of a common pea plant (*Pisum sativum*), as traced on a bell jar and transferred to paper with illumination coming from the direction of the bottom of the diagram. From Darwin (1865), p. 65, fig. 6.

glass, and at regular time intervals mark the position of the upper stem seen through the glass with painted dots. The dots were then traced onto a flat sheet of paper, much as stars in the apparent curved "dome" of the night sky can be represented on a flat star map. The result showed the back-and-forth elliptical movement of the stem.

The American geologist James D. Hague visited Down House in 1871, and described another experimental setup in an 1884 memoir of his visit:

> The work of the forenoon was the careful observation of a number of tender shoots that were growing in pots, each under a separate bell-glass, and all ranged on a table exposed to the morning sun. To the growing tip of each plant there had been attached by wax . . . one end of a straight piece of very finely drawn glass thread, in such a manner that its other end projected about two inches, horizontally or nearly so, from the point of attachment, where any revolutionary movement of the stem must be imparted to the glass thread, and cause it to turn like a radial arm from a central point. The ends of the glass thread were made conspicuous by little "blobs" of [paint], thus giving two easily distinguished points in one straight line. By marking then upon the outer surface of the bell-glass a third point in line with the two "blobs", the subsequent departure of the outer "blob" from that line, caused by the turning of the stem of the plant, very soon became distinctly visible.[10]

At some point in the late 1870s Charles and Frank realized that the bell jars weren't ideal for their circumnutation studies—maybe owing to the curvature of the glass, or the limited size of the plants the jars could accommodate. Charles had an idea: why not try Emma's and Etty's plant case? Back in the early 1860s the Darwins became the proud owners of a Miss Maling's Patent In-door Plant Case (a "handsome ornament for the Drawing Room or the Sitting Room"). Basically a large terrarium, about the size of a 50-gallon fish tank, Miss Maling's ingenious design included slots in the base for pans of hot water, to be

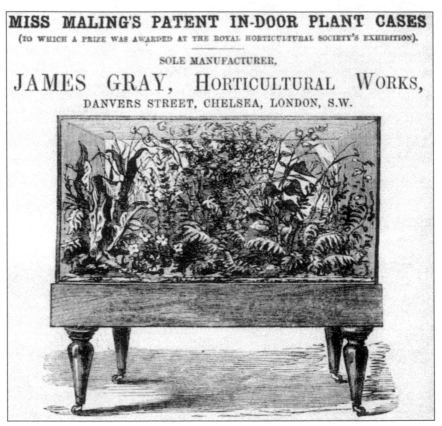

Advertisement for Miss Maling's Patent In-door Plant Case, an early terrarium. Darwin co-opted the family's case for circumnutation studies. From the *Gardeners' Chronicle and Agricultural Gazette*, January 17, 1863, p. 64.

replaced twice daily in order to warm the case and "preserve many of the most tender Exotic Plants through the severest winter." Darwin had originally bought the case for just that reason—as a hothouse for experiments—but it didn't quite live up to its billing.

His frustration was one reason he resolved to build the greenhouse and hothouse, abandoning Miss Maling's case to a delighted Etty. She and her mother kept it in the drawing room done up with hyacinths, azaleas, deutzias, and other plants. After Etty married Richard Litchfield in the summer of 1871, upkeep of the case fell to Emma. It was about this time that Darwin had the idea of using the top and side panes of the case for his circumnutation tracings—flat surfaces made tracing on them much easier than on the curved bell jars. Some of the species he documented were the very ones Emma most prized for the plant case, among them slender deutzia (*Deutzia gracilis*) and Indian azalea (*Azalea indica*), showy flowering shrubs from Japan. She surely didn't mind having the case co-opted for the cause of science.

Darwin's circumnutation studies burgeoned and blossomed like the plants themselves, growing into a multitude of circumnutation tracings—roots and stolons, seedlings and leaves, flower stalks and tendrils. All of these entailed variations on the theme of his basic experimental design, described so well by Hague: the attachment of a fine glass needle with a small ball of wax affixed on the end to the structure being traced. The wax ball was viewed through a glass plate—the roof or side window of the plant case—using a fixed point of reference, and the position of the ball in the line of sight marked on the glass.

It was tedious work, but in its way just what Frank needed to occupy his mind at a time of great sadness. Frank had studied medicine and physiology, and was married a few years after Etty, in 1874. He and his wife, Amy, had settled in the village of Downe, not far from Down House, and Frank would walk over almost daily to help his father with various experiments. Tragedy struck just 2 years later when Amy died in childbirth, though fortunately the baby survived. A devastated Frank and his newborn son Bernard moved back home with Charles and Emma. In his grief he threw himself into work and continued to help at Down House with all manner of experiments, especially plant movement, until his father's death in 1882.

For a while, the movement of plants became all-consuming for father and son, and one of their great insights was to realize that the gyratory movements Darwin had observed with climbing plants was actually a special case of a far more general tendency for nearly all plants and plant parts to revolve as they grow—roots, shoots, flowers, and leaves all are constantly but slowly, imperceptibly, tracing irregular and often narrow ellipses or ovals. Charles and Frank had hit upon a general principle of plant growth, one that if true could show how basic movements seemingly common to nearly all plants could be elaborated and modified in manifold ways. "Every growing part of every plant is continually circumnutating," Darwin later wrote, "though often on a small scale. . . . In this universally present movement we have the basis or groundwork for the acquirement, according to the requirements of the plant, of the most diversified movements."[11] The uncanny power of movement in climbers was likely a special, derived case of movement found in plants generally, movement that may be simply a by-product of the growth process.

Touchy Plants

There is more to the climbing habit than circumnutation, and touch perception of all kinds in climbers fascinated Charles. Frank, too, became ensnared by the "irritability" of tendrils and other structures. Recall that in the eighteenth and nineteenth centuries one definition of the words "irritable" and "irritability" referred to being highly responsive to stimuli, a definition that botanists soon applied to the movement of plants in response to touch. Darwin's climbers were plenty touchy, and he and Frank set about documenting when and under what circumstances. It's plain enough that climbers sense the object that they move toward and grow upon, but are all parts of the growing shoot or tendril irritable? Will the plant respond to a fleeting touch as well as prolonged contact? Darwin's approach was to lightly rub the petioles or internodes with a small stick and observe the response, or suspend a bit of string or thread for a very light touch. Unlike the lightning-fast action of plants like Venus flytrap, these climbers would respond in more usual "plant time"—taking hours. Darwin found that touch sensitivity resided in whatever organ was modified for climbing—petioles in the case of *Clematis*, for example, the last couple of internodes of the shoot for hops, and the entire tendril for grape vines. Of the many climbers he studied one of the most curious was the Mexican species *Lophospermum* (now *Maurandya*) *scandens*, a member of the plantain family, Plantaginaceae. This was a leaf-climber that used its petioles to grasp, but did so in an odd way: when the petiole bows and clasps a stick the adjacent internode is drawn to the stick too, and upon contacting the stick it also bends, trapping the stick between petiole and stem like pincers or forceps. Darwin tried 15 trials, each time rubbing the internodes two or three times and timing the result: most bent within 2 hours (all within 3), and then straightened themselves by the next day. He determined that all sides of the internodes were sensitive, and he was able to get them to bend first in one direction, then another. The response, he noted, was always toward the rubbed side.

Another curious expression of touch perception was found in the crossvine, *Bignonia capreolata,* of the southeastern United States, a vine celebrated for its lovely large, red-orange hummingbird-pollinated flowers. Each tendril is multiply branched, and each branch is divided into bifid or trifid hooked ends, like tiny grappling irons. What made them remarkable, however, was more their behavior than their appearance. Darwin was puzzled at first to find the tendrils had little sensitivity and seemed almost indifferent to the sticks he offered for climbing. Then he noticed that the stem curved away from the light, toward the darkest side of the room. He tried to trick them into climbing a tube, simulating a stick but blackened within, and found that the plant soon "recoiled, with what I can only call disgust, from these objects, and straightened themselves." In another experiment, Darwin placed a potted crossvine bearing six tendrils into a box open on one side only, and situated the opening to obliquely face a light source. Within two days he found that all of the tendrils were pointing toward the darkest corner of the box. This observation was all the more striking because the tendrils, being on different locations along the shoot, had to bend different degrees and directions to point that way. "Six wind-vanes could not have more truly shown the direction of the wind, than did these branched tendrils the course of the stream of light which entered the box,"[12] he marveled.

He began to connect the dots when he replaced the smooth slender stick with stouter posts, one with rough bark and one with fissures. The tiny tendril hooks soon found their niche, literally: "the points of the tendrils crawled into all the crevices in a beautiful manner." Serendipitously, he discovered what the crossvine was really after: when a tendril seized on a bit of wool left within reach it dawned on him that the plant sought out fibrous, textured surfaces. He wrapped flax, moss, and wool around a smooth stick and voila!—the tendrils grasped and the vine climbed with abandon. What's more, the little hooks at the end of the tendrils metamorphosed into tiny disks or balls that secreted an adhesive substance. Darwin had a hunch that this was related to the plant's native habitat. He wrote to Gray in May 1864:

> *Have you travelled south, & can you tell me, whether the trees, which* Bignonia capreolata *climbs, are covered with moss, or filamentous lichen or Tillandsia [Spanish moss]; I ask because its tendrils abhor a simple stick, do not much relish rough bark, but delight in wool or moss. They adhere in curious manner, by making little disks at end of each point, like the Ampelopsis [Virginia creeper]; when the disk sticks to bundle of fibres, these fibres grow between them & then unite, so that the fibres of wool end by being embedded in middle.*[13]

Gray replied that he had seen *B. capreolata* growing in the mountains of North Carolina and Tennessee. The vines, he said, climbed moist and shady trees, the trunks of which he didn't doubt were "well furnished with Lichens and Mosses." Gray remembered well: the southern Appalachian mountains are so wet that the vast deciduous forests of the region are distinctly green even when leafless in winter, so thick is the coating of epiphytic mosses, lichens, and ferns. The shade-seeking behavior of the tendrils thus makes sense: faced with a choice between light or shade, growing toward shade or shadow is more likely to pay off for young plants in terms of trees to climb, ultimately giving the vine access to light.

Darwin was doubly delighted by his discovery: "It is a highly remarkable fact," he declared, "that a leaf should become metamorphosed into a branched organ which turns from the light, and which can by its extremities either crawl like roots into crevices, or seize hold of minute projecting points."[14] This turning away from light belongs to a general class of plant movements called *tropisms*, from the Greek *tropos*, "turning." Tropism refers to turning movements in response to environmental stimuli, of which there are several types: hydrotropism for moisture, thermotropism for temperature, and so on. Tropisms are further described as "positive" for movement toward and "negative" for movement away from the stimulus. So, "negative phototropism" describes the darkness-seeking behavior of Darwin's *Bignonia* tendrils, an oddity in structures derived from light-loving leaves.

Of Roots and Shoots

Darwin came to suspect that insofar as all plant parts circumnutated, the rotary searching motion of climbers was an exaggerated expression. He embarked upon an investigation of circumnutation in a multitude of species and, crucially, at different stages of their growth. He found to his surprise that even the roots and shoots of seedlings gyrated. He was even more surprised, however, to find that in some cases an environmental stimulus overrode the gyration and influenced the direction of movement—certainly that was true of crossvine's flight from light. For most plants attraction *to* light is the norm, of course. But just how is this accomplished? Is the entire plant light-sensitive, or are there specialized cells or organs, analogous to eyes, that see the light and direct the plant?

Like the tendrils that John Horwood fancied could see, Darwin found that the growing tips of canary grass, *Phalacris canariensis*, also seemed to have eyes: "It can hardly fail to be of service to seedlings, by aiding them to find the shortest path from the buried seed to the light, on nearly the same principle that the eyes of most of the lower crawling animals are seated at the anterior ends of their bodies."[15] Grass shoots are encased in the seed leaf, the coleoptile. Father and son devised a series of experiments to show that the coleoptile has a distinct zone of sensitivity, one that signals other cells in the shoot about what direction to grow. Here is an example of their experimental approach:

- As a control, they topped the coleoptile tips of nine seedlings with transparent glass caps and exposed them to a bright southwest window on a bright day for 8 hours. All seedlings strongly curved toward the light as expected, but the experiment also showed that the glass tubes did not prevent shoot movement.

- They similarly enclosed 19 other coleoptiles in glass tubes, but these tubes were painted with black ink to block out the light. Five had to be rejected after the ink dried and cracked, permitting some light to get in. Of the remaining 14 seedlings, 7 remained upright and 7 were bowed, showing a light response. The Darwins suspected that the light response was due to leakage, or maybe reflected light from below the cap.

- In an effort to better control the light exposure to the tips, they next fitted 24 seedlings with tiny tinfoil caps, like diminutive botanical sunglasses, covering the last 0.15–0.2 inch of the coleoptile. Three seedlings still detected light and bent, but of the remaining ones, 17 stayed upright, while 4 only slightly bent. To ensure that the lack of movement wasn't due to injury from the caps themselves, they were removed and the plants fully exposed to light; all soon bowed toward the light, showing they were uninjured.

- Since tinfoil caps between 0.15 inch and 0.2 inch long proved to be efficient in preventing the seedlings from bending toward light, they fitted another 8 coleoptiles with even smaller caps, just 0.06 to 0.12 inch long. Of these, only 1 bent markedly toward the light; the lack of response from all but this one showed that the sensitive zone was very close to the tip of the shoot.

- Finally, the Darwins took another step toward localizing the zone of sensitivity by "bandaging" 8 shoots with strips of tinfoil about 0.2 inch wide, leaving the very tip of the coleoptile exposed. After 8 hours, the seedlings were curved toward the light. This was complemented with an experiment using the varnish-coated glass tubes again, only this time a tiny line or slit about 0.1–0.2 inch wide was etched into one side of the coating of each cap. While the lower half of 27 seedlings was left fully exposed to light, at the tips the only light exposure came from the etched slits. After 8 hours, half (14) of the seedlings remained vertical, while the other 13 were bowed—and not simply toward the window, but rather in the direction of the etched slits. Clearly, just a tiny amount of light was often all that was needed to induce the light response, and this contained directional information too.

Charles and Frank were certain that light exposure at the tip of the shoot somehow signaled the cells lower down to bend: the bowing that they observed took place not at the top of the shoot, but well below it. This means that the plant is capable of transmitting a signal from an area of stimulation—in this case the coleoptile—to elicit a response from cells elsewhere. It was a "striking fact," Darwin declared. "These results seem to imply the presence of some matter in the upper part which is acted on by light, and which transmits its effects to the lower part. It has been shown that this transmission is independent of the bending of the upper sensitive part."[16]

Indeed, there is a means by which plants transmit the effects of stimulation: here Darwin anticipated the discovery of plant growth hormones. Nearly 50 years later, in 1928, Dutch botanist Fritz Went isolated the first plant hormone from canary grass, the very same species that the Darwins studied. He called the substance "auxin" from the Greek *auxein*—to grow or increase. Its chemical structure was determined by British plant physiologist Kenneth V. Thimann, and in 1937 he and Went coauthored a book entitled *Phytohormones* on the subject.

Charles and Frank found a similar phenomenon with bean radicles. It was received wisdom that just as shoots and leaves grow toward light, so too do roots grow downward, toward the earth. If you pin or in some other way affix a germinated bean, say, so that the radicle is held out horizontally, it will soon bend downward. This was termed *geotropism* in Darwin's day, but is more commonly known as *gravitropism* today. The conventional view of the time was that the growing radicle was *pulled* by gravity, it did not *sense* and respond to gravity per se. Or did it? Just as the Darwins showed with canary grass that the part of the growing shoot that is light-sensitive is not the same as the part that responds by bending toward light, they argued that there was more than met the eye with growing root radicles: the tip of the root is sensitive to gravity, they showed, and transmits a signal to cells farther up the radicle to induce the downward growth.

A role for gravity in downward root growth was established in the early nineteenth century by the celebrated English experimental plant breeder Thomas Knight, whom we learned about in Chapter 6. In a paper presented to the Royal Society in 1806, Knight noted that if gravitation were the cause of the downward growth of the radicle, it must act either on the growing tissue during its formation, or on the motion of the sap and other liquids within the cells. Thus, Knight reasoned, "as gravitation could produce these effects only whilst the seed remained at rest, and in the same position relative to the attraction of the earth, I imagined that its operation would become suspended by constant and rapid change of the position of the germinating seed, and that it might become counteracted by the agency of centrifugal force."[17] On a small waterwheel along a stream in his garden, like that used to mill grain, Knight affixed a number of soaked beans to the wheel's circumference with their radicles pointing in all directions. He set the wheel spinning, and noted that no matter what direction the growing radicles initially pointed, they soon extended outwards away from the center of the wheel. Knight concluded that the radicles were influenced by centrifugal force, therefore suggesting that they must be similarly influenced by gravity when growing in their normal situation. Darwin did not comment on Knight's experiment, but agreed with his conclusion about the importance of gravity. Geotropism was real, but the important question for him was, did gravity act directly upon the radicle, as Knight seemed to suggest, or was this another case of a stimulus-response reaction like with canary grass? The title of the chapter on the subject in *Power of Movement* signals his answer: "Sensitiveness to gravitation, and its transmitted effects."

The key word is "transmitted." The definitive demonstration is straightforward: a controlled experiment testing for geotropic response with root tips intact versus root tips in some way removed or destroyed. It turned out that the Darwins did not think of this experiment first: just a few years earlier in 1872 a young plant physiologist in Poland, Theophil Ciesielski, published his doctoral dissertation on the downward curvature of roots. Ciesielski noted the difference between the region of cell growth and extension in the radicle and the zone of sensitivity at the tip influencing downward movement, and showed that removing the tip resulted in a loss of downward growth, or gravity sensitivity. The Darwins agreed, and in *Power of Movement* credited Ciesielski with this discovery. For the sake of completeness they replicated Ciesielski's tip-excision experiments in various ways, including amputation and killing the tip with chemical treatment. These measures effectively knocked out the geotropic response, but it could be restored if the tips were allowed to grow back. They concluded that although most authors thought radicles bent in response to gravity, in fact "we now know that it is the tip alone which is acted on, and that this part transmits some influence to the adjoining parts, causing them to curve downwards. Gravity does not appear to act in a more direct manner on a radicle,

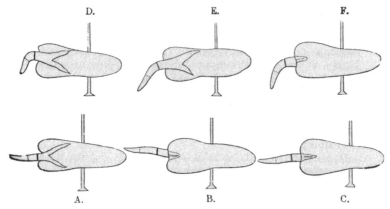

Common bean (*Vicia faba*) radicles grown horizontally for nearly 24h. (A, B, C) Radicle tips cauterized with silver nitrate, preventing a geotropic response. (D, E, F) Radicle tips left intact, showing normal geotropism. From Darwin (1880), p. 531.

than it does on any lowly organised animal, which moves away when it feels some weight or pressure."[18]

This conclusion put the Darwins on a collision course with one Julius von Sachs, distinguished professor of physiology at the University of Würzburg in Germany. At that time Germany was at the cutting edge of physiological research, and Sachs's laboratory was one of the finest. Sachs made his name with acclaimed work on transpiration, seed germination, and flowering. Frank, who was fluent in German, made two visits to Sachs's lab in Würzburg in the late 1870s to learn the latest techniques, and he also learned a great deal by reading German scientific papers and Sachs's own well-known 1868 text, the *Lehrbuch der Botanik*. His personal acquaintance with Sachs did not seem to help reduce friction between the German botanist and his father, however. In fact, Sachs eventually broke off his relationship with Frank after the perceived affront of the Darwins' suggesting that he was incorrect in his experimental conclusions over geotropism.

Sachs knew of Ciesielski's experiments and had one of his research assistants replicate them. Curiously, he failed to get the same result as Ciesielski and concluded that the Pole was incorrect: the radicle tips were *not* sensitive to gravity, he declared, and there were no so-called transmitted effects. Charles and Frank were puzzled by this, since their experiments clearly confirmed Ciesielski and they said so in print. Apparently their sin was not to simply disagree with Sachs, however, but to suggest there was a problem with his experimental technique. Adding insult to injury, in *The Power of Movement* Charles asserted that Ciesielski "[spoke] confidently of the obliteration of all geotropic action" with good reason, and that "it seems probable that Sachs unintentionally amputated the radicles on which he experimented, not strictly in a transverse direction."[19] In other words, if the tips were sliced off at a bit of an angle, some tiny piece of tip tissue might have remained, enough to show some geotropic response and lead Sachs astray. The German savant took this as a grave personal affront, priding himself as having exemplary experimental technique, with his laboratory the very embodiment of cutting-edge science.

There was more going on here than a bruised ego. In short, Sachs was contemptuous of the Darwins' work in large part because of the rough-and-ready style and home environment of their approach, for their failing to use the sophisticated instruments and controlled conditions of the German labs. The run-in with Sachs was thus symptomatic of a deeper issue, a sea change in the conduct of science in the late nineteenth century; historian Soraya de Chadarevian has aptly characterized their disagreement as a disconnect between old-fashioned "country-house" science and the emerging new laboratory-based science of the time. This was bound up with the increasing professionalization of science in the same period; Charles Darwin was of course very much in the tradition of the gentleman-naturalist, while Sachs was a brash, talented professional who had risen from humble beginnings to assume the botany chair at a fine university. He had won acclaim putting the latest high-tech equipment to good use, and had no use for Darwin's brand of drawing-room puttering with potted plants.

But ego surely played a role too. Sachs had a deserved reputation as a skilled experimentalist, which meant the suggestion of not only incorrect conclusions but sloppy technique was like pouring lemon juice on a cut. As excellent a scientist as Sachs may have been, unfortunately he was not as well known for his generosity of spirit. Already touchy about the term "circumnutation," which Charles and Frank coined to replace Sachs's own term "revolving nutation," things came to a head over geotropism. In his 1882 book on plant physiology Sachs was publicly dismissive of what he called the Darwins' "unskillfully made and improperly explained" experiments, and a more personal animus is evident in his private notebook: "Personal acquaintances often have their good side. I first became aware of the whole wre[t]chedness of Darwin's activities when Francis Darwin studied here in 1878 and 79; I had the opportunity to look behind the scenes and when the miserable book 'On Movements' appeared, I realized that here we are dealing with literary rascals."[20] Strong words from the esteemed professor! The Darwins thought there must be some mistake, as they in fact had high regard for Sachs and cited his excellent work in many places in *The*

Power of Movement. But it never occurred to them to be deferential to authority on any scientific point, let alone one with which their experiments disagreed. Frank traveled to Würzburg to see if he could make amends, but was curtly dismissed by Sachs and never saw him again.

As de Chadarevian concluded, if time has vindicated the Darwins' conclusions about root radicles and climbers, it has also vindicated the philosophy of Sachs: in the ensuing century, the science of country-house laboratories and gentleman-naturalists was replaced nearly wholesale by professional science conducted in specialized laboratories and with precision equipment. In the years following the death of his father in 1882, Frank himself was in the vanguard of this movement in Britain, becoming first Lecturer and then Reader in Botany at Cambridge, and was among an influential cohort of young scientists bringing to Britain the cutting-edge experimental techniques they learned in Germany. Geotropism remained a research interest, and when, around the turn of the century, the "statolith theory" of gravity-sensing was proposed, Frank tested and supported the theory in an ingenious manner.

Plant statoliths are starch granules (technically called amyloplasts) found in specialized oblong cells (statocytes) near the root tip. They function in the same way as otoliths, the tiny calcium carbonate granules of the inner ear that play a role in our sense of balance. (In fact, the term statolith was coined to correspond to otolith, as the mechanism of balance and orientation in animals was discovered earlier.) In both cases the granules can free-fall within their cells in response to gravity and pile up on whatever side of the cell is "down," where they trigger a signal. In the case of the otoliths in our ears the signal is triggered by touching and bending tiny hairs that line the inner walls of the cell; the nerve impulse is carried to our brain, which together with other input (from our eyes, for example) uses the information from these cells to gauge our body's position and keep us upright. In the case of the plants there are molecular chemical sensors in the cells rather than hairs. The plants lack nerves, despite Darwin's notion of his "sagacious" and "crafty" plants as animal-like, but the effect is the same: the signal is carried from the statocytes to the zone of active cell growth further up the radicle, and between the growth rate of different cells in that zone and changes in cell turgor pressure the root tip can be turned and guided in the right direction. Like us, the plants can tell which way is up and which is down.

Frank became fascinated by the statolith theory of geotropism, and pointed out that the experimental evidence for the theory was persuasive but incomplete. For example, he felt that experimental support for the theory based on destruction of the starch granules was suggestive but not definitive, since the demonstrated loss of geotropism in those cases could have been incidental to the loss of some other function simply as a result of eliminating the granules. Working with sorghum seedlings, he carried out experiments based on the destruction of the starch granules by heating. Finding that both sensitivity to gravity and to light seemed to be affected, and not suspecting that statoliths could also play a role in upward—"heliotropic"—shoot growth, he concluded that "these and other similar experiments showed us that we had no right to conclude that the loss of geotropic capacity depended on the absence of the special mechanism (statoliths), but rather that the loss of the starch may perhaps be no more than a symptom of exhaustion which shows itself both geo- and heliotropically."

He then devised a means of testing for the importance of statoliths by intensifying rather than removing their effects. His "tuning fork" method involved subjecting the seedlings to constant vibration, reasoning that if the response to gravity stems from "contact-irritability" by the statoliths within the cells, then the effect ought to be magnified if the statoliths could be made to continually jiggle and bounce on the cell wall they were resting on. He hypothesized that this would increase the stimulus, rather like increasing the stimulus to your ear by repeatedly ringing a door bell, pressing and removing your finger over and over in rapid succession. He rigged up a tuning fork to continually vibrate, and connected the tines of the fork to metal boxes in which he placed up to a half-dozen seedlings with their radicles growing horizontally. A similar number was set up in adjacent control boxes with no direct contact with the tuning fork. After vibrating the treatments for a time he placed both treatments and controls into an instrument called a klinostat—a device invented by Sachs to negate the effects of gravity by rotation, based on the waterwheel principle that Knight had used in his garden. A few hours later, Frank measured the curvature of the radicles, and lo and behold, those subjected to vibration showed considerably greater curvature than the controls. He repeated the process with several species and under various conditions; besides the geotropism of the radicle Frank also measured the upward curvature of the shoot (looking at effects on what he called "heliotropism"—shoot growth toward light, and so generally away from the direction of gravity's pull). Reporting his results to the Royal Society, Frank concluded that vibration did not affect the shoot response but clearly affected the radicles; he therefore supported the statolith theory of geotropism with his device. His father would have been proud of his son's experimental ingenuity.

Coming Full Circle

The experimental work on shoot, tendril, and root movement I discuss in this chapter is just a fraction of the many research questions explored in *The Power of Movement in Plants*; another research agenda of the Darwins focused on "sleep" in plants—both the diurnal movement of leaves and other organs and the touch sensitivity seen in the leaves of some species like mimosa or sensitive plant. We haven't space enough to delve into these ingenious experiments, too, but the underlying point was the same: to better understand plant sense perception and its myriad adaptations, qualities that at once underscore their common origin with animals and illustrate the evolutionary principles of diversification and co-optation. In his characteristically self-deprecating way, Darwin referred to *The Power of Movement in Plants* as "dry as dust" in a letter to his publisher. Fair enough, it's not exactly a page-turner. Okay, it's *far* from a page-turner—but of course it was not meant to be. But *The Movements and Habits of Climbing Plants*—which grew to over 200 pages by the second edition—and *The Power of Movement in Plants*—

running to 573 pages—together constituted a botanical tour-de-force, helping open up new research avenues in plant sense perception that remain vital today, and also helping to solidify Darwin's evolutionary vision of a truly universal Tree of Life.

That unitary vision was never far beneath the surface for Darwin. Consider his concluding remarks in both of these botanical works. In the last paragraphs of *Climbing Plants* he remarked that "The most interesting point in the natural history of climbing plants is their diverse powers of movement," and that "the most different organs—the stem, flower-peduncle, petiole, mid-ribs of the leaf or leaflets, and apparently aërial roots—all possess this power."[21] His deeper interest is evident in the last paragraph: "It has often been vaguely asserted," he says, "that plants are distinguished from animals by not having the power of movement. It should rather be said that plants acquire and display this power only when it is of some advantage to them." He points to the tendril-bearers, which provide a lesson in "how high in the scale of organization a plant may rise." It is worth looking closely at the language he uses in the final sentences:

> It first places its tendrils ready for action, as a polypus places its tentacula. If the tendril be displaced, it is acted on by the force of gravity and rights itself. It is acted on by the light, and bends towards or from it, or disregards it, whichever may be most advantageous. During several days the tendril or internodes, or both, spontaneously revolve with a steady motion. The tendril strikes some object, and quickly curls round and firmly grasps it. In the course of some hours it contracts into a spire, dragging up the stem, and forming an excellent spring. All movements now cease. By growth the tissues soon become wonderfully strong and durable.[22]

Note his carefully chosen words, expressing what can only be read as deliberate actions of the plant: it *places* its tendrils, just as a marine invertebrate polyp *places* its tentacles, and can *right itself*. It *bends toward* or *from* light, or *disregards* it. A tendril *spontaneously revolves*, and one *striking* an object *quickly curls* around and *firmly grasps* it. Over time it *contracts*, *dragging up* the stem of the climber.

Yes, tendrils are superb adaptations for seeking, grasping, and helping the plant climb—they move with intentionality relevant to their mode of life. Tendril-bearers certainly have ascended "high in the scale of organization," as Darwin put it—in a manner of speaking one might say they used those tendrils to climb the Tree of Life. He concludes, finally, that "the tendril has done its work, and done it in an admirable manner." Indeed, and the intentionality, the *behavior* of the climbers, is really of a piece with that found in the probing shoot or radicle. Hence the very last sentence in *The Power of Movement* echoes that of *Climbing Plants*: "It is hardly an exaggeration to say that the tip of the radicle thus endowed, and having the power of directing the movements of the adjoining parts, acts like the brain of one of the lower animals; the brain being seated within the anterior end of the body, receiving impressions from the sense-organs, and directing the several movements."[23] An animal in plant's clothing.

Experimentising: Seek and Ye Shall Find

Climbing plants were nearly as "sagacious" to Darwin as his beloved *Drosera*, illustrating for him several evolutionary principles. Seemingly animal-like in their powers of movement and sense perception, they too point to the fundamental underlying unity of plants and animals. Shoots that incessantly rotate, tendrils that probe and grasp, roots that can tell down from up . . . these responses to touch, light, and gravity may take place in slow motion, but they are "behaviors" nonetheless.

Obtaining Your Climbers

Try to find climbers from three or four groups below. Most should be readily available at garden centers and nurseries, or from online retailers, garden clubs, and websites dedicated to climbers.

Common ivy (*Hedera helix*)	Root-climber
Brewer's hops (*Humulus lupulus*)	Twiner
Virgin's bower (*Clematis virginiana*)	Leaf-climber
Sweet autumn clematis (*C. terniflora*)	Leaf-climber
Anemone clematis (*C. montana*)	Leaf-climber
Wild cucumber (*Echinocystis lobata*)	Tendril-climber
Bur cucumber (*Sicyos angulatus*)	Tendril-climber
Passionflower (*Passiflora* spp.)	Tendril-climber

Prairie Moon Nursery (www.prairiemoon.com) sells seeds of native North American climbing plants including *Clematis*, *Sicyos*, and *Echinocystis*.

NOTE: Depending on the species (and vendor) your climbers may be obtained as young plants, rhizomes or tubers, or may be grown from seed. Seeds may require stratification (some period of exposure to cold and moisture) before germinating; follow instructions on the seed packet. Vines and other climbers can get, of course, inconveniently long. Work with shoots or young plants of a manageable stature, planted in pots or a suitable outdoor garden with a small trellis or pole.

I. Climbing Styles and Behavior

Darwin was struck by how different plant groups climb with different organs, modified in various ways—some climb with their leading shoot, others with modified roots, leaf petioles, midribs, flower peduncles, or specialized tendrils, a nice illustration of evolutionary variations on a theme.

A. Materials
- English ivy, brewer's hops, clematis, passionflower, and bur or wild cucumber, potted or planted in the ground with trellis
- Hand lens or dissecting microscope
- Bamboo chopstick, wood skewer, or wood dowel, 4 in. (10 cm) in length, for probe

B. Procedure

Observation often goes hand in hand with experimentation, and as we've seen Darwin learned a great deal about his subjects simply by looking closely.

1. *Root-climbing.* Common or English ivy (*Hedera helix*): Native to Europe and Eurasia, this vine sports aerial roots along its stem. Root-climbers didn't hold much interest for Darwin, since their climbing organs show few of the active "behaviors" of other climbers. Still, for comparative purposes it is worth taking a look. Root-climbers can ascend even smooth walls by producing a kind of glue. Tugging on a vine growing up a vertical surface, note how strongly it adheres. Inspect the rootlets for adhesive disks at their tips, which secrete a glue-like substance. The "glue" includes nanoparticles just 60–85 nm in diameter, which were first described in 2008. They are being explored for applications in medicine, cosmetics, and cleaning agents.

2. *Twining.* Hops (*Humulus lupulus*) vines are also capable of climbing smooth, unbranched supports. Lightly draw your finger up a stem to learn how: the rough sandpapery feel comes from thousands of tiny downward-pointing hooks. Inspect the stem with a magnifying glass or, better, a dissecting microscope, and you will see that the surface is covered with these hooks. Note that they are pointed at two ends, with a pivot point in the center. From the side they look like tiny anvils. These are modified trichomes, or plant hairs, that function as miniature grappling hooks. As the stem winds around its support the hooks catch on the surface, increasing stability; with the combination of the spiral twining and tiny hooks, hops is remarkably resistant to slippage, and stability actually increases the larger the vine grows. This is because the downward-pulling weight of the vine creates tension that helps the spiral stem tighten. The principle at work is similar to that

of the bamboo "finger trap" toy, with helical bands of bamboo that tighten with stretching (tension) but loosen and open with inward force (compression).

3. *Leaf-climbing.* Virgin's bower (*Clematis*) vines provide a good example of a leaf-climber that uses its petioles to ascend. Gently stroke one side of an actively growing leaf petiole with a small twig or even your finger for a minute or two every hour, and within as little as 3–4 hours the petiole will have curled around on itself. You can uncurl it by repeating the process and stroking the opposite side of the petiole. The touch-sensitive petioles bend and clasp supporting stems or branches, or trellises in the garden. Clematis is "positively thigmotropic," bending toward an object making physical contact.

4. *Tendril-climbing.* Just as there are different ways to be a climber, among tendril-climbers we see the same principle of diversification at work: there are different ways to make a tendril. Tendrils, branched or unbranched, are variously specialized stems or leaves that may be derived from shoots, leaf petioles, or branches, depending on lineage.

   a. Bur cucumber (*Sicyos angulatus*) and wild cucumber (*Echinocystis lobata*) are North American climbers of the squash family, Curcurbitaceae. It was Asa Gray's observations of the tendrils of bur cucumber that initially got Darwin interested in climbing plants. Using either of these species, try Gray's demonstration of tendril sensitivity, from his 1858 paper:

   > A tendril which was straight, except a slight hook at the tip, on being gently touched once or twice with a piece of wood on the upper side, coiled at the end into 2½–3 turns within a minute and a half.

   Following Gray's lead, locate a straight tendril and gently stroke the upper surface a few times with the probe, then remove the probe. Observe and time the response. How long did it take for coiling to initiate, and how long did it last? Gray found that if the coiled tendril is left alone it will straighten itself. Time this: how long does it take your tendril to become straight again?

NOTE: Use actively growing straight tendrils, not ones that are coiled up. In many tendril-bearers the tendrils contract into a tight coil and hang limply after a few days (longer in some species) if they fail to come into contact with an object to grab. When that happens, the tendril has become more or less insensitive and won't respond to prodding, and so is useless for our purposes.

b. Observe older bur or wild cucumber vines with well-developed tendrils that have grasped a support to find examples of coiling and countercoiling, or coil "perversion"—the formation of coils that go in opposite directions, functioning as shock absorbers (see illustration below).

c. Passionflowers (*Passiflora* spp.) are a large and mostly tropical group native to the Americas, and many species are now available as garden plants prized for their unusual and complex flowers. Many people who grow passionflowers in their garden don't give much thought to the plant's tendrils. Not us! Note that the tendrils are unbranched, and extend from the axil of each leaf along the stem. Plants that are shaded produce a long leading or searcher tendril that extends toward sunlight like a long finger pointing the way the plant is yearning to go. You can see how readily this "finger" will grasp by placing a wood dowel or stick in contact with the lower surface of a leader ten-

A coiling and counter-coiling tendril.
Drawing by Leslie C. Costa.

dril held out horizontally. Observe how the tendril goes from being extended firmly to seemingly going limp at the point of contact with the dowel. After the tendril has begun its coiling, remove the dowel, and see how the angle remains in the tendril. What happens? Will it continue to coil, or straighten itself over time?

II. Circumnutation

Up-and-coming climbers get ahead by constantly probing in a circular or elliptical motion, a phenomenon that the Darwins dubbed circumnutation (Latin *circum*, circle, and *nutatio*, nodding or swaying). Circumnutation is slow, but plants vary quite a lot in just *how* slow, from the glacial to just on the edge of perceptibility by watchful humans. It's fun to try to gauge the movement with some circumnutation speed demons (bearing in mind that qualifying as a speed demon is a relative thing in the plant world). All you need is a point of reference, and that's where our handy "circumnutometer" comes in. Try this with brewer's hops, one of the speediest circumnutators around.

A. Materials

- Hops (*Humulus lupulus*) vine. Rootstock is readily available commercially, or from home-brewer friends who grow their own hops. Use a young plant with a single actively growing shoot perhaps a foot long, planted in the center of a circular planter about 10 in. (25 cm) in diameter.
- Paper plate with a diameter a bit smaller than the planter diameter so that it fits comfortably within the planter
- Ruler
- Scissors
- Markers
- Pins
- Watch or stopwatch

B. Procedure

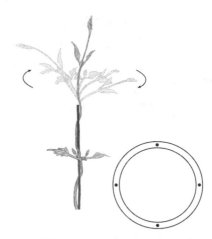

1. Cut out the center of the paper plate, creating a ring like the example at right. Mark the cardinal directions or hours of a clock (or only the 12, 3, 6, and 9 o'clock points) around the ring using the ruler and marker.

2. Carefully guide the ring over the hops plant and central support, centering the ring inside the planter laying it on the soil surface; secure if necessary with pins. This is your circumnutometer.

Potted hops (*Humulus lupulus*) plant with rotating shoot. "Circumnutometer" disk at right can be labeled with cardinal directions or hours like a clock, and fitted over the hops atop the planter. Observed from above, the progress of the rotating shoot can be timed. Drawing by Leslie C. Costa.

3. Looking down on the planter with 12 o'clock at the top, record at about what "time" or direction the hops shoot is pointing.

4. Check the shoot periodically, perhaps at hourly intervals, each time recording the "time" or direction indicated on the circumnutometer. Which way is the shoot revolving? Darwin's hops took about 2½ hours to complete one revolution. Is your hops revolving faster or slower than what Darwin observed? This may be expressed in CPH (circles per hour).

III. Geotropism: Downwardly Mobile Beans

If shoots are up-and-coming, roots are downwardly mobile. Here we show, like Darwin, that the expanding young roots (radicles) of sprouting beans sense gravity at their tips.

A. Materials

- 15–20 Dried pinto, black, or kidney beans
- 20 Paper towels
- 4 Quart-sized plastic sealable bags
- 4 Widemouthed jars (canning, peanut butter, mayonnaise, etc. jars; glass or plastic)
- Thick cardboard (e.g., from a corrugated cardboard box), cut into squares a bit wider than the mouth of the jars
- Waxed paper, cut into squares matching the size of the cardboard squares
- Cotton balls
- Straight pins (long)
- Craft knife or razor blade
- Small bowl and water
- Protractor and ruler

B. Procedure

1. First germinate your beans:
   a. Divide beans into 3–4 groups of up to five beans each.
   b. Place five paper towels on top of each other and wet them so they are saturated, but not dripping wet.
   c. Place up to five beans in the center of the paper towels about one in. (2.5 cm) apart, and fold the paper towels.
   d. Place the paper towels with beans inside the quart-sized sealable bag and seal. Follow this same procedure for the remaining groups of seeds.
   e. Allow the beans to germinate and the radicles develop, about 2–3 days.
2. The germinating radicles will often point in various directions, even doing loop-the-loops. Select four beans with well-developed radicles extending more or less horizontally, straight out from the bean.

3. With the knife or razor blade carefully cut the last 1–2 mm of the radicles off two of the beans (treatments), and leave the other two intact (controls).

4. Fold a saturated paper towel and place it in the bottom of a jar, or pour in a small amount of water.

5. Place a pin through a bean, and then carefully through a wet cotton ball.

6. Place a waxed-paper square on a cardboard square. Next pass the pin with the bean and cotton ball through the waxed paper and cardboard square. Secure the pin with a dab of glue if necessary. Place the cardboard on the opening of a jar, with the bean inside the jar to maintain humidity and prevent the bean from drying out.

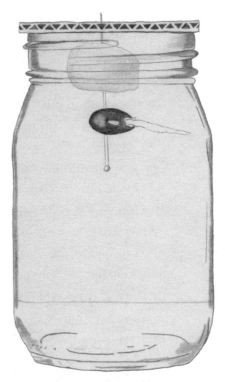

7. Repeat with the other three beans and jars, and label the jars according to whether they contain treatment or control beans.

8. Place jars in a dark space like a cabinet or closet, or cover with black construction paper, and allow radicles to further develop over a few days.

9. Make daily observations: the pinned radicles can be removed from the jars for measurement. Use a protractor to measure the

Pinned bean gravitropism experiment, with radicle initially held horizontally in jar. Different jars can be set up with treatment (excised) and control (intact) radicle tips to test for response to gravity. Drawing by Leslie C. Costa.

angle (deviation) of each radicle from the horizontal, and a ruler to measure radicle length.

10. Graph your data to obtain growth rate of the radicles and rate of change in angle of deviation over successive days. How rapidly did the intact radicles respond to gravity? Did the growth rate of the intact and excised radicles differ? By how much did the radicles deviate from the horizontal? Graph the deviation in degrees on successive days. The radicles with the tips removed should no longer sense gravity, and so should continue to grow more or less horizontally. The intact radicles, on the other hand, should strongly dip as they develop.

See also:

"Climbing Plants" and "Power of Movement in Plants" at the Darwin Correspondence Project: www.darwinproject.ac.uk/learning/universities/getting-know-darwins-science/climbing-plants and www.darwinproject.ac.uk/learning/universities/getting-know-darwins-science/power-movement-plants.

10

Earthworm Serenade

In the summer of 1880 Emma Darwin commented dryly on her husband's latest hobbyhorse. In a letter to their son Leonard, she wrote that his father "has taken to training earthworms but does not make much progress, as they can neither see nor hear." They were a source of amusement, however, spending hours "seizing hold of the edge of a cabbage leaf and trying in vain to pull it into their holes."[1] Most people simply *assumed* that earthworms could neither see nor hear, but Emma had firsthand experience: her husband had experimentally demonstrated this, and she even helped with some of the experiments. Emma was joking about Darwin trying to "train" the earthworms, but Leo, then 30 years old and teaching chemistry and photography at the School of Military Engineering in Chatham, surely chuckled on reading her letter—he knew all about his father's worm-wrangling, with his flowerpot "wormeries" in the study, midnight worm-watching excursions on the sandwalk, and experiments designed to test earthworm IQ. He had helped with his share of experiments too over the years, most recently procuring colored glass at the request of his father to use as light filters in other earthworm investigations.

For the past few years, even as Darwin struggled to complete his sixth and final botanical book (*Power of Movement in Plants*, published in 1880), earthworms had taken center stage at Down House. He was fascinated by them, even smitten. But this latest hobbyhorse was not new: it was in fact a return to one of the first studies he had undertaken after the *Beagle* voyage, namely, earthworms as a geological force. He had long thought that, despite being dismissed as mere fish bait, earthworms taught object lessons in Lyellian uniformity. Now he was sure there was even more to them than that—they may seem senseless, but they were animals, after all, cryptic but exquisitely attuned to their environment. They had a certain wormy intelligence, he came to appreciate: problem solvers, with personality. Darwin's fascination with the industriousness and psychology of worms might seem like the obsession of an eccentric on the surface of it, but by now you surely appreciate what a mistake it is to judge Darwin on surface appearances. No, beneath the surface his mind was always churning, turning out remarkable insights from the grist of simple observation and experiment, like those worms continually turning the soil, unseen yet remaking the world.

The Maer Hypothesis

Darwin was making quite a mark in London geological circles in 1837. He was just 2 months back from his voyage around the world, and the new year barely opened when he read his first paper, on the elevation of the Chilean coast. Geology was a grand subject, awe-inspiring in its scope, from the vast sweep of time to the power of its forces to the inexorability of its changes. He followed that paper with another on the extinct mammals of South America, and then another on his theory of coral reef formation. The cycle of time and slow geological building up and wearing down were on his mind: the coastal elevation and coral reef papers were all about uplift, growth, and construction, while the one treating long extinct and fossilized mammals was all about erosion, subsidence, and burial. Visiting his Wedgwood relatives at Maer, as he often did, Darwin's ever-observant Uncle Jos (Josiah II) showed him a curious thing: some years earlier he had spread over different fields an early form of fertilizer containing granular lime, cinders, and burnt marl (reddish nodules of clay mixed with calcareous material like shell bits). Now, he showed his nephew, they were all buried in a recognizable layer a few inches beneath the surface. Farmers had long noticed that objects spread on their fields were buried over time, but they simply assumed these things somehow just "worked their way down," through the action of rain and gravity. Josiah Wedgwood thought differently: he opined to his nephew that the explanation lay in the constant churning of the soil by earthworms.

Earthworms are segmented (annelid) worms of the class Oligochaeta, a group of terrestrial and aquatic species with a simple and mostly smooth, tubular body form. They are hermaphroditic (bearing both male and female sexual organs), and deposit their eggs in a sac produced by the glandular clitellum, the thickened body section resembling a wormy Band-Aid. There are some 4,100 or so described earthworm species worldwide, and their distribution often bears the signature of historical geologic or climatic conditions; in North America, for example, all earthworms in northerly ice-covered areas of the last glacial period were extirpated, and the northern extent of native earthworms on the continent today still largely mirrors the glacial front. Despite their name some earthworms have taken to an aquatic or semiaquatic lifestyle, but the majority are terrestrial. These make their living in rather different ways in the soil, reflecting the kind of divergence and niche partitioning that Darwin explored with plants (Chapter 3). There are smallish pigmented *epigeic* worms that feed on decaying organic matter, living in the area where the soil and leaf litter interface, under logs, or in compost; topsoil-burrowing *endogeic* worms that make and live in horizontal burrows and rarely surface; and the often large-bodied and deeper-dwelling *anecic* worms that make and live in permanent vertical burrows from which they surface to forage for leaves.

The soil-burrowing endogeic and anecic earthworms literally eat their way through the earth, grinding with their muscular gizzards the tiny particles they ingest into tinier particles still, digesting the organic matter as they go. They often expel the indigestible soil with their waste products at the surface in the form of small mounds called castings. On closer inspection the mounds can be seen to consist of long, convoluted tubes of extremely fine particles—like extruding the contents of a very narrow-mouthed tube of toothpaste into a pile. Or, better, think of spraying a little mounded pile of that scariest of processed foods, aerosol cheese in a can. You get the picture. The tubular shape is that of the worm's intestine. Uncle Jos posited that by ingesting soil below and defecating it at the surface, earthworms in their multitudes are constantly turning the soil, in the process undermining and slowly burying surface objects and helping create the organic humus or topsoil layer (called "mould" at the time, to use the British spelling). Darwin's imagination was fired immediately by the idea.

He examined several fields with his uncle, one of which had been plowed, harrowed, and covered in cinders and marl 15 years earlier, and left undisturbed since. They carefully dug a number of test pits and trenches, confirming that the sunken material occurred in a distinct layer—not haphazardly at varying depths. Different-sized objects seemed to have sunk at the same rate, too, which would not be expected if rainfall and gravity were responsible. No, clearly something was acting upon the scattered granules and cinders simultaneously, and that something was humble earthworms. It was precisely the kind of explanation that most excited the Lyellian in Darwin: a mundane and modest force constantly applied that added up to great effects over time. Not only soil turnover and burial of objects, but also their intestinal processing of coarse soil particles, grinding them into ever-finer particles: why, these worms were a veritable erosional force, unrecognized though literally beneath our feet.

Consulting with his uncle, Darwin prepared a paper for the Geological Society. It was read on November 1, 1837, and published in the society's *Proceedings* the following year. He described the observations from the Staffordshire fields and pointed out similar observations, such as the elevated marine shells he noticed high in the mountains of Chile, covered by a shallow layer of earth in areas where rain was unlikely to have washed it over the shells. "The explanation of these circumstances," he told the Society, "which occurred to Mr. Wedgwood, although it may at first appear trivial," was the digestive process of worms. Darwin proceeded to speak to the abundance of worms and their soil-eating habit. The quantity of soil moved is evidenced by their ubiquitous castings—there was "scarcely a space of two inches square without the cylindrical castings of the worms" throughout his uncle's fields. Their digestive process is a veritable geological power, he urged them to realize.[2] With coral reefs fresh in his mind he pointed to a similar process going on in the oceans of the world. In reefs the incessant excavations of boring and tunneling mollusks and marine worms is helped along by coral-crunching fish like seabream (genus *Sparus*), converting great quantities of calcareous coral into fine mud. The accumulated calcium-rich sedimentary mud they produce eventually becomes limestone. He thus argued that a significant proportion of the chalky limestone formations of Europe was produced from coral, "by the digestive action of marine animals, in the same manner as mould has been prepared by the earth-worm"[3] from long-disintegrated rock. Today, scientists believe that the calcium carbonate shells (called *tests*) of near-microscopic foraminifera—amoeboid protists occurring in untold scintillions, which incessantly accumulate on the sea floor as they expire and drift down like a never-ending calcareous snowstorm—are the single most important source of limey ocean sediments (and therefore limestone). Darwin's coral-crunching fish and burrowing marine worms do make a contribution, albeit a minor one. In any case, the image of great limestone formations owing their existence to the accumulated action of humble organisms, whether foraminifera or marine worms, is certainly evocative. Returning to the terrestrial world, he declared it probable "that every particle of earth in old pasture land has passed through the intestines of worms."

The paper caused a stir—no less a figure than the eccentric and distinguished geologist William Buckland of Oxford (Charles Lyell's old professor) strongly urged that the Geological Society publish Darwin's paper not just in the meeting proceedings, but also in the more prestigious *Transactions* of the society, with illustrations to boot. Darwin had described nothing less than "a new Geological Power," Buckland

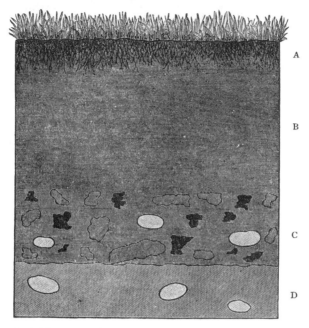

Cross-section of a field at Maer, showing the stratum of cinders, burnt marl, and quartz pebbles buried by earthworms. From Darwin (1840), p. 506.

declared.[4] Darwin was delighted, though probably not with Buckland's recommendation to strike his hypothesis about the creatures that made the chalk formations. The earthworm argument was so compelling, Buckland reasoned, why distract from it with a disputable conjecture that, if it ends up being well supported, would deserve a paper in its own right? It was dropped from the *Transactions* paper, but Darwin stubbornly included it in later writings. His family cheered him on. Cousin Elizabeth Wedgwood wrote to say that Uncle Jos "desires me to tell you he is very much struck with your hypothesis of chalk being made by fishes—if fish made Chalk Hill I don't see why worms may not make a meadow."[5] Buckland's reservations proved prescient, however. A few years later, Berlin-based naturalist Christian Gottfried Ehrenberg astounded the scientific community with a series of masterful microscopic analyses of various kinds of rocks and sediments, including the chalk. Ehrenberg showed that microorganisms were the more likely source of the chalk and other limestone formations, and coined the name foraminifera, "hole-bearers," a reference to the perforated tests of these ubiquitous marine protists.

Darwin's earthworm ideas fared better, and back home his extended family shared his delight at the buzz he created in the London geological scene. "We are very much obliged to you for sending us a copy of ye 'Maer Hypothesis,' which you so handsomely give my father all the credit of," Elizabeth wrote. "We shall all be exceedingly glad when you can come & philosophize again with us."[6] Darwin decided to expand the study, and asked Uncle Jos and Elizabeth to gather information on different fields in the parish for him. Others, too, volunteered their observations. When Charles Lyell sent a copy of Darwin's paper to friends in Edinburgh it caught the attention of industrialist and farmer William Fullerton Lindsay-Carnegie, of Forfarshire, who eagerly wrote Lyell to say that not only was Darwin's theory correct, but that beyond its geological insights "the discovery may do much in an economical sense."[7] He related how farmers in his district had misinterpreted how lime and other fertilizing materials spread on their unplowed fields had ended up below ground, thinking that the lime simply had a tendency to sink. Assuming therefore that these things would sink even faster in plowed fields, they held off applying the fertilizer until very late, "thus, losing much of the advantages to be derived from a thorough intermixture." When Lindsay-Carnegie bucked convention and laid down lime and plowed immediately, the farmers grumbled that he had made a grave error. But the results said otherwise, and Darwin's theory proved the error of their ways. "By means of Mr Darwin's observation I think the prejudice will be entirely removed," he concluded. Darwin was able to incorporate this account and more from Lindsay-Carnegie as well as fresh information from the Wedgwoods into the *Transactions* paper, which came out in 1840.

By then much had transpired: Darwin had married, read a host of papers at the learned societies, begun serving as editor of the *Beagle* voyage zoological volumes, published his *Beagle* travel memoir to some acclaim, was hard at work on his coral reef book, *and* he harbored a burning secret, having become a closet transmutationist. A rising star in the British geological firmament, in some ways Darwin was a victim of his own success at that busy time: parrying an invitation to become secretary of the Geological Society in 1837, he succumbed the following year while also accepting the post of vice-president of the new Entomological Society of London. Earthworms and the "Maer hypothesis" were put on the back burner, but were never taken off the stove altogether. When the March 16, 1844, issue of the *Gardeners' Chronicle* reprinted in commentary form his Geological Society earthworm paper—making the point that "the poor despised 'worm of the earth'" is in fact "one of the farmer's best friends"—he took advantage of the opportunity to correct an error in the original paper and give refined estimates for the worms' rate of burial. Published in the April 6th issue, Darwin's note reveals that he had continued to work on worms, at least occasionally: he checked on the progress of Uncle Jos's flints and cinders, and interviewed the local farmer, William Dabbs, whose nearby croft was the subject of data sent by his Wedgwood cousins in 1837. He wanted to verify when the Dabbs's field was initially treated, and dug holes and trenches to measure the layer of marl, cinders, and broken pottery beneath the surface. The rate of burial in the field was nearly 3½ inches per year, much faster than the rate of ¼ inch per year found in his uncle's boggier field. He also found a large clod of earth "penetrated by eight upright, cylindrical worm-holes, nearly as large as swanquills, so that I could see through them."[8] Those cylindrical wormholes were, to Darwin, so many barrels of a smoking gun: where is all that soil missing from the long tunnels? Why, removed by the worms and deposited on the surface.

Then, all was quiet on the earthworm front for over 20 years, at least as far as the documentary record goes. All the while Darwin was as industrious as his worms, constantly churning ideas day in and day out. Despite his sometimes-debilitating illness and through the joys and tragedies of his family life, we've seen how he was incessantly probing, writing, experimentising as reflected in the multitude of "research programmes" he undertook: the barnacle years of the late 1840s and early 1850s (see Chapter 2); his insights into the nature of competition and diversity (Chapter 3); the mystery of bees' cells (Chapter 4); understanding the "grand game of chess" that is geographical distribution (Chapter 5); plumbing the secrets of pollination and flower morphs (Chapter 6); and the masterful case study of orchids (Chapter 7). Along the way his barnacle monographs appeared in this period, followed by the *Origin*, his orchid book, and his two-volume treatise on domestication. Then, too, carnivorous plants caught his eye (Chapter 8) and climbing plants beckoned (Chapter 9)—the stuff of books yet to come.

And earthworms? They may have been out of sight, figuratively and literally, but they were not altogether out of mind. For one thing they played a role in some of Darwin's other investigations, such as buried seeds: one February day in 1856 he collected some earth along the sandwalk consisting of worm castings and planted them under bell jars in his study. Three plants came up in one, and two in the other. Earthworms moved seeds in two ways, he realized: carrying them in their gut they moved seeds deep underground or up to the surface. Also, they buried seeds by constantly turning the soil. On balance they probably buried more than they brought to the surface, since two processes carried seeds down, while just one carried them up. Darwin was struck by just how deeply seeds can be moved underground, speaking from experience when he later described "how easily a botanist might be deceived who wished to learn how long deeply buried seeds remained alive, if he were to collect earth from a considerable depth, on the supposition that it could contain only seeds which had long lain buried."[9] The movement of seeds is important for understanding the struggles that plants endure, too: those seeds "in right climate & soil," and "which got planted at right depth" by earthworms and other agents, will germinate. Then, however, comes the "battle for life with slugs & insects & other plants."[10] There seems to be no further mention of worms for the next dozen years after his seed burial observations; he was hot on the trail of other investigations.

Fishing for Clues with Worms

Worms made an appearance again, cropping up suddenly in Darwin's notes like castings in a field after an autumn rain. His nieces Lucy, Margaret, and Sophy Wedgwood had begun helping their Uncle Charles with field collections and observations of flower morphs beginning in 1862. They were the daughters of his sister Caroline. She had married Josiah (Joe) Wedgwood III in 1833 (another Darwin-Wedgwood cousin liaison), and were now living at Leith Hill Place in Surrey. Their career as their uncle's field assistants came about in typical Darwin fashion. Upon hearing that his sister's family was heading to north Wales for a holiday, Darwin asked his nieces to check local purple loosestrife populations, and let him know how many of each of its three flower morphs they counted (see Chapter 6) so that he could compare the ratios found in different localities. (They did, and he included their data in the paper that eventually resulted.) Of the three young Wedgwood ladies, Lucy had the naturalist's bent. It was likely on one of the Darwin family visits at Leith Hill that Lucy was introduced to the finer points of earthworms. She was soon sending her uncle questions and observations, one of which he forwarded to the *Gardeners' Chronicle*: "As gardeners have much to do with worms," he wrote the editor, "I think that you will find the enclosed little communication, written by my niece, worth insertion in Gard. Chronicle. I can vouch for her remarkable accuracy."[11] Lucy had noticed small stones arranged atop and around the entrance to wormholes. On several occasions she removed the stones and came out at night to see how the worms replaced them. She noticed the worms would hold their bodies in the holes with their tails and extend themselves until they felt one and dragged it to the hole with their mouth. Her query: What was the worms' reason to make stone heaps over their holes?

Lucy's investigations got Darwin thinking—he long admired worms as irrepressible earthmovers, but now he was beginning to view their behavior as more sophisticated than he thought. To most people these blind and mute creatures wriggling around gardens have no behavior to speak of. He knew better, but Lucy's experiments were intriguing. What was it about a worm's life that would result in so curious a behavior as to gather stones? There was more to worms than even *he* realized, but just at the moment he was swamped with other projects. Having completed his domestication volumes and a paper on flower morphs, he had resolved to publish a book giving his views on human evolution, a topic he had glossed over in the *Origin*. In the 1860s Lyell and Huxley in Britain and Haeckel in Germany all came out with books bearing on human origins. Darwin had mixed feelings about their efforts: unsurprisingly, he had his own ideas about human evolution, and he simply couldn't put off publishing them any longer. (This would become *The Descent of Man*, appearing just a few years later in 1871. But an awful lot of work lay ahead to get there.)

On February 4, 1868, Darwin recorded that he "Began on Man & Sexual Selection."[12] But there was a month's break in March to visit his brother Erasmus and then Emma's older sister Elizabeth. They returned home on April 1st; it was quiet now, with just Etty (then 25) and Bessy (21) living at home. Lenny and Horace were off at boarding school, and Frank and George were at Cambridge. Darwin managed to do some work on *Descent*, but there were always the letters: in April 1868 alone they take up just over 100 pages in the collected Darwin correspondence volume for that year (albeit with endnotes).

Darwin was unwell much of the summer, so the family took a month-long holiday on the Isle of Wight, renting the home of the noted portrait photographer Julia Cameron. There they met the poets Tennyson and Longfellow (a very distant Darwin relation) through Cameron's good offices. It was restful, but he was impatient. "We have been here for 5 weeks for a change," Darwin wrote in one letter, "& it has done me some little good; but I have been forced to live the life of a drone, & for a month before leaving home, I was unable to do anything & had to stop all work. We return to Down tomorrow."[13] That was late August, and they stopped in Southampton on the way home to visit William, now well established in banking. A few weeks later Darwin was cheered by a visit from Alfred Russsel Wallace and his wife Annie, along with ornithologist John Jenner Weir and all-around zoologist Edward Blyth, just back from Calcutta. The visit was a great success, notwithstanding that Blyth was "a dreadful bore" according to Bessy.[14] That welcome visit was followed by another: his American botanist friend Asa Gray and his wife Jane and several nieces came to see the Darwins in October, on their way to southern Italy and Egypt.

That visit occasioned an opportunity to recruit the Grays for yet another of Darwin's ongoing investigations: as part of his researches into human evolution he became keenly interested in the physical expression of emotions. That is, the muscles involved in expressing happiness, anger, or grief, for example, as well as gestures and body language. By now you realize that seemingly odd investigations like this were no tangent for Darwin. He was nothing if not consistent, and his investigations were always undertaken to extend and reinforce his theory of evolution by natural selection. In this case, it was about the universality of expressing similar emotions in similar ways, by contracting the same sets of muscles. This line of research was consistent in another way too: it had deep roots with Darwin, dating back at least to 1839 when Willy was born and Darwin set about closely observing his behavior even as a tiny infant—the natural history of babies. The story of these investigations is beyond our scope here, so I will mention only that by 1868, with renewed interest in the subject, he had drafted and printed a questionnaire regarding facial expression and gestures in peoples of other races and cultures, which he distributed among friends and colleagues.

His "Queries About Expression" consisted of 17 questions, remarkable for their scope and detail: Is astonishment expressed by the eyes and mouth being opened wide, and by the eyebrows being raised? When low in spirits, are the corners of the mouth depressed? Do the children pout when sulky? How is fear expressed? It covered the shrugging of shoulders, nodding of the head, uplifted palms and raised eyebrows. Darwin would not only seek to establish a universal bond among humans but throughout the animal kingdom, as reflected in the title of his eventual book on the subject: *Expression of the Emo-*

tions in Man and Animals (1872). The Grays furnished detailed observations from Italy and Egypt, sent near the end of their tour in May of 1869. Jane even made observations of Renaissance paintings while she was at it: "I thought you would have been interested in seeing an old picture here of Fra Angelico's of the deposition from the cross," she wrote Darwin from Florence; "The Madonna has the distress muscles very carefully painted."[15]

Jane Gray's letter from Italy was written on the very same day—May 8, 1869—as another letter that brought earthworms to the surface of Darwin's mind once again. Gardener and horticultural writer David Taylor Fish had reservations about Darwin's theory of the efficacy of worms in generating "vegetable mould"—what we call topsoil. He thought topsoil was simply decayed plant matter, and published an open letter in the *Gardeners' Chronicle* saying so. Surely so much plant matter decomposing "annually, and hourly, in fact, over the surface of the world" provides "sufficient energy, force, and persistency to cover the earth with vegetable mould." This, Fish asserted, "would account for all the elevations of surface so graphically described in the admirable paper of Mr. Darwin." Furthermore, he wrote the editor, he was sure that "it would be gratifying to many of your readers to know if this distinguished geologist [yes, geologist] still adheres to his original estimate of the usefulness and power of worms as earth-lifters and surface-elevators." You bet he did. Darwin immediately wrote the magazine: "As Mr. Fish asks me in so obliging a manner whether I continue of the same opinion as formerly in regard to the efficiency of worms in bringing up within their intestines fine soil from below, I must answer in the affirmative." Here was an opportunity to provide some updated information: he had not made actual measurements of late, he said, but had

> watched during the last 25 years the gradual, and at last complete, disappearance of innumerable large flints on the surface of a field with very poor soil after it had been laid down as pasture. I have also purposely covered a few yards square of a grass-field with fine

chalk, so as to observe the worms burrowing up through it, and leaving their castings on the surface, which were soon spread out by the rain. The Regent's Park in early autumn is a capital place to observe the wonderful amount of work effected under favourable circumstances by worms, even in the course of a week or two.[16]

Regent's Park, home of Emma's sister, was regularly visited by the Darwins, so his mention of the locale here underscores his incessant observing. You can imagine him stealing away for solitary walks in the park, or perhaps taking after-dinner strolls arm-in-arm with Emma, to admire the industrious worms of Regent's Park. His observations told him that Fish could not be correct. Judging from how quickly he observed the surface to become covered with fine soil, he reasoned that if mere decay of grass was responsible, soil many feet in thickness would form in the course of just a few centuries. On the contrary, "In ordinary soils the worms do not burrow down to great depths, consequently fine vegetable soil is not accumulated to any inordinate thickness."[17]

Rising to the defense of worms that May was probably a welcome diversion for Darwin: the year already had had its ups and downs, and it wasn't even half over. For starters it took him a month and a half to complete the editing for the latest (fifth) edition of his "everlasting old *Origin*"; some of the continuing comments and criticisms were well taken while others were maddeningly out of left field. "I begin to understand your sufferings over the *Origin*," Huxley wrote him on March 17th; "a good book is comparable to a piece of meat & fools are as flies, who swarm to it, each for the purpose of depositing and hatching his own particular maggot of an idea."[18] Huxley did not suffer fools gladly, but Darwin acknowledged that some of his critics, notably Swiss botanist Carl Nägeli and Scottish engineer Fleeming Jenkin, made good points that he had to address.

That same month, too, Wallace managed to elicit both deep gratitude and profound shock and disappointment in Darwin: he might have made himself a case study for his own Queries on Expression. Wallace's book *The Malay Archipelago* came out in early March to much acclaim, bearing a warm and heartfelt dedication to Darwin. "The dedication is a thing for my children's children to be proud of," Darwin wrote appreciatively to Wallace.[19] But Wallace soon wrote him a letter that mentioned, incidentally, that he would soon come out with an article of which he feared Darwin would disapprove. Darwin was apprehensive. Wallace's bombshell of a paper, dropped just a few weeks later, included an about-face over human evolution, arguing that natural selection could not explain the evolution of the human brain after all. Late March 1869 found a shocked and dismayed Darwin bitterly declaring to Wallace that he hoped his friend had not "murdered too completely your own & my child." He followed up with another letter, still upset: "As you expected I differ grievously from you, & I am very sorry for it."[20] Right about then Darwin was thrown from his horse, injuring his back and leg. While incapacitated, Horace helpfully pulled him along on a wagon so he could make the rounds of his botanical observations in the greenhouse. He was still not quite recovered by June, and he and Emma decided it was time to get away. They headed to northern Wales, spending most of June and all of July near Barmouth, amid the salubrious mountain scenery and atmosphere. While there they visited with family friends Lawrence and Mary Ann Ruck, the parents of Frank's future wife Amy, who would become another one of Darwin's earthworm field assistants. At the moment, however, her interest lay more with Horace than Frank.

Wales did Darwin some good, but he was slow to recover. He wrote Wallace about his condition, their friendship strained but still strong: "I can say nothing good about my health, & am so weak that I can hardly crawl half a mile from the House . . . Oh dear what [would] I not give for a little more strength to get on with my work."[21] Somehow he found the strength, and then some: all through 1870 he worked at *Descent*, which came out at last in February 1871. That January he started on *Expression* and, in a tremendous burst of writing, finished the manuscript in April . . . only to take on the editing for what he resolved would be the final edition of the *Origin*. That project took another half year, punctuated by visits with family in London and Southampton, and the happy occasion of Etty's marriage to Richard Litchfield in August of 1871.

The Earthworm Returneth

Earthworms cropped up again in 1870 and 1871, this time with more staying power. Visiting his sister's family at Leith Hill Place in October of 1870, Darwin and Lucy, his ever-ready research assistant, got to speculating on the vast amount of earth the worms must bring to the surface annually. They had an idea: why not quantify it? They soon set up two square-yard plots, one in a partially shaded grassy spot near her home, and the other on unenclosed common land near Leith Hill Tower (an eighteenth-century Gothic-style landmark atop Leith Hill). The plan was for Lucy to collect worm castings over the course of a year, and they would weigh them at the end. She was diligent in her collecting, and when the Darwins came for a visit the following November she had two bags full of castings for her uncle. When he got home Darwin excitedly dried the castings and wasted no time getting hold of the most accurate scale he could find, which he borrowed from the local pharmacist.[22] The more shady plot yielded exactly 3.5 pounds of castings, scaling up to 7.56 tons of soil per acre, while the open hilltop one yielded 7.45 pounds for a whopping 16.1 tons per acre, close to the quantity he had guessed for the site. That's a remarkable quantity of earth moved annually by such humble organisms. Darwin sent the results to a delighted Lucy. "My dear Uncle Charles," she replied, "Thank you very much indeed for sending me such a full account of the worm castings which I was deeply interested to read. What a wonderfully accurate guess of yours about the amount! The 16 tons an acre is the most astonishing part." (The rest of her letter, true to form, concerns observations she was just making on male turkey displays.)[23]

Around this time Darwin resolved to produce a thorough study of the effects of worms, but he couldn't do it alone, or quickly—carnivorous and climbing plants were competing for his attention at precisely this time, not to mention correcting page proofs for both the sixth edition of the *Origin* and his new book *Expression*. When he wrote Gray for information about North American worms, his friend admonished him: "Now, pray don't run off on some other track till you have worked out and published about Drosera and Dionaea"[24]—the carnivorous plants he had been working on since 1860 (see Chapter 7). Yes, he knew he had to buckle down, but he had a hard time working on just one project at a time, and the worms were growing more insistent. He took a page from their play book, a uniformitarian strategy that Lyell would appreciate: marshal his field assistants, enlist friends, family, and colleagues, and gather data little by little. As with the worms' castings, it adds up.

Darwin—and what became a small army of assistants and informants—proceeded with worms on several fronts through the 1870s. As with his Queries on Expression, Darwin drafted instructions to distribute among his contacts explaining what he was looking for. He wanted to extend his analysis of the quantity of soil brought to the surface by worms, document their effects in terms of the burial of objects and landform erosion, and, more and more, he wanted to better understand their mental qualities.

On the erosion front, he sought help from far and near. He wrote Archibald Geikie, newly appointed Professor of Geology and Mineralogy at the University of Edinburgh, who not long before had written on "denudation," the nineteenth-century term for erosion. He had to first acknowledge his Glen Roy blunder of some 30 years prior, when he assumed that turf-covered slopes were resistant to erosion even over thousands of years. "This I now see as a complete error," he admitted.[25] Not only was such erosion possible, but it was also aided by worms. Based on the fineness of the mineral soil defecated by the worms, Darwin was sure that they played a role in breaking down bedrock, not by directly attacking it, but by ingesting and grinding down small stone particles. It was well known even then that worms, like certain birds, have muscular gizzards in their alimentary tract that function to mechanically break down tough food items as an aid to digestion. They augment the action of the muscle by ingesting small stones as grinders. Naturally occurring acids from decaying vegetation help break down the rock substrate, but once rock particles reached a certain size the earthworms could ingest them along with soil, further wearing them down.

The end result, literally and figuratively, was very finely triturated mineral soil mixed with organic waste. Geikie agreed, and pointed out that his fellow Scot John Playfair, the eighteenth-century Edinburgh geologist who communicated James Hutton's uniformitarianism to the geological world, had written that the permanence of the soil furnishes "a demonstrative proof of the continual destruction of the rocks."[26] His idea was that despite continuous erosion by rivers and streams the soil seems never to diminish, and concluded that this was because it was being constantly renewed by the disintegration of bedrock. The role that worms might play in this process was unappreciated until Darwin posited it. Darwin related his and Lucy's worm-casting study to Geikie, and his interest in the persistence of ridges and furrows in old pastures plowed in the distant past. He enclosed a copy of his questionnaire, worth reproducing here in full to appreciate how minutely he was thinking about the action of worms:[27]

I have seen (some 45 years ago!) in North Wales land, showing by ridges & furrows that it had been ploughed apparently centuries ago. Do such exist in your neighbourhood; or can you find a field with ridges & furrows which has been laid down in grass for about half a century or more. In all such cases if the surface slopes I want much to learn whether the ridges & furrows run straight or obliquely down the slope, or transversely across it. In the former case it [would] be of the greatest service to me if you [would] observe whether the ridges & furrows are as distinct near the base of the slope as at its top. By far the best way to observe [would] be to stretch a string in 2 or 3 places from ridge to ridge & measure the depths of the furrow in the upper & in the lower part of the slope. If the ridges & furrows runs transversely across the slope, the only point to observe [would] be whether the two slopes of the ridges have become different to

what they are in a recent stubble field.— In any field observed
please to notice whether there are many worm-castings about.—
        My object is to trace the manner of gradual obliteration of
ridge & furrow in old pasture-land, as far as such obliteration is
ever effected | C. Darwin

The Scotsman readily assisted, sending details of furrowed fields
and old fortifications in the Edinburgh area. Closer to home Darwin's
sons George and Horace were enlisted. The two were at Trinity College, Cambridge, where George was a Fellow and Horace was a student. In early December the younger Darwin passed his first exam for
the B.S. degree—the Litte Go as it was called—an event that occasioned some reflection on his father's part. "I have been speculating last
night," Darwin mused in a congratulatory letter, "what makes a man a
discoverer of undiscovered things, & a most perplexing problem it is.—
Many men who are very clever—much cleverer than discoverers,—
never originate anything. As far as I can conjecture, the art consists
in habitually searching for the causes or meaning of everything which
occurs. This implies sharp observation & requires as much knowledge
as possible of the subjects investigated."[28] That's the voice of experience
speaking, though *doggedly* might have been a better term than "habitually" searching for causes or meaning, as persistence certainly was a
hallmark of Darwin's working method—and procuring assistance.

Just after Christmas, George and Horace made observations of a
field near Biggin Hill, a couple of miles down the road from Down
House (a locale later famous for its RAF air field and the heroic role it
played in defending London during World War II). They measured the
depth of soil in furrows and ridges running transversely across the hill
by probing with a stick. "The furrows appeared slightly more filled up
on the hill side than on the flat," George reported. "There were a good
many earth castings about tho' I should say that they were by no means
very numerous."[29] On New Year's Eve William took more precise measurements of a field in Wiltshire, near Stonehenge, braving the winter
cold to measure the slope and the furrow depths at different points

along a steep hill. His father's idea was that worms hasten the disappearance of old ridges and furrows, especially on hillsides, by helping to erode the former and fill in the latter. Frank sent further observations, and his future wife Amy was even initiated into the family fieldwork. In early February 1872 she sent observations on earthworm activity in old hillside furrows looking for little ledges on the sides of steep, grass-covered slopes which, Darwin suspected, resulted from the washing down of worm castings. Her future father-in-law dubbed her his "geologist in chief for N. Wales."[30] Of course, Lucy was in on it, too, sending observations of a cow-trodden hill while on holiday in the Swiss Alps.

The idea of worms hastening the erosion of ridges and furrows along hillsides may sound reasonable to us today, but for his skeptical contemporaries, unaccustomed to appreciating the finer points of uniformitarian worms, it was a stretch. He was determined to marshal irrefutable data. No detail was to go unexamined, as when he asked Lucy to probe wormholes with a knitting needle on steep, grassy slopes to determine whether or not they come out at right angles to the slope. "We have no steep grass-covered slopes here," he lamented, and he acknowledged that "It is not easy to probe the holes." Lucy was not deterred. Two weeks later, she sent her findings: "Out of 25 worm-holes probed with a blunt wire between Jan 6 & 14 on different slopes for a few inches, 8 came to the surface nearly vertical to the slope; the rest at various angles." She promised to probe some more, and to dig trenches for data about furrows. Her uncle was thrilled: "My dear Lucy," he wrote right back, "You are worth your weight in Gold." He shared what he was after: "If worms would be so good as to come up generally at right angles to slope, it would bring the earth down grandly." His idea was that since earthworms leave their castings piled up at the mouths of their tunnels, if the tunnels are at right angles to the sloping surface then gravity and rain will more easily draw the castings downslope.[31]

Darwin was fixated on gaining enough data to prove his theory. Later that year he tested the idea while on one of his enforced holidays. In late October 1872, people strolling at beautiful Knole Park, the 1000-acre deer park south of London, were met with a curious sight:

there was Mr. Darwin flat on his belly, carefully observing earthworm castings on the steepest hillsides. After several wet days he found that "almost all the many castings were considerably elongated in the line of the slope." At the mouths of worms' tunnels where earth had been ejected, there was always more earth below (downslope) than above the tunnel. He continued his observations back home, seeking suitable hills in the neighborhood. At Holwood Park, outside nearby Bromley, he measured slopes between 8 and 11 degrees where "castings abounded in extraordinary numbers," creating a uniform surface coating of fine material extending downslope.[32]

It was time to get more quantitative. He selected 11 castings on a hillside, and carefully measured their length in the direction of the slope (average: 2.03 inches). He then took a sharp knife and methodically cut each in two right across the mouth of the worm tunnel. Weighing the above-burrow and below-burrow portions, he found that the average weight of those occurring above (upslope) of the tunnels was 103 grains (1 grain = 64.8 mg), while the average weight of the downslope portion was twice that, at 205 grains. The idea was that fresh castings, while still a bit wet and fluid, slump a bit on the downslope side owing to gravity. George the mathematician then took over the calculations. Given this finding that $2/3$ of the cast soil is found on the downslope side of worms' burrows and $1/3$ on the upslope side, George extrapolated to calculate the weight of the soil ejected by worms annually flowing down a slope with a mean inclination of 9° 26'. To be conservative he halved the downslope $2/3$ of the worm casts. Based on an average casting size of 2.03 inches and using Lucy's Leith Hill data (7.45 pounds of earth brought to the surface annually per square yard), he estimated that each year 1.1 ounces of dry earth would flow across each linear yard running across the slope, or nearly 7 pounds across a 100-yard long line drawn across the slope.[33] Rain and wind would eventually erode these hills, but the worms help out by substantially accelerating the process, it appeared. Buckland hit the nail on the head back in 1837 when he declared that Darwin's worms were "a new Geological Power."

The Archaeologist's Friend

If the worms were the geologist's friend by helping create sedimentary strata, they were also the archaeologist's friend by helping preserve ancient remains, as with the first observation that drew Darwin
into the world of worms: those buried flints in Uncle Jos's fields. Over
the next decade Darwin's field-workers—most notably sons William,
Frank, George, and Horace—visited site after archaeological site
armed with trowels, probes, and measuring rules. The list of sites, from
druidical monuments and Roman villas to ancient abbeys and grand
old manor houses, was impressive: Wroxeter, Stonehenge, Chedworth,
Chirencester, Brading, Silchester, the Cloisters of Neville, Beaulieu
Abbey, Gravetye Manor, and more. He did his own archaeological
work too. His appetite for fieldwork was whetted in June 1877 when,
on a visit to Leith Hill, he found a pair of large worked stones at the
bottom of a hill—the remains of an old lime kiln that had been torn
down 35 years prior, it turned out. He hired a local to help dig them
out so he could inspect the nature of the soil beneath. Emma looked
after him, running out with an umbrella one day, and fretting in a
letter to Etty that she thought he would get sunstroke the next. Then
Emma excitedly announced that he had bigger archaeological fish to
fry: they would be going to Stonehenge, the awe-inspiring 5,000-year-
old monumental stone circle rising from the Salisbury Plain. "I am
afraid it will half kill [Father]—two hours' rail and a twenty-four mile
drive—but he is bent on going, chiefly for the worms, but also he has
always wished to see it."[34]

Five years before, William had visited Stonehenge to measure the
depth of the humus at different spots and assess how deeply some of
the smaller fallen blocks had sunk. There was no getting under the
gigantic blocks, but by digging along the edges of several and measuring with a trowel and skewer he was able to gather some useful data.

George met them at Salisbury Station the next day and off they went
to the site by carriage. Emma told Etty, who was away with her husband
on holiday in Germany, that they were met by Stonehenge's owner, Sir

Edward Antrobus, who was "agreeable to any amount of digging, but sometimes visitors came who were troublesome, and once a man came with a sledgehammer who was very difficult to manage. . . .'That was English all over,' said he."[35] Some things never change—relic hunters are as destructive today as they were in Darwin's day, though not so much at Stonehenge anymore, now a UNESCO World Heritage site well safeguarded by English Heritage. Charles and George dug trenches near some of the larger fallen stone blocks, recording the depth of the topsoil. With a bubble-level and rod they measured the contours of the land around the huge embedded blocks, determined how deeply they were buried, and measured the depth of the narrow sloping shoulders of turf surrounding the blocks. They concluded that worms, not simply the blocks' enormous weight, were responsible for the sinking of the stones.

Later that summer Darwin received exciting news from one of his correspondents, Thomas Henry Farrer, who had an estate called Abinger Hall in Surrey, not far from Leith Hill Place. Farrer had been in touch with Darwin since the late 1860s, at first regarding their mutual interest in orchids. Soon, Farrer was drawn into making observations on cross-pollination and, with Darwin's encouragement, published on the subject. He even became a relation by marriage when, in 1873, he wed Emma's niece Katherine Euphemia Wedgwood. (In a few years he was to become a double relation by marriage, when his daughter Ida married Horace.)

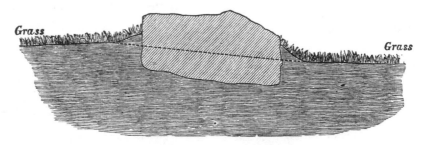

Section through a nearly 8-foot fallen stone at Stonehenge, measured by Charles Darwin and his son George in 1877, showing its burial depth (given as 4" below the general ground level, for scale) and the narrow sloping shoulders at the rock's edges indicative of worm activity. From Darwin (1881), p. 156.

Hearing of Darwin's recent interest in worms and ancient structures, Farrer invited him to his estate to inspect the excavation of a Roman villa that one of his farm hands had recently discovered in a field adjacent to the house. Charles and Emma spent most of August 1877 at Abinger. Darwin and Farrer measured the slope of the field, and each day Darwin climbed into the pits to take measurements on the depth of worm activity and the various strata of humus and underlying materials. They departed for home on the 20th, but Darwin, curious about just how many worms lived under the newly exposed ancient mosaic floor, asked Farrer if he would continue to make observations for him. Farrer kept up the good worm work, sending Darwin his "worm journal" the following month with counts of worm castings and burrows over the next month or so, destroying the castings and burrows each day to count the new ones. On September 22nd, George came from Cambridge to help Farrer: "43 holes: defaced all" reads the worm journal. Day to day, Farrer counted as many as 61 fresh burrow openings in the 14 x 9-foot space.[36] Once back at Cambridge, Emma kept George up to date: "[Father] is highly pleased about the worms. T. H. [Farrer] writes him word today that he counted 40 worm holes thro' the absolute wall after heavy rain—I think it will come to be a wonder why ruins do not sink more."[37] A wonder indeed. Using the Roman coins found at the site to peg an approximate age, with George's help Darwin estimated the rate of soil deposition by the worms over the centuries at about an inch per dozen years—worms had slowly but surely buried the Roman structure.

In early September, soon after Charles and Emma returned home, Horace presented his father with a device he invented to measure precisely the rate of stone subsidence. The soon-to-be cofounder of the Cambridge Scientific Instrument Company was at his inventive best. He unveiled his "wormstone," a precision micrometer attached to a millstone of known weight and dimensions, with a metal bar driven down to bedrock through the center hole. His father eagerly helped him select a site, close to the house under a large Spanish chestnut tree. Later, a second one was installed at Leith Hill Place. Horace recorded

data with the wormstone long after his father's death, until it was accidentally dismantled in 1896—no doubt to Horace's chagrin. (A reconstruction can be seen in approximately the original location at Down
House today.) The device was sensitive enough to record the expansion and contraction of the earth owing to rain and temperature, and
Horace even quantified this by testing with "artificial rain" (watering)
in one experiment. In 1901 he published his data from 1878 to 1896;
taking seasonal and weather factors into account he found that the
stone sank 2.22 mm per year in that period, a bit less than 1 inch in 10
years. His father's data from an adjacent open field gave an estimated
rate of 2.2 inches in 10 years, a difference that could be attributed to
field conditions (Darwin later lamented siting it under the tree, where
he began to suspect were fewer worms than in open fields), and the
smaller-sized objects Darwin had used in the field, which sink faster
than large stones.

Randal Keynes explaining the wormstone to Addison (right) and Eli (left) Costa in
2011. Photograph by the author. Inset: Horace Darwin's "wormstone" micrometer,
from H. Darwin (1901), p. 255.

Darwin must have been tickled by his son's ingenuity. That fall also brought good news twice over: Darwin was to be awarded an honorary degree from Cambridge, and the bachelor William suddenly announced his engagement to Sara Sedgwick, originally from the *other* Cambridge, in Massachusetts. Both happy events took place in November 1877. The conferral of Darwin's honorary Doctor of Laws degree at Cambridge was met with much fanfare, including the appearance of a stuffed monkey in robes and mortar board dangled from the gallery by mischievous undergraduates. Still, Emma "felt very grand walking about with my LL.D. in his silk gown," she wrote William with pride.[38] William and Sara were married on the 29th. When Thomas Farrer sent a note of congratulations, Emma, helping with the correspondence as she often did, relayed that Charles joked that the cream of his letter lay in the P.S. he appended about "the beloved worms, and not in any such trifles as marrying, &c."[39] They really were beloved worms, and not just to Darwin. Writing to his niece Sophy Wedgwood a few years later Darwin mentioned her sister Lucy's "well-known affection for worms." "I am also becoming deeply attached to worms," he owned.[40]

Earthworm Serenade

It may sound like Darwin worked on little else besides earthworms through the late 1870s, but in fact they were a sideline—his books on cross- and self-fertilization in plants and the different forms of flowers came out in 1876 and 1877, respectively, and he completed his manuscript for *Power of Movement in Plants* in the spring of 1880. He then began his book on worms in the fall of 1880, which would appear the following year as *The Formation of Vegetable Mould, Through the Action of Worms, With Observations On Their Habits*, usually abbreviated simply as *Earthworms*. It was to be Darwin's last book.

The subject of the last part of his title, the *habits* of worms, was also his last set of investigations, and they were just gearing up in 1880. Since Lucy's observations of her pebble-gathering worms, he was intrigued by their behavior. By then Lucy's question about what they were doing had been answered: Darwin and others, having excavated more than a few worm domiciles, knew the annelids often lined their burrows with pebbles and leaves, creating a wormy version of a snug nest. They expertly plugged the burrow's mouth with leaves as well, but would resort to using the pebbles for this too if no leaves were at hand, so to speak. No mean feat for an animal with no eyes, no limbs, and not much of a mouth. How did they manipulate items, and evaluate size, shape, and suitability for their purposes? Their ability to do these things convinced Darwin that earthworms possess *une petit dose de raison*—a small dose of reason, as he quoted the celebrated Swiss entomologist Pierre Huber. Anything bearing on the intelligence and curious behavior of worms were subjects of great import to him, so it's unsurprising that he put his findings front and center in the opening two chapters of his book.

But first, he took to bringing his worms indoors, housing them in glass-covered flower pots, eventually converting the billiards room into a wormery. They were probably *Lumbricus terrestris*, the common garden earthworm. Frank was his right-hand experimentiser now. As previously mentioned, Frank's wife Amy had died in childbirth in 1876, and he moved in at Down. Their son, Bernard, survived, and was doted on by his grandparents—and was an occasional lab hand for worm experiments. The Darwins, father, son, and sometimes grandson, would sneak up on the worms to see what they were up to in their wormeries. The worms often lolled about on the surface, but always with the tip of their tail just inside their burrows, ever-ready to beat a hasty retreat. What were their daily rhythms? And what of their sensory world—they have no eyes, but can they see? The Darwins tested the worms' reaction to bright light, dim light, sudden flashes of light, even colored light using red and blue glass filters that Leo procured for them. The worms proved indifferent to all except bright flashes, which induced them to zip into their burrows, "like a rabbit," Bernard marveled.[41] Darwin concluded that their thin skin admitted enough light to be perceived directly by their nervous system.

What other sense perceptions do worms have? The Darwins tested various stimuli. Radiant heat? Darwin waved a red-hot glowing poker over their pots, but the response varied. They reacted more decisively to concentrated heat from candlelight focused with a lens. Air movement? Gentle breathing had no effect, but strong puffs got the worms' attention. What of odors? Darwin stuck cotton balls soaked in vinegar, perfume, or tobacco juice in his mouth (not a Darwin experiment advisable to replicate) and let the odors waft over the pots as he breathed on them. The worms were unfazed. Could they hear? Darwin resolved to serenade the earthworms to find out. The assembled musicians included 5-year-old Bernard on penny whistle, Frank on bassoon, and the accomplished pianist Emma on piano. It was a natural test to try, but may have been partly inspired by somewhat similar experiments carried out with ants a couple of years earlier by his neighbor and sometime protégé, John Lubbock, with Darwin's cousin Francis Galton. Lubbock and Galton used a whistle that could produce very high-pitched sounds with forced oxygen or hydrogen, thinking maybe insects hear only ultrahigh frequencies. Having befriended Alexander Graham Bell, the following year Lubbock even borrowed a prototype telephone to test the ants' ability to communicate via sound, holding a receiver and mouthpiece between different colonies in a nearby field. They riled up one colony to see if sounds transmitted by phone would agitate the other. In a notebook entry for November 24, 1879, Lubbock recorded that he made "some experiments on Ants with the telephone, but they had no result."[42] Darwin doesn't appear to have tried the telephone, but in any case the worms seemed as indifferent to sound as the ants, being wholly unappreciative of the Down House Trio:

> Worms do not possess any sense of hearing. They took not the least notice of the shrill notes from a metal whistle, which was repeatedly sounded near them; nor did they of the deepest and loudest tones of a bassoon. They were indifferent to shouts, if care was taken that the breath did not strike them. When placed on a table

close to the keys of a piano, which was played as loudly as possible, they remained perfectly quiet.[43]

However, Darwin noted, when pots with two worms that had been "quite indifferent" to Emma's playing were placed directly on the piano, they reacted: when "the note C in the bass clef was struck, both instantly retreated into their burrows," he recorded. "After a time they emerged, and when G above the line in the treble clef was struck they again retreated." It was the vibration, of course. He experimented further with the piano, creeping into the drawing room at night and suddenly striking a high note; one worm "dashed into its burrow" but another stayed until "C in the treble clef" was played. Darwin mentioned only these specific notes: did he try the entire scale? Or would they react to any note being played resonantly enough to send vibrations through their wormery? He noted that the worms were not touching the sides of the pots, so they were sensitive to dampened vibration coming through the soil.[44]

Although not all worm species show vibrational sensitivity, the phenomenon is widespread. In the southern United States there is a long tradition of "worm grunting" for fish-bait, inducing *Diplocardia* earthworms to surface by sending vibrations through the soil using a wooden stake driven into the ground, over which another object, often metal, is stroked rhythmically back and forth. As for *why* the worms surface in response to vibration, there are various hypotheses. Darwin thought that it might be to escape from burrowing predators like moles, which send telltale vibrations through the soil as they tunnel, alerting the worms to their approach. If so, there may be a cruel irony in the way some other predators apparently use the vibrational flight response of worms to their own advantage, including foot-stomping herring gulls (a behavior described by the distinguished ethologist Niko Tinbergen in 1960) and wood turtles (described in the 1980s) that induce worms to surface to their doom.

These are adaptations that would have impressed but not surprised

Darwin. What did surprise him was the intelligence of worms. Who would have thought that these humble creatures have much of a brain? Then again, he reflected, consider the remarkable repertoire of ants and bees, with a far smaller brain. But how much is instinct and how much "reason" with the worms? He tested their problem-solving abilities, giving them access to variously shaped leaves and other materials to see how and what they select for their burrows. There seemed to be an urgency with the worms in lining and, perhaps more importantly, plugging their burrows—maybe to keep predators or parasites out. They are choosy, and carefully select and position materials they deem suitable. In one slightly mean investigation he pinned the leaves to the soil of their pots, allowing him to make extended observations as the worms seemed to puzzle and struggle to move the leaves without success. They grabbed leaves with their mouths, but would latch on here and there as they maneuvered objects into place. It's astounding, if you think about it—not seemingly trial and error, but rather more deliberate in grabbing hold after feeling the leaves all over. He made observations of the kinds of leaves found in burrows in the vicinity of the house, discovering that worms loved the lime tree leaves by the veranda, but were not as keen on rhododendron leaves. Long, narrow objects would be retrieved as well: they readily drew in pine needles and thin leaf petioles. Fixated, he set out to test further.

Commandeering some of Bernard's coloring paper, Darwin cut out paper diamonds, triangles, and discs, varying the size and the acuteness of the angles, and then sprinkled them on the soil surface in the pots for the worms to evaluate. "The worms are drawing in my triangular card nicely," he reported in one letter to William; "but enough have not yet been drawn in for any conclusion. I think that I shall try feathers."[45] Feathers were not a success, and he stuck with the paper cutouts. He could not only tell which shapes the worms preferred, but by excavating their paper-lined burrows he could also see the orientation of the cutouts. Most had been pulled in by their narrowest, most pointed ends—easiest to grab hold of, he surmised, but they managed to draw some of them in by the middle too. Flexible leaves were often drawn in from the center of the midrib, neatly folding each leaf as it was pulled into the burrow, making a more effective plug. Darwin was sure that the worms somehow acquire a "notion" of shape, most likely by feeling the objects all over. He likened them to a blind person with a heightened tactile sense, and felt they had some learning ability. Curious about their food preferences, he also did taste tests: cabbage, turnip, and onion leaves were worm favorites, but they weren't keen on kitchen herbs. Darwin came to appreciate that his worms had personality, describing how they enjoy "the pleasure of eating" (based on their eagerness for certain leaves) and their sexual passions ("strong enough to overcome . . . their dread of light"), and even commenting on their feelings: "They perhaps have a trace of social feeling, for they are not disturbed by crawling over each other's bodies, and they sometimes lie in contact." He cited Werner Hoffmeister's 1845 account of worms passing the winter "either singly or rolled up with others into a ball at the bottom of their burrows"—an incongruous image, perhaps, of cuddly worms snugly coiled in their burrows together.[46]

The Formation of Vegetable Mould, through the Action of Worms was published on October 10, 1881, to instant acclaim—3,500 copies sold within a month. "I must own I had always looked on worms as amongst the most helpless and unintelligent members of the creation," his old friend Hooker wrote, declaring that he was now "amazed to find that they have a domestic life and public duties!"[47] Lady Derby, who had long corresponded with Darwin on scientific subjects, was equally impressed: "I have read your book with the greatest interest. You said once, laughing—that you were finding that "Worms" could revolutionise the world—you have succeeded in proving the greatness of their power."[48]

Darwin came full circle with his worms: they were the subject of one of his first published studies and represented his earliest scientific interest, geology. Now, 40 years later he completed the chapter he opened with them by writing his last book on worms. He would die 6 months later, on April 19, 1882.

Caricature of Darwin as the sage of worms, by Linley Sambourne for *Punch* (December 6, 1881).

Earthworm Reprise

Formation of Vegetable Mould presaged if not prompted several lines of research that came into their own in the century to come. Pedology, soil ecology, bioturbation, ethology, and even erosion and archaeological and forensic stratigraphy all owe something to Darwin's little book, though recognition of the significance of worms has sometimes been slow. This brings to mind Darwin's admonishment to David Fish, the Doubting Thomas of his worm gospel: "Here we have an instance of that inability to sum up the effects of a recurrent cause, which has often retarded the progress of science, as formerly in the case of geology, and more recently in the principle of evolution."[49] Note his rhetorical device of drawing a parallel with the early resistance to Lyell's geological vision, and to the cumulative effects of selection acting over vast time periods. You will come to see the error of your ways, he is saying. But there have always been those who, like Lady Derby, appreciated that Darwin well and truly succeeded in proving the greatness of the power of worms.

Some researchers have returned to Darwin's own landscape laboratory over the years to revisit the effects of worms there—descendants, no doubt, of the very worms that Darwin knew and loved. Sir Arthur Keith, Scottish anthropologist, revisited Darwin's fields in the early 1940s, digging 70 trenches to see how far Darwin's flints had sunk. Publishing his findings in *Nature* in 1942, Keith reported the important observation that objects are buried only to a point, where they collect in a layer—often on a denser stratum of clay or rock below which the worms can't go. Just as a stone, thrown into a pond, comes to rest on the bottom, so too do objects on the surface come to sink to a point where they cannot sink any farther. Understanding more precisely the dynamics of object burial by earthworms is as important for forensic investigations as archaeology.

The most recent (and ongoing) study was initiated in 2007 by ecologist Kevin Butt, who heads the Earthworm Research Group at the University of Lancashire in the UK, in collaboration with US Forest Service soil ecologists (one of whom happens to be my brother-in-law, Mac Callaham), English Heritage, Down House staff, and dedicated volunteers. The team set up a long-term field experiment to replicate Darwin's 1842 flint experiment in Great Pucklands Meadow— Darwin's own replication of his Uncle Jos's observations at Maer, and the same meadow in which he later undertook his biodiversity studies with Miss Thorley (see Chapter 3). This time an experimental design informed by modern statistical approaches was implemented: a randomized 2 × 2 factorial design, with "large" and "small" flint size categories and "low" and "high" flint densities distributed among 16 1-square-meter (m^2) plots. At the same time 10 0.1-m^2 plots were established nearby for quantifying worm-casts, and a parallel study revealed 19 earthworm species in the greater area, 9 at Down House alone. The star of the show proved to be *Aporrectodea longa*, which belongs to the largest and most widespread temperate zone earthworm genus. A copious caster, *A. longa* abounds in Great Pucklands Meadow, and is the main species responsible for transforming Great Pucklands from the "stony field" remembered by the Darwin children to the stone-free and well-vegetated meadow it became in Darwin's later years, and remains today.

The plots were revisited in 2013. In the intervening 6 years the flints were buried at an average rate of 0.96 cm per year—a good deal faster than Darwin's estimate of 0.21 cm per year. The greatest proportion of buried flints were small ones at high density, but the larger flints at both densities were buried deeper—over 5 cm on average. This differs too from Darwin's finding that smaller stones tend to get buried more quickly than larger ones, but it's worth bearing in mind that the difference in depth between small and large flints in the contemporary study amounts to a centimeter or less on average. It will be interesting to see how well the worms of Great Pucklands Meadow get on with their uniformitarian work in the years to come; the plots are next scheduled to be revisited in 2019. Reflecting on the countless earthworms that are incessantly working away unseen, can we possibly look at the landscape in the same way again? As Darwin so eloquently put it in the final paragraph of his earthworm book:

> When we behold a wide, turf-covered expanse, we should remember that its smoothness, on which so much of its beauty depends, is mainly due to all the inequalities having been slowly levelled by worms. It is a marvellous reflection that the whole of the superficial mould over any such expanse has passed, and will again pass, every few years through the bodies of worms. The plough is one of the most ancient and most valuable of mans inventions; but long before he existed the land was in fact regularly ploughed, and still continues to be thus ploughed by earth-worms. It may be doubted whether there are many other animals which have played so important a part in the history of the world, as have these lowly organised creatures.[50]

———

Darwin's fields and meadows are a microcosm of the wider world. As with all of the naturalist's investigations over the 40 years he lived in Downe, what he learned locally—in his study and greenhouse, field and garden, and woodland and lawn—had global implications. It is resonant that Darwin's researches on worms spanned his entire life at Down House; he worked away above ground as persistently as they reworked the soil beneath. That 40-year earthworm odyssey was classic Darwin: seeing common things with new eyes, intuiting new significance; asking with childlike wonder *why* or *how;* always *experimentising* to find an answer. He was as revolutionary as his worms, with his knack for asking questions and the determination to follow out and reinforce his evolutionary ideas through subject after subject, book after book. Those books were like worm castings, slowly burying received wisdom and old prejudices, making the world anew.

Experimentising: Get Thee to a Wormery

With the publication of his earthworm book, Darwin had come full circle. He not only established these humble organisms as a uniformitarian geological force, but as animals with some measure of sense and sensibility. Seen in this light, the sensibilities of worms are not far removed from the sense perception of climbing plants, or the appetites of carnivorous plants. Here we'll follow Darwin's lead in testing for choosiness in worms.

Obtaining Your Worms

Collect or purchase a dozen or more common night crawlers, *Lumbricus terrestris*, available from most live-bait shops. Avoid the common "red wiggler" worms often associated with compost piles; these do not exhibit the burrowing and surface-foraging behavior that we are interested in.

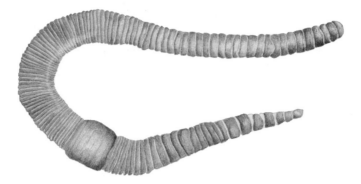

Mighty earthmover: *Lumbricus terrestris*, the common earthworm.
Drawing by Leslie C. Costa.

A. Materials

- 6–8 *Lumbricus terrestris* worms
- Terra-cotta flowerpot with drainage hole (medium-sized, large enough to hold 3–4 L of soil), OR 2-L clear plastic bottle (a soft-drink bottle with the label removed will work), cut near the top at its widest diameter and with a drainage hole cut in the base.
- Garden trowel
- Soil, locally dug and preferably not organically enriched
- Clear plastic wrap
- Rubber band, large enough to stretch the width of the flower-pot or bottle
- Masking tape
- Ruler or tape measure
- Spray bottle with water (preferably bottled, rain, or well water)
- Lard, butter, or oil
- Construction paper (black)
- Plain white paper
- Scissors
- Spoon (teaspoon size)

- Assorted fresh leaves (e.g., oak, maple, dandelion, pine needle, lettuce, spinach)
- Corn meal, shredded cabbage, and fresh grass clippings for food

B. Procedure

You have the option of making a terra-cotta flowerpot wormery, much as Darwin did, or to make one out of a clear plastic bottle, which has the advantage of permitting burrow observations should any worms construct a burrow against the bottle wall.

1a. Flowerpot wormery: Fill your flowerpot ¾ full of moistened (not saturated) potting soil and place two or three small to medium 1½–2 in. (4–5 cm) *L. terrestris* worms on the surface. Sprinkle a bit of cabbage and corn meal over the surface for food, spritz with water, and secure with plastic wrap and rubber band. Allow your worms to settle in for a week, giving a daily spritz of water and replacing cabbage bits and corn meal as they disappear.

1b. Plastic bottle wormery: Carefully cut the top off of the 2-liter plastic bottle, leaving most of the length of the cylindrical bottle. Cover the cut edge with masking tape. If the bottle has a label, remove it if possible so that you can easily observe what's inside, and punch a few small holes in the bottom for drainage. Fill the bottle with ~1.5 L of soil, and place one or two *L. terrestris* on the surface. Add food, water, and cover as above, and tape black construction paper around the bottle in such a way that you can easily remove it later for observations. Place this wormery in a cool and dark place (although covered with black paper, some light will get in and disturb the worms).

2. Note the number and size of wormholes at the soil surface of your wormery. These mark burrow entrances. Are the holes surrounded by mounds or other materials?

3. If you made a plastic bottle wormery, remove the black construction paper and see if any burrows were constructed along

the wall, permitting you to see into the burrow. If so, what is its length and diameter? Do you see a worm in the burrow? Is there any evidence of cabbage or corn meal in the burrow?

4. Using the white paper, cut out paper triangles, squares, disks, and diamonds, each about 3 cm in length. Darwin rubbed lard on his paper cutouts to repel water and prevent them from

A flowerpot "wormery" like those Darwin used to house his earthworms. Drawing by Leslie C. Costa.

becoming soggy and limp on the soil surface. If necessary, do this with lard, butter, or oil.

5. Place two of each shape (eight total) randomly onto the soil surface of your wormery, and re-cover it with plastic wrap.

6. Check the wormery twice daily and note the position of the paper cutouts. Are they moved between observations?

7. After 3 or 4 days, how many cutouts remain on the surface? Were any plugging the mouth of a burrow? Any unaccounted for were drawn into the burrows, out of sight. Did the worms prefer some shapes over others? Why do you think this was the case, or not?

8. Repeat with three kinds of cutout triangles: (1) equilateral triangles, 1 cm to a side; (2) isosceles triangles, with 1-cm base and 2-cm sides; and (3) right triangles, with 1-cm sides meeting perpendicularly. Prepare and place three of each on the wormery surface. Check daily, and note position, type, and number of visible triangles.

9. After a few days of observations, try carefully excavating a burrow to note the orientation of the paper cutouts within. Use a spoon to remove soil around the burrow until you come to the paper cutouts. Note the orientation of each triangle you excavate.

10. Did the worms prefer some triangles over others in drawing

them into their burrows? Of those drawn in, how were they ori-
ented? Darwin noted that nearly $^2/_3$ of the triangles he provided
to his worms were drawn in by one of the pointy ends.

11. Test the worms' preference for different natural materials: small
    leaves, pine needles, leaf petioles, and so on. Do the worms
    exhibit any preferences? Darwin also did taste tests with the
    leaves of garden vegetables. Try cabbage, tomato, turnip, and
    green onion leaves and see which ones they have an appetite for.

See also:

"Earthworms" at the Darwin Correspondence Project: www.darwin
project.ac.uk/learning/universities/getting-know-darwins-science
/earthworms.

ACKNOWLEDGMENTS

"I cannot express too strongly my obligations to the many persons who have assisted me." So wrote Charles Darwin in the first volume of *The Variation of Animals and Plants Under Domestication*, words of gratitude that I heartily echo. A number of friends, colleagues, and family members have directly or indirectly assisted me with this project, beginning with the very direct contributions of my wife, Leslie Costa. I am deeply appreciative of Leslie's keen editorial eye, and her deft artistic hand with the illustrations.

It has been my good fortune to have learned a great deal from David Kohn and Randal Keynes, who share an infectious enthusiasm for Darwin's experiments. The Darwin Manuscripts Project, directed by David and hosted by the American Museum of Natural History, is an invaluable treasure trove of Darwin manuscript material, and I am grateful to David for facilitating my use of this wonderful resource. Randal has curated another invaluable treasure trove: thematic compilations of letters, manuscripts, and other documents bearing on Darwin's varied research interests. Deepest thanks to Randal for so generously sharing his compilations, and, indeed, more—it is difficult to adequately express my gratitude to Randal and Zelfa Hourani for their warm hospitality during my visits to London. The *Wissenschaftskolleg zu Berlin*, where I spent a blissful fellowship year in 2012–2013, also helped advance our experiments project by kindly supporting Randal's and Zelfa's visit to Berlin, enlivening those short winter days with Darwin-inspired brainstorming.

Exploring Darwin's "backyard"—Down House and environs, managed by English Heritage—is always inspirational. I thank the Down House manager and staff for kindly hosting our meetings, and especially gardeners Rowan Blaik and Kristyna Slivova for sharing their enthusiasm and knowledge of Darwin's garden and grounds, and presenting Darwin's experiments. One year my family and I spent a magical day with Randal and Irene and John Palmer exploring verdant Orchis Bank, dotted with beautiful orchids descended from those that so enthralled Darwin. Icing on the cake that day was a slow worm caught by the kids—also descended, I fancied, from ones that Darwin surely rejoiced over when his own kids turned one up. Many thanks to Irene, warden with the Kent Wildlife Trust, for sharing her deep knowledge of orchids and Orchis Bank.

I have been gratified by the enthusiasm and encouragement of several friends and colleagues for the idea of the pedagogical—and inspirational—value of Darwin's experiments. Kefyn Catley, Louise Mead, and I co-taught the Evolution in the Blue Ridge course for science educators at Highlands Biological Station in the summer of 2011, in which we used Darwin's experiments as a new approach to teaching and learning about Darwin and his method. We learned as much from the teachers participating in the course as they learned from us: thank you Megan Cassidy, Tom Copeland, Alyssa Fuller, Jena Gladden, Hannah Grimm, Paula LaPoint, Mark Lonac, Randi Neff, Gloria Painter, Leslie Schoof, and Susan Steiner.

Special thanks to Leslie Costa for her comments, corrections, and suggestions on multiple earlier drafts of the book; our children Addison and Eli for conducting some of the experiments; Mac Callaham for assistance with the earthworm chapter; Nuala Caomhanach for sharing ideas and enthusiastically developing the first *Darwin's Dangerous Experiments* program at the AMNH; and Richard Milner for kindly providing the *Beagle* illustration. The wonderful librarians at Western Carolina University's Hunter Library were ever-ready to assist with all manner of literature-hunting, and I would be remiss if I failed to acknowledge too the tremendously useful online resources of the Darwin Correspondence Project, Wallace Correspondence Project, and Biodiversity Heritage Library.

I had the pleasure of working with a truly outstanding team at W. W. Norton to bring this book to fruition, starting with my editor, Amy Cherry, who deftly wields a veritable scimitar of an editorial pen, and her able editorial assistant Remy Cawley—thanks to you both! I thank, too, the book's designer, Helene Berinsky, and copyeditor Charlotte Kelchner for their fine efforts, and Norton's contracts manager Jesse Fox for advice and assistance. Last but far from least, I owe a debt of gratitude to filmmaker Erica Rothman of Nightlight Productions for expertly documenting our Evolution in the Blue Ridge Darwin-inspired experiments in Highlands, and for introducing me to Wendy Strothman and Lauren McLeod of the Strothman Agency. Many thanks, Wendy, for your advice and for championing this project from start to finish, and Lauren for all of your kind assistance along the way. I could not have asked for finer agents and advocates.

NOTES

Abbreviations

People

AG	Asa Gray
ARW	Alfred Russel Wallace
CD	Charles Darwin
CL	Charles Lyell
ED	Emma Darwin
EAD	Erasmus Alvey Darwin (CD's brother)
ErD	Erasmus Darwin (CD's grandfather)
FD	Francis (Frank) Darwin
GD	George Darwin
HD	Horace Darwin
HL	Henrietta (Etty) Litchfield née Darwin
JDH	Joseph Dalton Hooker
JL	John Lubbock
JSH	John Stevens Henslow
LD	Leonard (Lenny or Leo) Darwin
THH	Thomas Henry Huxley
WD	William (Willy) Darwin
WDF	William Darwin Fox

Works

Annotated Origin C. R. Darwin and J. T. Costa, *The Annotated Origin: A Facsimile of the First Edition of* On the Origin of Species (Cambridge, Mass.: Harvard University Press, 2009)

Autobiography	C. R. Darwin, *The Autobiography of Charles Darwin 1809–1882*, ed. N. Barlow (London: Collins, 1958)
BD	C. R. Darwin, Beagle *Diary*, ed. R. D. Keynes (Cambridge: Cambridge University Press, 1988)
CCD	C. R. Darwin, *The Correspondence of Charles Darwin*, ed. F. Burkhardt et al., 24 vols. (Cambridge: Cambridge University Press)
CDSP	C. R. Darwin, *Charles Darwin's Shorter Publications 1829–1883*, ed. J. van Wyhe (Cambridge: Cambridge University Press, 2009)
DCP	Darwin Correspondence Project (www.darwinproject .ac.uk/)
DED	E. Darwin, Diaries of Emma Darwin, DAR 242.1-60 (Cambridge University Library)
Climbing Plants	C. R. Darwin, "On the Movements and Habits of Climbing Plants," *Journal of the Linnean Society of London* (*Botany*) 9 (1865): 1–118.
Descent	C. R. Darwin, *The Descent of Man, and Selection in Relation to Sex* (London: John Murray, 1871)
EDFL	E. Darwin, *Emma Darwin: A Century of Family Letters*, ed. H. Litchfield (London: John Murray, 1915)
ExB	C. R. Darwin, Experiment book, DAR 157a (Cambridge University Library)
Fertilisation	C. R. Darwin, *The Effects of Cross and Self Fertilisation in the Vegetable Kingdom* (London: John Murray, 1876)
Forms of Flowers	C. R. Darwin, *The Different Forms of Flowers on Plants of the Same Species* (London: John Murray, 1877)
Foundations	C. R. Darwin, *The Foundations of the Origin of Species*, ed. F. Darwin (Cambridge: Cambridge University Press, 1909)
Insectivorous Plants	C. R. Darwin, *Insectivorous Plants* (London: John Murray, 1875)
Journal	C. R. Darwin, Darwin's *'Journal'* (1809–1881), DAR 158.1-76 (Cambridge University Library)
LL	C. R. Darwin, *The Life and Letters of Charles Darwin*, ed. F. Darwin (London: John Murray, 1887)

Movement	C. R. Darwin, *The Power of Movement in Plants* (London: John Murray, 1880)
Natural Selection	C. R. Darwin, *Charles Darwin's* Natural Selection *Manuscript*, ed. R. C. Stauffer (Cambridge: Cambridge University Press, 1975)
Orchids	C. R. Darwin, *On the Various Contrivances By Which British and Foreign Orchids Are Fertilised by Insects* (London: John Murray, 1862)
Origin	C. R. Darwin, *On the Origin of Species* (London: John Murray, 1859)
Principles	C. Lyell, *Principles of Geology* (London: John Murray, 1830–1833)
Q&E	C. R. Darwin, Questions & Experiments Notebook, ed. Barrett et al. (Ithaca, NY: Cornell University Press, 1987)
Researches	C. R. Darwin, *Journal of Researches* (London: Colburn, 1839; 2nd ed. London: John Murray, 1845)
RN	C. R. Darwin, Red Notebook, ed. P. H. Barrett et al. (Ithaca, NY: Cornell University Press, 1987)
TAN	C. R. Darwin, Torn Apart Notebook, ed. P. H. Barrett et al. (Ithaca, NY: Cornell University Press, 1987)
TN-A, -B, -C, -D, -E	C. R. Darwin, Transmutation Notebooks A–E, ed. Barrett et al. (Ithaca, NY: Cornell University Press, 1987)
Variation	C. R. Darwin, *Variation of Animals and Plants Under Domestication* (London: John Murray, 2 vols., 1868)
WCP	Wallace Correspondence Project, ed. G. Beccaloni (www.nhm.ac.uk/research-curation/scientific-resources /collections/library-collections/wallace-letters-online)
Worms	C. R. Darwin, *The Formation of Vegetable Mould Through the Action of Worms* (London: John Murray, 1881)

Chapter 1. Origins of an *Experimentiser*

1   *Autobiography*, 46. See Janet Browne, *Charles Darwin: Voyaging* (Princeton, NJ: Princeton University Press, 1995), chapter 1, for an outstanding account of Darwin's early life.

2   For lucid treatments of the spirit of the times, see J. Uglow, *The Lunar*

Men (London: Faber and Faber, 2002), and R. Holmes, *The Age of Wonder* (New York: Pantheon, 2008).

3   *Autobiography*, 27.

4   *Autobiography*, 46.

5   ErD, *The Temple of Nature* (London: J. Johnson, 1803), Canto I, part V.

6   See, for example, A. Desmond, *The Politics of Evolution* (Chicago: University of Chicago Press, 1989).

7   P. R. Sloan, "Darwin's Invertebrate Program," in *The Darwinian Heritage*, ed. D. Kohn (Princeton: Princeton University Press and Nova Pacifica, 1985), 86.

8   Grant is quoted in Sloan (1985), 84.

9   *Autobiography*, 49.

10   *Autobiography*, 59.

11   *Autobiography*, 62.

12   CD to Leonard Jenyns, 17 October 1846; *CCD* 3:354.

13   CD to JL, September 1854; *CCD* 5:211.

14   CD to WDF, 15 February 1831; *CCD* 1:118.

15   W. Whewell, *The Philosophy of the Inductive Sciences* (London: John W. Parker, 1840), 1:cxiii.

16   CD to WDF, 7 April 1831; *CCD* 1:120.

17   CD to Caroline Darwin, 28 April 1831; *CCD* 1:122.

18   John Medows Rodwell to FD, 8 July 1882 (DAR 112:94v); quoted in *CCD* 1:125, n. 2.

19   CD to WDF, 9 July 1831; *CCD* 1:124.

20   M. Roberts, "Darwin at Llanymynech: The Evolution of a Geologist," *British Journal for the History of Science* 29 (1996), 471.

21   *Autobiography*, 69–70.

22   *Autobiography*, 139.

23   See the wonderfully detailed accounts of the voyage by Darwin biographers Adrian Desmond and James Moore, *Darwin: The Life of a Tormented Evolutionist* (New York: Warner Books, 1991), and Janet Browne, *Charles Darwin: Voyaging* (Princeton, NJ: Princeton University Press, 1995). See also Darwin's own account, initially published as the *Journal of Researches* and later reissued as *The Voyage of the Beagle* (*Researches*; Darwin 1839a, 1845).

24   R. Fitzroy, *Proceedings of the Second Expedition, 1831–1836*. 3 vol. (London: Henry Colburn, 1839), 2:49.

25   CD to JSH, 18 May–16 June 1832; *CCD* 1:236.

26　*Autobiography*, 77.

27　C. Lyell, *Principles of Geology* (London: John Murray, 1830), 1:74.

28　CD to JSH, 18 May–16 June 1832; *CCD* 1: 237.

29　*BD* (11 January 1832), 22.

30　R. L. Keynes, ed., *Charles Darwin's Zoology Notes & Specimen Lists from H.M.S.* Beagle (Cambridge: Cambridge University Press, 2000), 232; DAR 31.1:262, Cambridge University Library.

31　CD to Catherine Darwin, 20 July 1834; *CCD* 1:391.

32　CD to JSH, 24 July 1834; *CCD* 1:399–400.

33　Quotations taken from the transcription by Sloan (1985), 105 [DAR 31.1:279v], 107 [DAR 5:99], and 109 [DAR 31.2:360].

34　*BD* (19 January 1836), 402.

35　N. Barlow, ed., "Darwin's Ornithological Notes," *Bulletin of the British Museum (Natural History) Historical Series* 2, no. 7 (1963), 262a.

36　*Researches*, 31.

37　*Researches*, 222–223.

38　See Keynes (2000), xxvii, n. 49 and *CCD* 4:29.

39　*Researches*, 384, 387.

40　*Researches*, 37.

41　*Researches*, 44.

42　CD to Alexander von Humboldt, 1 November 1839; *CCD* 2:239.

43　Indeed, Darwin's model of coral atoll evolution still stands today, and therein lies a fascinating story in itself—see D. Dobbs, *Reef Madness: Charles Darwin, Alexander Agassiz, and the Meaning of Coral* (New York: Pantheon, 2005).

44　CD to Leonard Horner, 29 August 1844; *CCD* 3:55.

45　*Researches*, 368–369.

46　*Researches*, 376.

47　*Researches*, 379.

48　*Origin*, 366.

49　CD to Henry Fawcett, 18 September 1861; *CCD* 9:269.

50　TN-C, 76–77.

51　TN-A, 180, and TN-B, 125e and 248.

Chapter 2. Barnacles to Barbs

1　*BD* (8–14 and 15–16 January 1835), 279.

2　R. L. Keynes, ed., *Charles Darwin's Zoology Notes & Specimen Lists from*

H.M.S. Beagle (Cambridge: Cambridge University Press, 2000), 274–276; DAR 31.305–307, Cambridge University Library.

3   *BD* (19 January 1835), 280.

4   CD to WDF, 15 February 1836; *CCD* 1:491.

5   *BD* (1 October 1836), 446.

6   C. Kinglsey, *Glaucus, or, The Wonders of the Shore* (London: Macmillan & Co., 1855), 87.

7   For a list of books aboard the *Beagle* see http://test.darwin-online.org.uk /BeagleLibrary/Beagle_Library_Introduction.htm.

8   A transcription of the zoology notes from the Edinburgh notebook can be found in P. H. Barrett et al., eds., *Charles Darwin's Notebooks, 1836–1844* (Ithaca, NY: Cornell University Press, 1987), 475–486.

9   Richard Owen, Hunterian Lecture no. 4, 9 May 1837. See P. R. Sloan, ed., *The Hunterian Lectures in Comparative Anatomy, May and June 1837* (Chicago: University of Chicago Press, 1992), 193.

10   TN-B, 44.

11   This marginal note appears on p. 442 of volume two of Darwin's copy of the *Principles of Geology* (fifth edition, 1837)—the very last page. The Biodiversity Heritage Library and Cambridge University Libraries have partnered to make Darwin's personal library available online, fully digitized and with transcribed marginal notes. To view this page go to: www .biodiversitylibrary.org/item/105893#page/451/mode/1up.

12   Quoted in T. A. Appel, *The Cuvier-Geoffroy Debate: French Biology in the Decades Before Darwin* (Oxford: Oxford University Press, 1987), 1.

13   TN-B, 111.

14   See Barrett et al. (1987), 197, notes 111-1–112-5.

15   Barrett et al. (1987), 242.

16   J. S. Sebright, *The Art of Improving the Breeds of Domestic Animals, in a Letter Addressed to the Right Hon. Sir Joseph Banks, K.B.* (London: John Harding, 1809), 5, 24–25.

17   Sebright (1809), 26.

18   J. Wilkinson, *Remarks on the Improvement of Cattle, etc. In a Letter to Sir John Saunders Sebright Bart. M.D,* 3rd ed. (Nottingham: H. Barnett, 1820), 4–5.

19   TN-C, 133.

20   TN-D, 20, 44.

21   TN-D, 118.

22   TN-E, 118.

23   A transcription of the Questions and Experiments (Q&E) notebook can be found in Barrett et al. (1987), 487–516.

24   The mystery of Darwin's illness has inspired much speculation and medical sleuthing. See R. O. Pasnau, "Darwin's Illness: A Biophychosocial Perspective," *Psychosomatics* 31 (1990) 121–1281; F. P. Smith et al., "Darwin's Illness," *Lancet* 336 (1990), 1139–1140; J. Adler et al., "The Dueling Diagnoses of Darwin," *Journal of the American Medical Association* 277 (1997), 1275–1277; R. Colp, "The Dueling Diagnoses of Darwin," *Journal of the American Medical Association* 277 (1997), 1275–1276; R. Colp, "More on Darwin's Illness," *History of Science* 38 (2000), 219–236; and A. K. Campbell and S. B. Matthews, "Darwin's Illness Revealed," *Postgraduate Medical Journal* 81 (2005), 248–251, to review a diversity of hypotheses.

25   See *Foundations* for published versions of the 1842 "*Sketch*" and 1844 "*Essay*," edited by Francis Darwin.

26   *Autobiography*, 68.

27   CD to JDH, 11 January 1844; *CCD* 3:2.

28   JDH to CD, 4–9 Sept 1845; *CCD* 3:250–251.

29   CD to JDH, 10 September 1845; *CCD* 3:253.

30   See letters of 10 and 14 September 1845; *CCD* 3:253 and 254.

31   J. Browne, *Charles Darwin: Voyaging* (Princeton, NJ: Princeton University Press, 1995), 471.

32   CD to JDH, 6 and 12 November 1846; *CCD* 3:363 and 365.

33   CD to JSH, 1 April 1848; *CCD* 4:128.

34   CD to JDH, 10 May 1848; *CCD* 4:140.

35   CD to JDH, 10 May 1848; *CCD* 4:140.

36   CD to CL, 2 September 1849; *CCD* 4:253.

37   See *CCD* volume 4, 388–409 (Appendix II) for a concise overview of Darwin's barnacle studies. See also the engaging treatment by R. Stott, *Darwin and the Barnacle* (London: Faber and Faber, 2003).

38   CD to JDH, 13 June 1850; *CCD* 4:344.

39   This anecdote, attributed to the Darwin's friend and neighbor John Lubbock (Lord Avebury), was reported by Francis Darwin, *More Letters of Charles Darwin* (London: John Murray, 1903), 1:38.

40   CD to WDF, 27 March 1855; *CCD* 5:294.

41   TN-E, 136.

42   D. J. Browne, *The American Bird Fancier; Considered with Reference to the Breeding, Rearing, Feeding, Management, and Peculiarities of Cage and House Birds* . . . (New York: C. M. Saxton, 1850), 83.

43   CD to WD, 29 November 1855; *CCD* 5:508.

44   CD to WD, 26 February 1856; *CCD* 6:45.

45   CD to William B. Tegetmeier, 18 May 1857; *CCD* 6:397.

46   CD to Thomas Eyton, 3 & 9 December 1855; *CCD* 5:513, 522.

47   ARW to Samuel Stevens, 21 August 1856 (WCP letter 1703). Wallace also corresponded with Darwin directly; in a letter to Wallace dated 1 May 1857 (*CCD* 6:387) Darwin mentioned receiving one from Wallace dated 10 October 1856, but this letter has been lost.

48   CD to ARW, 1 May 1857; *CCD* 6:387; WCP letter 2086.

49   *Journal*, 36v. A transcription of Darwin's *Journal* for 1858 can be found in *CCD* 7:503 (Appendix II).

50   Whitwell Elwin to John Murray, 3 May 1859; *CCD* 7:290.

51   CD to THH, 16 December 1859; *CCD* 7:434.

Chapter 3. Untangling the Bank

1   *Origin*, 489.

2   *Origin*, 489.

3   *Origin*, 490.

4   TN-E, 114.

5   TN-D, 134–135.

6   *Foundations*, 8, 90.

7   See *Annotated Origin*, 67, n. 1, for a brief account of Darwin's famous wedge metaphor.

8   *Origin*, 62, 64.

9   H. Milne-Edwards, *Introduction á la zoologie générale* (Paris: Victor Masson, 1851), 9.

10   DAR 205.5, 149.

11   CD to JDH, 5 June 1855; *CCD* 5:343.

12   CD to JDH, 15 June 1855; *CCD* 5:354.

13   See BBC news reports about the repeat of Darwin's and Miss Thorley's experiment at http://news.bbc.co.uk/2/hi/sci/tech/4607037.stm and http://news.bbc.co.uk/2/hi/science/nature/5201816.stm.

14   DAR 205.2:119r and 119v.

15   DAR 205.5:157; note dated 19 August 1855.

16   *Natural Selection*, 230.

17   *Natural Selection*, 228.

18   G. Sinclair, "On Cultivating a Collection of Grasses in Pleasure-Grounds or Flower-Gardens [and on the Utility of Studying the Gramineae]," *The*

Gardener's Magazine and Register of Rural & Domestic Improvement 1 (1826), 113.

19 E. M. Fries, "A monograph of the *Hieracia*; being an Abstract of Prof. Fries's *'Symbolae ad Historiam Hieraciorum,'* translated and abridged," *Botanical Gazette* 2 (1850), 188.

20 CD to JDH, 11 March 1858; *CCD* 7:47.

21 DAR 205.5; Darwin elaborated on this in *Origin*, p. 126: "Looking to the future, we can predict that the groups of organic beings which are now large and triumphant, and which are least broken up, that is, which as yet have suffered least extinction, will for a long period continue to increase."

22 ExB, 24–25.

23 See *Origin*, 67.

24 DAR 46.1:37r; note dated 24 April 1857.

25 *Journal*, 36r; entry dated 22 April 1857.

26 CD to JDH, 2 June 1857; *CCD* 6:404.

27 CD to JL, 14 July 1857; *CCD* 6:430.

28 CD to JDH, 14 July 1857; *CCD* 6:429.

29 CD to JL, 14 July 1857; *CCD* 6:430.

30 See *Natural Selection*, 149–154, for a summary of Darwin's calculations.

31 CD to JDH, 22 August 1857; *CCD* 6:443.

32 CD to AG, 5 September 1857; *CCD* 6:448.

33 *Origin*, 62.

34 Anecdotes of Hutchinson's childhood antics are recounted in N. G. Slack, *G. Evelyn Hutchinson and the Invention of Modern Ecology* (New Haven: Yale University Press, 2010), chapter 2.

Chapter 4. Buzzing Places

1 CD to JL, September 1854; *CCD* 5:211.

2 GD on "humble bees," DAR 112.B18 ff.

3 R. B. Freeman, "Charles Darwin on the Routes of Male Humble Bees," *Bulletin of the British Museum (Natural History), Historical Series* 3, no. 6 (1968), 182.

4 Freeman (1968), 182.

5 LD, "Memories of Down House," *The Nineteenth Century* 106 (1929), 118.

6 Freeman (1968), 183.

7 LD (1929), 119.

8 This statement comes from Kirby's Bridgewater Treatise, first published in 1835. See W. Kirby, *On the Power, Wisdom, and Goodness of God as*

Manifested in the Creation of Animals, and in Their History, Habits, and Instincts, new edition, ed. Thomas Rymer Jones, 2 vols. (London: Henry G. Bohn, 1852), 2:246.

9   Indeed, the "honeycomb conjecture" (as mathematicians know it) dates back more than 2,000 years to the Roman scholar Marcus Terentius Varro, who wrote in 36 BC that honeycomb consisting of hexagons is more space-efficient than other shapes. Varro was proven correct only as recently as 1999, by University of Michigan mathematician Thomas Hales; mathematically inclined readers can check out his proof at the archive http://lanl.arxiv.org, "The Honeybee Conjecture," paper ID 9906042.

10   G. R. Waterhouse, "Bees," *Penny Cyclopaedia of the Society for the Diffusion of Useful Knowledge,* 4 (1835), 153.

11   CD to JDH, 11 and 28 February 1858; *CCD* 7:30–31 and 39–40.

12   Erasmus Darwin helped his brother with the geometry of bees' cells in May and June 1858; see *CCD* 7:86, 99–100, 100–102, 104, and 108–112.

13   Gray's comments were given at the 5 July 1858 meeting of the Entomological Society of London; see *The Zoologist* vol. 16, p. 6189.

14   The discoverer was renowned Hungarian mathematician László Fejes Tóth (1915–2005); see L. F. Tóth, "What the Bees Know and What They Do Not Know," *Bulletin of the American Mathematical Society* 70 (1964), 468–481, and T. C. Hales, "Cannonballs and Honeycombs," *Notices of the American Mathematical Society* 47, no. 4 (2000), 440–449.

15   *Origin,* 235.

16   *Origin,* 51.

17   Thoreau's stirring account of the raid, which appears in chapter 12 of *Walden,* was much reprinted as an essay entitled *"The Battle of the Ants."*

18   CD to Frederick Smith, [Before 9 March] 1858; *CCD* 7:44.

19   CD to ED, 28 April 1858; *CCD* 7:84.

20   CD to ED, 25 April 1858; *CCD* 7:80.

21   DAR 205.11 (2):106. Darwin wrote "Ch. 10" on this note, intending to include it in the chapter of *Natural Selection* treating instincts; see *Natural Selection,* 470–471 and 574–575.

22   CD to WD, 26 April 1858; *CCD* 7:81.

23   CD to JDH, 13 July 1858; *CCD* 7:129.

24   DAR 205.11:94r. The full notebook entry reads:
      "July. 58 Took specimens of F. Rufa near Sandown, with remarkable differences in size of workers— very many smallest workers in nest.—
      I saw July 24th a migration & only the largest & medium sized were

carrying pupae; indeed I doubt if the last [could] have carried them; so,«there is» little difference in ‹function› «Habits», almost compulsory."

25   *Origin*, 223–224.

26   *Origin*, 243–244.

27   See J. Moore, *Darwin's Sacred Cause: How a Hatred of Slavery Shaped Darwin's Views on Human Evolution* (Boston: Houghton Mifflin Harcourt, 2009) for discussion of the interplay between Darwin's science and his abolitionist convictions.

Chapter 5. A Grand Game of Chess

1   CD to WDF, 7 May 1855; *CCD* 5:326.

2   See D. Pauly's *Darwin's Fishes: An Encyclopedia of Ichthyology, Ecology, and Evolution* (Cambridge: Cambridge University Press, 2007), 70–72, for a summary of Darwin's fish-related experiments.

3   CD to C. J. F. Bunbury, 21 April 1856; *CCD* 6:80.

4   *BD* (19 January 1836), 402–403.

5   *Principles*, 2:124.

6   CD to JDH, 10 February 1845; *CCD* 3:140.

7   *Principles*, 2:159.

8   *Foundations*, 31.

9   *Autobiography*, 124–125.

10   E. Forbes, "On the connexion between the distribution of the existing fauna and flora of the British Isles, and the geological changes which have affected their area, especially during the epoch of the Northern Drift," *Memoirs of the Geological Survey of England, and of the Museum of Economic Geology in London* 1 (1846), 337.

11   *Foundations*, 31 (text and note 1).

12   *Foundations*, 169.

13   CD to JDH, 17–18 June 1856; *CCD* 6:147.

14   CD to CL, 16 June 1856; *CCD* 6:143.

15   JDH to CD, 16 June 1847; *CCD* 4:50.

16   J. D. Hooker, *Introductory Essay to the Flora of New Zealand* (London: Lovell Reeve, 1853), xxi.

17   CD to JDH, 13 April 1855; *CCD* 5:305.

18   CD to JSH, 2 July 1855; *CCD* 5:365.

19   CD to JDH, 15 May 1855; *CCD* 5:329–330.

20   CD to JDH, 10 October 1855; *CCD* 5:477.

21 CD to JDH, 12 April 1857; *CCD* 6:371–372.

22 CD to JDH, 10 December 1856; *CCD* 6:305.

23 James Tenant to CD, 27 March 1857; *CCD* 6:364.

24 Q&E, [5]a.

25 ExB, 17.

26 ExB, 8.

27 CD to JDH, 10 December 1856; *CCD* 6:305.

28 Davy's paper was read before the Royal Society and published in 1856: "On the Ova of Salmon," *Philosophical Transactions of the Royal Society*, London 146:21–29.

29 *Origin*, 385.

30 C. R. Darwin, "Transplantation of shells," *Nature* 18 (1878), 121; *CDSP*, 422.

31 C. R. Darwin, "On the Dispersal of Freshwater Bivalves," *Nature* 25 (1882), 529–530; *CDSP*, 486–487.

32 CL to CD, 1–2 May 1856; *CCD* 6:89.

33 G. Nelson, "From Candolle to Croizat: Comments on the History of Biogeography," *Journal of the History of Biology* 11, no. 2 (1978), 289.

34 See Censky et al., "Over-water Dispersal of Lizards Due to Hurricanes," *Nature* 395 (1998), 556–557.

35 See the excellent treatments by Alan de Queiroz: "The Resurrection of Oceanic Dispersal in Historical Biogeography," *Trends in Ecology and Evolution* 20 (2005), 68–73; *The Monkey's Voyage: How Improbable Journeys Shaped the History of Life* (New York: Basic Books, 2014).

36 P. Sebastian et al., "Darwin's Galapagos Gourd: Providing New Insights 175 Years after his Visit," *Journal of Biogeography* 37 (2010), 975–980.

37 *Origin*, 388.

Chapter 6. The Sex Lives of Plants

1 J. S. Sebright, *The Art of Improving the Breeds of Domestic Animals, in a Letter Addressed to the Right Hon. Sir Joseph Banks, K.B.* (London: John Harding, 1809), 26.

2 TN-C, 133.

3 CD to William Herbert, 26 June 1839; *CCD* 2: 201–202.

4 TN-E, 150–151.

5 T. A. Knight, "An Account of Some Experiments on the Fecundation of Vegetables," *Philosophical Transactions of the Royal Society* 89 (1799), 200.

6 Knight (1799), 202.

7　Knight (1799), 202–203.

8　*Foundations*, 70–71. See also *Natural Selection*, an annotated edition of the manuscript edited by R. C. Stauffer (Cambridge: Cambridge University Press, 1975).

9　Darwin's marginal note is found on p. 18 of his copy of Sprengel's book. View this and other works from Darwin's personal library at the Biodiversity Heritage Library's virtual collection: http://biodiversitylibrary.org /collection/darwinlibrary.

10　CD to AG, 19 January 1863; *CCD* 11:57.

11　*LL*, 3:258.

12　*EDFL*, 1:51.

13　TAN, 135.

14　Ruricola, "Humble-bees," *The Gardeners' Chronicle* 34 (1841), 485.

15　C. R. Darwin, "Humble-bees," *Gardeners' Chronicle* 34 (1841), 550; *CDSP*, 134–136.

16　*Natural Selection*, 54 (ms. pp. 34–35).

17　*Natural Selection*, 53 (ms. p. 33).

18　*Fertilisation*, 417–418.

19　*Fertilisation*, 82–83.

20　CD to JDH, 30 May 1862; *CCD* 10: 226.

21　C. R. Darwin, "Bees and the Fertilisation of Kidney Beans," *Gardeners' Chronicle and Agricultural Gazette* 43 (1857), 725; *CDSP*, 267.

22　*Origin*, 73–74.

23　*Fertilisation*, 361.

24　CD to JL, 2 September 1862; *CCD* 10:388.

25　CD to JL, 2 September 1862; *CCD* 10:392.

26　*Natural Selection*, 52 (ms. p. 32).

27　CD to JDH, 27 April 1860; *CCD* 8:169.

28　CD to JDH, 7 May 1860; *CCD* 8:191–192.

29　ExB, 55.

30　AG to CD, 11 October 1861; *CCD* 9:299–300.

31　C. R. Darwin, "On the Sexual Relations of the Three Forms of *Lythrum salicaria*," *Journal of the Linnean Society of London (Botany)* 8 (1864), 173; *CDSB*, 345.

32　CD to AG, 9 August 1862; *CCD* 10:362.

33　CD to WD, 9 July 1862; *CCD* 10:309.

34　CD to AG, 28 July 1862; *CCD* 10:341.

35　Margaret Susan Wedgwood to CD, before 4 August 1862; *CCD* 10:351.

36 CD to Katherine Elizabeth Sophy, Lucy Caroline, and Margaret Susan Wedgwood, 4 August 1862; *CCD* 10:355.

37 CD to WD, 2–3 August 1862; *CCD* 10:349–350.

38 CD to AG, 16 October 1862; *CCD* 10:470.

39 CD to AG, 4 August 1863; *CCD* 11:581.

40 Quoted in Baker's foreword to the facsimile edition of Darwin's *The Different Forms of Flowers on Plants of the Same Species* published by the University of Chicago Press (1986), ix.

41 Henry David Thoreau, journal entry for 22 October 1839; quoted in O. Shepard, ed., *The Heart of Thoreau's Journals* (New York: Dover Publications, 1961), 9.

Chapter 7. It Bears on Design

1 CD to CL, 6 June 1860; *CCD* 8:243.

2 ExB, 29.

3 *EDFL*, 2:376.

4 J. Ruskin, *The Complete Works of John Ruskin*, library ed. (London: George Allen/New York: Longmans, Green, and Co., 1906), 391.

5 Ruskin (1906), 342.

6 CD to JDH, 5 June 1860; *CCD* 8:238.

7 C. R. Darwin, "Fertilisation of British Orchids by Insect Agency," *Gardeners' Chronicle and Agricultural Gazette* 23 (1860), 528; *CDSP*, 301.

8 CD to Alexander Goodman More, 17 July 1861; *CCD* 9:207.

9 CD to JDH, 12 July 1860; *CCD* 8:286.

10 *Orchids*, 215.

11 CD to CL, 12 September 1860; *CCD* 8:356.

12 Darwin was invited by the naturalist and clergyman Leonard Jenyns, Henslow's brother-in-law, to contribute a memorial to Henslow. This was included in Jenyns' *Memoir of the Rev. John Stevens Henslow* (London: Van Voorst, 1862), 51–55.

13 CD to JDH, 13 July 1861; *CCD* 9:202.

14 CD to JDH, 27 July 1861; *CCD* 9:220.

15 CD to JDH, 27 July 1861; *CCD* 9:220.

16 *Orchids*, 160.

17 CD to JDH, 28 July–10 August 1861; *CCD* 9:222.

18 CD to John Murray, 21 September 1861; *CCD* 9:273.

19 CD to CL, 1 October 1861; *CCD* 9:291.

20 CD to JDH, 28 September 1861; *CCD* 9:284.

21   CD to JDH, 11 October 1861; *CCD* 9:301.

22   J. Lubbock, *On British Wild Flowers Considered in Relation to Insects* (London: Macmillan and Co., 1875), 173–174.

23   R. Schomburgk, "On the Identity of Three Supposed Genera of Orchidaceous Epiphytes, in a Letter to A. B. Lambert," *Transactions of the Linnean Society of London* 17 (1837), 552.

24   J. Lindley, *The Vegetable Kingdom, or, The Structure, Classification, and Uses of Plants, Illustrated Upon the Natural System* (London: Bradbury and Evans, 1846), 178.

25   C. R. Darwin, "On the Three Remarkable Sexual Forms of *Catasetum tridentatum*, an Orchid in the Possession of the Linnean Society," *Proceedings of the Linnean Society of London (Botany)* 6 (1862), 151–157.

26   George Bentham to CD, 15 May 1862; *CCD* 10:194.

27   *Orchids*, 244.

28   *Origin*, 453.

29   *Origin*, 453.

30   For treatments of the Gray-Agassiz debates see D. N. Livingstone, *Darwin's Forgotten Defenders: The Encounter Between Evangelical Theology and Evolutionary Thought* (Vancouver: Regent College Publishing, 1984), 57–64; A. Ward, "Evolution and Creation Debates," in *A Companion to the History of American Science*, ed. G. M. Montgomery and M. A. Largent, (Chichester: John Wiley & Sons, 2016), 361–373.

31   *Orchids*, 2.

32   ARW to CD, 23 May 1862; *CCD* 10:217.

33   AG to CD, 2–3 July 1862; *CCD* 10:292.

34   CD to AG, 23–24 July 1862; *CCD* 10:331.

35   AG to CD, 22 September 1862; *CCD* 10:428.

36   CD to JDH, 25 & 26 January 1862; *CCD* 10:48.

37   Duke of Argyll (G. D. Campbell), *The Reign of Law* (London: Alexander Strahan, 1867), 46.

38   A. R. Wallace, "Creation by Law [Review of *The Reign of Law* by the Duke of Argyll, 1867]," *Quarterly Journal of Science* 4 (1867), 475.

39   Wallace (1867), n. 2.

40   *Orchids*, 2nd ed. (1877), 230.

Chapter 8. Plants with Volition

1   J. Browne, *Charles Darwin: Power of Place* (Princeton, NJ: Princeton University Press, 2003), 46.

2   *EDFL*, 2:177.

3   *Origin*, 484.

4   CD to AG, 26 September 1860; *CCD* 8:389.

5   T. Slaughter, ed., *Bartram: Travels and Other Writings* (New York: Library of America, 1996), 17. The quotation is found on pp. xx–xxi in the original 1791 edition of Bartram's *Travels*. An electronic transcription of the 1791 text can be found in the *Documenting the American South* collection of the University of North Carolina–Chapel Hill Libraries: docsouth.unc .edu/nc/bartram/bartram.html.

6   *Origin*, 187.

7   C. R. Darwin, "Irritability of *Drosera*," *Gardeners' Chronicle and Agricultural Gazette* 38 (1860), 853; *CDSP*, 303.

8   CD to JL, 18 November 1860; *CCD* 8:477.

9   CD to JDH, 11 September 1862; *CCD* 10:402.

10 CD to AG, 3–4 September 1862; *CCD* 10:390–391.

11 DAR 54:29 (Cambridge University Library).

12 *Insectivorous Plants*, 4.

13 CD to AG, 4 August 1863; *CCD* 11:582.

14 *Journal*, 50v and 51v.

15 *Insectivorous Plants*, chapter 6.

16 *Insectivorous Plants*, 96.

17 F. Darwin, "Experiments on the nutrition of *Drosera rotundifolia*," *Journal of the Linnean Society (Botany)* 17 (1880), 23.

18 See note 5.

19 *Insectivorous Plants*, 286.

20 CD to Daniel Oliver, 29 September 1860; *CCD* 8:398.

21 CD to AG, 8 August 1867; *CCD* 15:343.

22 CD to AG, 8 January 1873; *CCD* 21:31.

23 CD to AG, 22 October 1872; *CCD* 20:454.

24 J. Burdon-Sanderson, "On the Electrical Phenomena which Accompany the Contractions of the Leaf of *Dionaea muscipula*," *Report of the British Association for the Advancement of Science* 43 (1873), 133.

25 *Insectivorous Plants*, 286.

26 *Insectivorous Plants*, 331.

Chapter 9. Crafty and Sagacious Climbers

1   CD to JDH, 25 June 1863; *CCD* 11:506.

2   AG to CD, 24 November 1862; *CCD* 10:554.

3   CD to AG, 26 June 1863; *CCD* 11:507–508.

4   JDH to CD, 21 July 1863; *CCD* 11:554.

5   AG to CD, 1 September 1863; *CCD* 11:614.

6   J. Browne, *Charles Darwin: Power of Place* (Princeton, NJ: Princeton University Press, 2003), 241.

7   *Climbing Plants*, 7.

8   *Climbing Plants*, 7–8.

9   *Climbing Plants*, 2.

10 J. D. Hague, "A reminiscence of Mr Darwin," *Harper's New Monthly Magazine* 69 (1884), 763.

11 *Power of Movement*, 3.

12 *Climbing Plants* (2nd ed.; 1875), 98–99.

13 CD to AG, 28 May 1864; *CCD* 12:211.

14 *Climbing Plants*, 59.

15 *Power of Movement*, 484.

16 *Power of Movement*, 486.

17 T. A. Knight, "On the Direction of the Radicle and Germen during the Vegetation of Seeds," *Philosophical Transactions of the Royal Society of London* 96 (1806), 100.

18 *Power of Movement*, 545.

19 *Power of Movement*, 529.

20 Sachs is quoted in P. Ayres, *The Aliveness of Plants* (London: Pickering and Chatto, 2008), 106.

21 *Climbing Plants*, 115.

22 *Climbing Plants*, 117–118.

23 *Power of Movement*, 573.

Chapter 10. Earthworm Serenade

1   ED to LD; *EDFL* 2:241.

2   C. R. Darwin, "On the Formation of Mould," *Proceedings of the Geological Society of London* 2 (1838), 575.

3   Darwin (1838), 576.

4   William Buckland to Geological Society of London, 9 March 1839; *CCD* 2:76.

5   Sarah Elizabeth Wedgwood and Josiah Wedgwood II to CD, 10 November [1837]; *CCD* 2:55.

6   Ibid.

7   W. F. Lindsay-Carnegie to CL, 14 February 1838; *CCD* 2:72.

8   C. R. Darwin, "On the Origin of Mould," *Gardeners' Chronicle and Agricultural Gazette* 14 (1844), 218; *CDSP*, 173.

9   *Worms*, 115.

10 ExB, 28.

11 CD to M. T. Masters, 21 March 1868; *CCD* 16(1):290. Lucy's observations were published in the March 28, 1868 issue of the *Gardeners' Chronicle*.

12 *Journal*, 46v and 47r.

13 CD to ARW, 19 August 1868; *CCD* 16(2):688.

14 J. Browne, *Charles Darwin: Power of Place* (Princeton, NJ: Princeton University Press, 2003), 479.

15 AG and Jane Loring Gray to CD, 8 and 9 May 1869; *CCD* 17:218.

16 CD to *Gardeners' Chronicle*, 9 May 1869 (published 15 May 1869); *CCD* 17:222.

17 Ibid.

18 THH to CD, 17 March 1869; *CCD* 17:136.

19 CD to ARW, 5 March 1869; *CCD* 17:111.

20 CD to ARW, 27 March 1869 (*CCD* 17:157) and 14 April 1869 (*CCD* 17:175).

21 CD to ARW, 25 June 1869; *CCD* 17:289–290.

22 W. W. Baxter to CD, 13 November 1871; *CCD* 19:682.

23 Lucy Wedgwood to CD, 20 November 1871; *CCD* 19:694.

24 AG to CD, 2 February 1872; *CCD* 20:61.

25 CD to Archibald Geikie, 27 December 1871; *CCD* 19:738.

26 Archibald Geikie to CD, 29 December 1871; *CCD* 19:743.

27 CD to Archibald Geikie, 30 December 1871; *CCD* 19:746–747.

28 CD to HD, 15 December 1871; *CCD* 19:721.

29 GD to CD, 30 December 1871; *CCD* 19:745.

30 CD to Amy Ruck, 24 February 1872; *CCD* 20:82.

31 CD to Lucy Wedgwood, 5 January 1872, *CCD* 20:12; Lucy Wedgwood to CD, 20 January 1872, *CCD* 20:32; CD to Lucy Wedgwood, 21 January 1872, *CCD* 20:34.

32 *Worms*, 262, 263.

33 *Worms*, 264–268.

34 ED to HL, *EDFL* 2:225–226.

35 ED to HL, *EDFL* 2:227.

36 T. H. Farrer to CD, 23 September 1877; DCP, letter 11150.

37 ED to GD, Down House Collection, letters 77.24 and 77.37.

38 ED to WD; *EDFL* 2:231.

39 ED to T. H. Farrer, 4 December 1877; DCP letter 11268.

40   CD to Sophy Wedgwood, 8 October 1880; DCP, letter 12745.

41   A. Desmond and J. Moore, *Darwin: Life of a Tormented Evolutionist* (New York: Warner Books, 1991), 649.

42   JL, *Diary*, 1879–1882, Supplementary Avebury Papers, Add. MSS 62682, fol. 14 (British Library); quoted in J. F. M. Clark, "'The Ants Were Duly Visited': Making Sense of John Lubbock, Scientific Naturalism and the Senses of Social Insects," *British Journal for the History of Science* 30, no. 2 (1997): 197 n. 71.

43   *Worms*, 26.

44   *Worms*, 27–28.

45   CD to WD, 5 February 1881; DCP, letter 13037.

46   *Worms*, 34.

47   JDH to CD, 23 or 30 October 1881; DCP, letter 13424.

48   Lady Derby [M. C. Stanley] to CD, 16 October 1881; DCP, letter 13406.

49   *Worms*, 6.

50   *Worms*, 313.

FURTHER READING AND RESOURCES

Resources for teaching and learning about Darwin through his writings, method, and experiments and other investigations.

Electronic

Charles Darwin Trust

http://www.charlesdarwintrust.org/content/19/darwin-inspired-learning

Central to the mission of the Charles Darwin Trust is "Darwin-inspired" teaching and learning that builds upon Darwin's life and work.

Darwin Correspondence Project

https://www.darwinproject.ac.uk/learning/7-11, https://www.darwinproject.ac .uk/learning/11-14, and https://www.darwinproject.ac.uk/learning/universities

The site that brings you the treasure trove of Darwin letters online provides these fine learning resources, drawing upon the Darwin correspondence.

Darwin's Pigeons

http://darwinspigeons.com

Undertake a virtual exploration of the pigeon breeds that so captivated Darwin at this website dedicated to Darwin's pigeons curated by pigeon fancier, breeder, and judge John Ross. Randal Keynes joins John in describing six of Darwin's pigeon breeds at www.youtube.com/watch?v=VFVueCs3gFI. This video was filmed at the Albemarle Street, London, location of Darwin's publisher, John Murray.

Down House

http://www.english-heritage.org.uk/visit/places/home-of-charles-darwin -down-house/

The home of Charles Dawin and family, now maintained by English Heritage. The house is maintained essentially as it was when the Darwins lived there. The gardens, greenhouses, and grounds have been restored to this period as well, and include presentations of several of Darwin's experiments.

English Heritage Blog

http://blog.english-heritage.org.uk/10-ways-experiment-like-darwin/

S. Kinchin-Smith's article on the English Heritage blog, "10 Ways to Experiment Like Darwin." Posted 22 August 2014.

Evolution: Education and Outreach [e-journal]

https://evolution-outreach.springeropen.com/

A SpringerOpen journal dedicated to the understanding, teaching, and learning of evolutionary theory for K–16 students, teachers, and scientists.

Linnean Society of London

https://www.linnean.org/education-resources/secondary-resources/darwin-inspired-learning

The LSL's Education Resources include outstanding Darwin-inspired modules by Drs. Carol Boulter, Emma Newall, and Dawn Sanders, developed in cooperation with the Charles Darwin Trust.

National Academies of Sciences, Engineering, and Medicine USA

http://www.nas.edu/evolution/

In the Light of Evolution book series, outstanding educational resources provided by the US National Academies, including free electronic editions.

National Center for Science Education

https://ncse.com/

Defending the integrity of science education against ideological interference, the NCSE "works with teachers, parents, scientists, and concerned citizens at the local, state, and national levels to ensure that topics including evolution and climate change are taught accurately, honestly, and confidently."

Nature's Evolution Gems

www.nature.com/nature/newspdf/evolutiongems.pdf

H. Gee, R. Howlett, and P. Campbell. 2009. "15 Evolutionary Gems: A Resource from *Nature* for those Wishing to Spread Awareness for Evolution by Natural Selection."

Print

Ayala, F. 2009. "Darwin and the Scientific Method." *Proceedings of the National Academy of Sciences USA* 106 (supp. 1): 10033–10039.

Boulter, C., D. Sanders, and M. Reiss, eds. 2015. *Darwin-Inspired Learning*. Rotterdam and Boston: Sense Publishers.

Catley, K. M. 2006. "Darwin's Missing Link—A Novel Paradigm for Evolution Education." *Science Education* 90 (5): 767–783.

Costa, J. T. 2003. "Teaching Darwin with Darwin." *BioScience* 53: 1030–1031.

Dennison, R. 1993. "Using Darwin's Experimental Work to Teach the Nature of Science." *American Biology Teacher* 55: 50–52.

Ellis, R. J. 2010. *How Science Works: Evolution. A Student Primer*. New York: Springer.

Grace, M., P. Hanley, and S. Johnson. 2008. "'Darwin-Inspired' Science: Teachers' Views, Approaches and Needs." *School Science Review* 90 (331): 71–77.

Hernández Laille, M., and C. A. Soler. 2014. *Charles Darwin and Lucia Sapiens: Lessons on the Origin and Evolution of Species*. Translated by N. Stapleton. Madrid: Universidad Nacional de Educación a Distancia.

Johnson, S. 2008. "Teaching Science Out-of-Doors." *School Science Review* (90) 331: 65–70.

Keynes, R. 2009. "Darwin's Ways of Working—the Opportunity for Education." *Journal of Biological Education* 43 (3): 101–103.

Lennox, J. G. 1991. "Darwinian Thought Experiments: A Function for Just-So Stories." In *Thought Experiments in Science and Philosophy*, edited by T. Horowitz and G. J. Massey, 223–246. Savage, MD: Rowman and Littlefield.

Love, A. C. 2010. "Darwin's 'Imaginary Illustrations:' Creatively Teaching Evolutionary Concepts and the Nature of Science." *American Biology Teacher* 72 (2): 82–89.

Morris, J. R., J. T. Costa, and A. Berry. 2015. "Adaptations: Using Darwin's *Origin* to Teach Biology and Writing." *Evolution* 69: 2556–2560.

National Biodiversity Network. 2009. *The Darwin Guide to Recording Wildlife*. Nottingham, UK: NBN Trust.

Slingsby, D. 2009. "Charles Darwin, Biological Education and Diversity: Past Present and Future." *Journal of Biological Education* 43 (3): 99–100.

BIBLIOGRAPHY

Electronic/Online Bibliographic Resources

Biodiversity Heritage Library

biodiversitylibrary.org/

An unprecedented consortium of major natural history, botanical, and research libraries cooperate to digitize and make freely accessible the historical literature held in their collections. Hosted by the Smithsonian Institution.

Darwin Correspondence Project

www.darwinproject.ac.uk/

University of Cambridge–based digital archive of the approximately 15,000 letters of Charles Darwin. Directed by Jim Secord et al., this archive forms the basis of both the printed *Correspondence of Charles Darwin* volumes (Cambridge University Press) and the annotated letters of the Darwin Correspondence Project website, currently available through 1876.

Darwin Manuscripts Project

www.amnh.org/our-research/darwin-manuscripts-project/

Based at the American Museum of Natural History (NY) and directed by David Kohn of the AMNH and Drexel University, this site presents extensive historical and textual editions of Darwin's scientific manuscripts built upon the DARBASE database at Cambridge University Library, featuring high-resolution color images and full transcriptions of Darwin manuscripts.

Darwin Online

darwin-online.org.uk/contents.html

The most extensive scholarly website on Darwin, featuring complete transcriptions and images of manuscripts, published works, private papers, and other documents. Made possible by a consortium of universities, museums, libraries, and other institutions and directed by John van Wyhe of the National University of Singapore.

Manuscripts

The following Darwin manuscripts were utilized extensively in this work:

Catalogue of Down Specimens (EH 88202576).
Notebook in the collection of Darwin's papers and other material transferred by his family through the Pilgrim Trust to Down House in 1942. Transcribed by Randal Keynes.
Darwin's *Experiment Book* (DAR157a).
Notebook in the collection of Cambridge University Library containing records of many experiments and related investigations undertaken by Darwin between 1855 and 1867.
Darwin's *Journal* (DAR158).
General record of activities kept by Darwin between August 1838 and December 1881, in the collection of Cambridge University Library. Transcribed and edited by John van Wyhe.

Print Sources and References

Agassiz, L. 1841. "On Glaciers, and the Evidence of Their Once Having Existed in Scotland, Ireland and England." *Proceedings of the Geological Society of London* 3: 327–332.

Alcock, J. 2006. *An Enthusiasm for Orchids: Sex and Deception in Plant Evolution*. Oxford: Oxford University Press.

Alder, J., R. Colp Jr., G. M. FitzGibbon, A. G. Gordon, T. J. Barloon, and R. Noyes Jr. 1997. "The Dueling Diagnoses of Darwin." *Journal of the American Medical Association* 277: 1275–1277.

Allen, M. 1977. *Darwin and His Flowers: The Key to Natural Selection*. London: Faber and Faber.

Appel, T. A. 1987. *The Cuvier-Geoffroy Debate: French Biology in the Decades Before Darwin*. Oxford: Oxford University Press.

Arditti, J., J. Elliott, I. J. Kitching, and L. T. Wasserthal. 2012. "'Good Heavens What Insect Can Suck It'—Charles Darwin, *Angraecum sesquipedale* and

Xanthopan morganii praedicta." *Botanical Journal of the Linnean Society* 169: 403–432.

Argyll, Duke of (George Douglas Campbell). 1867. *The Reign of Law.* London: Alexander Strahan.

Armstrong, P. 1991. *Under the Blue Vault of Heaven: A Study of Charles Darwin's Sojourn in the Cocos (Keeling) Islands.* Nedlands: Indian Ocean Centre for Peace Studies.

Athauda, S. B. P., K. Matsumoto, S. Rajapakshe, M. Kuribayashi, M. Kojima, N. Kubomura-Yoshida, A. Iwamatsu, C. Shibata, H. Inoue, and K. Takahashi. 2004. "Enzymic and Structural Characterization of Nepenthesin, a Unique Member of a Novel Subfamily of Aspartic Proteinases." *Biochemical Journal* 381: 295–306.

Ayres, P. 2008. *The Aliveness of Plants.* London: Pickering and Chatto.

Barlow, N., ed. 1958. *The Autobiography of Charles Darwin 1809–1882: With Original Omissions Restored* (edited by and with appendix and notes by his granddaughter Nora Barlow). London: Collins.

Barlow, N., ed. 1963. "Darwin's Ornithological Notes." *Bulletin of the British Museum (Natural History) Historical Series* 2 (7): 201–278.

Barrett, P. H. 1974. "The Sedgwick-Darwin Geologic Tour of North Wales." *Proceedings of the American Philosophical Society* 118: 146–164.

Barrett, P. H., P. J. Gautrey, S. Herbert, D. Kohn, and S. Smith, eds. 1987. *Charles Darwin's Notebooks, 1836–1844.* Ithaca, NY: Cornell University Press.

de Beer, G., ed. 1959. "Darwin's Journal." *Bulletin of the British Museum (Natural History), Historical Series* 2: 1–21.

de Beer, G. 2009. "Darwin's *Questions about the Breeding of Animals* [1840]." In *Darwin in the Archives* edited by E. C. Nelson and D. M. Porter, 154–164. Edinburgh: Edinburgh University Press.

Berra, T. 2013. *Darwin and His Children: His Other Legacy.* Oxford: Oxford University Press.

Boulter, C. J., M. J. Reiss, and D. L. Sanders, eds. 2015. *Darwin-Inspired Learning.* Rotterdam and Boston: Sense Publishers.

Brougham, H. 1839. *Dissertations on Subjects of Science Connected with Natural Theology: Being the Concluding Volumes of the New Edition of Paley's Work.* 2 vols. London: C. Knight and Co.

Brown, G. G., C. Feller, E. Blanchart, P. Deleporte, and S. S. Chernyanskii. 2003. "With Darwin, Earthworms Turn Intelligent and Become Human Friends." *Pedobiologia* 47 (5–6): 924–933.

Brown, R. 1833. "On the Organs and Mode of Fecundation in Orchideae and Asclepiadeae." *Transactions of the Linnean Society of London* 16: 685–746.

Browne, D. J. 1850. *The American Bird Fancier; Considered with Reference to the Breeding, Rearing, Feeding, Management, and Peculiarities of Cage and House Birds* . . . New York: C. M. Saxton.

Browne, J. 1995. *Charles Darwin Voyaging: A Biography*. Princeton, NJ: Princeton University Press.

Browne, J. 2003. *Charles Darwin: The Power of Place*. Princeton, NJ: Princeton University Press.

Bruford, M. W., D. G. Bradley, and G. Luikart. 2003. "DNA Markers Reveal the Complexity of Livestock Domestication." *Nature Reviews Genetics* 4: 900–910.

Bujok, B., M. Kleinhenz, S. Fuchs, and J. Tautz. 2002. "Hot Spots in the Bee Hive." *Naturwissenschaften* 89: 299–301.

Burdon-Sanderson, J. 1873a. "On the Electrical Phenomena which Accompany the Contractions of the Leaf of *Dionaea muscipula*." *Report of the British Association for the Advancement of Science* 43 (1873): 133.

Burdon-Sanderson, J. 1873b. "Note on the Electrical Phenomena which Accompany Irritation of the Leaf of *Dionaea muscipula*." *Proceedings of the Royal Society of London* 21 (20 November 1873): 495–496.

Burdon-Sanderson, J. 1874. "Venus's Fly-Trap (*Dionaea muscipula*)." *Nature* 10 (11 & 18 June 1874): 105–107, 127–128.

Burdon-Sanderson, J. 1882. "On the Electromotive Properties of the Leaf of *Dionaea* in the Excited and Unexcited States." *Journal of the Philosophical Transactions of the Royal Society of London* 173: 1–55.

Burkhardt, F., S. Smith, J. Secord, et al., eds. 1985–2016. *The Correspondence of Charles Darwin*, vols. 1–26. Cambridge: Cambridge University Press.

Butt, K. R., M. A. Callaham Jr., E. L. Loudermilk, and R. Blaik. 2015. "Action of Earthworms on Flint Burial: A Return to Darwin's Estate." *Applied Soil Ecology*. doi:dx.doi.org/10.1016/j.apsoil.2015.04.002.

Butt, K. R., C. N. Lowe, T. Beasley, I. Hanson, and R. Keynes. 2008. "Darwin's Earthworms Revisited." *European Journal of Soil Biology* 44: 255–259.

Campbell, A. K., and S. B. Matthews. 2005. "Darwin's Illness Revealed." *Postgraduate Medical Journal* 81: 248–251.

Canby, W. M. 1868. "Notes on *Dionaea muscipula* Ellis." *Gardener's Monthly & Horticultural Advertiser* 10: 229–232.

de Candolle, A. P. 1820. *Essai Élémentaire de géographie botanique*. Paris and Strasbourg.

Canti, M. G. 2003. "Earthworm Activity and Archaeological Stratigraphy: A Review of Products and Processes." *Journal of Archaeological Science* 30: 135–148.

Carreck, N., T. Beasley, and R. Keynes. 2009. "Charles Darwin, Cats, Mice, Bumblebees and Clover." *Bee Craft* 91 (February 2009): 4–6.

Censky, E. J., K. Hodge, and J. Dudley. 1998. "Over-Water Dispersal of Lizards Due to Hurricanes." *Nature* 395: 556–557.

Ciesielski, T. 1872. "Untersuchungen uber die Abwartskrummung der Wurzel." *Beitrage zur Biologie der Pflanzen* 1: 1–30.

Clark, J. F. M. 1997. "'The Ants Were Duly Visited': Making Sense of John Lubbock, Scientific Naturalism and the Senses of Social Insects." *British Journal for the History of Science* 30 (2): 151–176.

Colp, R. 1997. "The Dueling Diagnoses of Darwin." *Journal of the American Medical Association* 277: 1275–1276.

Colp, R. 2000. "More on Darwin's Illness." *History of Science* 38: 219–236.

Conlin, J. 2014. *Evolution and the Victorians: Science, Culture and Politics in Darwin's Britain*. London: Bloomsbury.

Correvon, H., and M. Pouyanne. 1916. "Un curieux cas de mimétisme chez les Ophrydées." *Journal de la Société Nationale d'Horticulture de France*, ser. 4. 17: 29–47.

Correvon, H., and M. Pouyanne. 1923. "Nouvelles observations sur le mimétisme et la fécondation chez le *Ophrys speculum* et *lutea*." *Journal de la Société Nationale d'Horticulture de France*, ser. 4. 24: 372–377.

Corsi, P. 1978. "The Importance of French Transformist Ideas for the Second Volume of Lyell's *Principles of Geology*." *British Journal for the History of Science* 11: 221–244.

Costa, J. T. 2009. *The Annotated Origin: A Facsimile of Charles Darwin's On the Origin of Species*. Cambridge, MA: Harvard University Press.

Culley, T. M., and M. R. Klooster. 2007. "The Cleistogamous Breeding System: A Review of its Frequency, Evolution, and Ecology in Angiosperms." *Botanical Review* 73: 1–30.

Curtis, M. A. 1834. "Enumeration of Plants Growing Spontaneously around Wilmington, North Carolina, with Remarks on Some New and Obscure Species." *Boston Journal of Natural History* 1: 82–141.

Darwin, C. R. 1838. "On the Formation of Mould." *Proceedings of the Geological Society of London* 2: 574–576.

Darwin, C. R. 1839a. *Journal of Researches into the Geology and Natural History of the Various Countries Visited by H.M.S.* Beagle. London: Colburn.

Darwin, C. R. 1839b. "Observations on the Parallel Roads of Glen Roy, and of Other Parts of Lochaber in Scotland, with an Attempt to Prove That They Are of Marine Origin." [Read 7 February] *Philosophical Transactions of the Royal Society* 129: 39–81.

Darwin, C. R. 1840. "On the Formation of Mould." *Transactions of the Geological Society of London* 1840: 505–509.

Darwin, C. R. 1841. "Humble-bees." *Gardeners' Chronicle* 34 (21 August): 550.

Darwin, C. R. 1842. "Notes on the Effects Produced by the Ancient Glaciers of Caernarvonshire, and on the Boulders Transported by Floating Ice." *Philosophical Magazine* 21: 180–188.

Darwin, C. R. 1844. "On the Origin of Mould." *Gardeners' Chronicle and Agricultural Gazette* 14 (6 April): 218.

Darwin, C. R. 1845. *Journal of Researches into the Natural History and Geology of the Countries Visited During the Voyage of H.M.S.* Beagle *Round the World.* 2nd ed. [*Voyage of the Beagle*] London: John Murray.

Darwin, C. R. 1851. *A Monograph of the Sub-class Cirripedia, with Figures of All the Species. Volume 1: The Lepadidae; or, Pedunculated Cirripedes.* London: The Ray Society.

Darwin, C. R. 1854. *A Monograph on the Sub-class Cirripedia, with Figures of All the Species. Volume 2: The Balanidae, (or Sessile Cirripedes); the Verrucidae, etc. etc. etc.* London: The Ray Society.

Darwin, C. R. 1855a. "Does Sea-Water Kill Seeds?" *Gardeners' Chronicle and Agricultural Gazette* 15 (14 April): 242.

Darwin, C. R. 1855b. "Does Sea-Water Kill Seeds?" *Gardeners' Chronicle and Agricultural Gazette* 21 (26 May): 356–357.

Darwin, C. R. 1855c. "Effect of Salt-Water on the Germination of Seeds." *Gardeners' Chronicle and Agricultural Gazette* 47 (24 November): 773.

Darwin, C. R. 1855d. "Effect of Salt-Water on the Germination of Seeds." *Gardeners' Chronicle and Agricultural Gazette* 48 (1 December): 789.

Darwin, C. R. 1857a. "On the Action of Sea-Water on the Germination of Seeds." *Journal of the Proceedings of the Linnean Society (Botany)* 1: 130–140.

Darwin, C. R. 1857b. "Bees and the Fertilisation of Kidney Beans." *Gardeners' Chronicle and Agricultural Gazette* 43 (24 October): 725.

Darwin, C. R. 1858. "On the Agency of Bees in the Fertilisation of Papilionaceous Flowers, and on the Crossing of Kidney Beans." *Gardeners' Chronicle and Agricultural Gazette* 46 (13 November): 828–829.

Darwin, C. R. 1860a. "Fertilisation of British Orchids by Insect Agency." *Gardeners' Chronicle and Agricultural Gazette* 23 (19 June): 528.

Darwin, C. R. 1860b. "Irritability of *Drosera*." *Gardeners' Chronicle and Agricultural Gazette* 38 (22 September): 853.

Darwin, C. R. 1861. "Fertilisation of British Orchids by Insect Agency." *Gardeners' Chronicle and Agricultural Gazette* 6 (9 February): 122.

Darwin, C. R. 1862a. *On the Various Contrivances by which British and Foreign Orchids Are Fertilised by Insects*. London: John Murray.

Darwin, C. R. 1862b. "On the Three Remarkable Sexual Forms of *Catasetum tridentatum*, an Orchid in the Possession of the Linnean Society." [Read 3 April] *Proceedings of the Linnean Society of London (Botany)* 6: 151–157.

Darwin, C. R. 1862c. "On the Two Forms, or Dimorphic Condition, in the Species of *Primula*, and on Their Remarkable Sexual Relations." [Read 21 November 1861] *Journal of the Proceedings of the Linnean Society of London (Botany)* 6: 77–96.

Darwin, C. R. 1863a. "On the Existence of Two Forms, and on Their Reciprocal Sexual Relation, in Several Species of the Genus *Linum*." [Read 5 February 1863] *Journal of the Proceedings of the Linnean Society (Botany)* 7: 69–83.

Darwin, C. R. 1863b. "Fertilisation of Orchids." *Journal of Horticulture and Cottage Gardener* (31 March): 237.

Darwin, C. R. 1864. "On the Sexual Relations of the Three Forms of *Lythrum salicaria*." [Read 16 June 1864] *Journal of the Linnean Society of London (Botany)* 8: 169–196.

Darwin, C. R. 1865. "On the Movements and Habits of Climbing Plants." [Read 2 Feb.] *Journal of the Linnean Society of London (Botany)* 9: 1–118.

Darwin, C. R. 1867. "Queries about Expression." Charles Darwin's *Queries about Expression*, edited by R. B. Freeman and P. J. Gautrey. *Bulletin of the British Museum of Natural History* (historical series) 4 (1972): 205–219.

Darwin, C. R. 1868. *The Variation of Animals and Plants under Domestication*. 2 vol. London: John Murray.

Darwin, C. R. 1869. "The Formation of Mould by Worms." *Gardener's Chronicle and Agricultural Gazette* 20 (15 May): 530.

Darwin, C. R. 1871. *The Descent of Man, and Selection in Relation to Sex*. 2 vol. London: John Murray.

Darwin, C. R. 1872. *On the Expression of the Emotions in Man and Animals*. London: John Murray.

Darwin, C. R. 1875a. *Insectivorous Plants*. London: John Murray.

Darwin, C. R. 1875b. *The Movements and Habits of Climbing Plants*. 2nd ed. London: John Murray.

Darwin, C. R. 1876. *The Effects of Cross and Self Fertilisation in the Vegetable Kingdom.* London: John Murray.

Darwin, C. R. 1877a. *On the Various Contrivances by which British and Foreign Orchids are Fertilised by Insects.* 2nd ed. London: John Murray.

Darwin, C. R. 1877b. *The Different Forms of Flowers on Plants of the Same Species.* London: John Murray.

Darwin, C. R. 1878. "Transplantation of Shells." *Nature* 18 (30 May): 120–121.

Darwin, C. R. 1880. *The Power of Movement in Plants.* London: John Murray.

Darwin, C. R. 1881. *The Formation of Vegetable Mould, through the Action of Worms, with Observations on Their Habits.* London: John Murray.

Darwin, C. R. 1882. "On the Dispersal of Freshwater Bivalves." *Nature* 25 (6 April): 529–530.

Darwin, C. R. 1986. *The Different Forms of Flowers on Plants of the Same Species.* Facsimile of the 1888 edition, with a foreword by Herbert G. Baker. Chicago: University of Chicago Press.

Darwin, E. 1803. *The Temple of Nature, Or, The Origin of Society.* London: J. Johnson.

Darwin, F. 1880. "Experiments on the Nutrition of *Drosera rotundifolia*." *Journal of the Linnean Society. Botany.* 17: 17–32.

Darwin, F., ed. 1887. *Life and Letters of Charles Darwin, With an Autobiographical Chapter.* 3 vols. London: John Murray.

Darwin, F. 1903. "The Statolith-Theory of Geotropism." *Proceedings of the Royal Society of London* 71: 362–373.

Darwin, F. 1904. "On the Perception of the Force of Gravity by Plants." *Nature* 70: 466–473.

Darwin, F., and A. C. Seward, eds. 1903. *More Letters of Charles Darwin: A Record of His Work in a Series of Hitherto Unpublished Letters.* 2 vols. London: John Murray.

Darwin, H. 1901. "On the Small Vertical Movements of a Stone Laid on the Surface of the Ground." *Proceedings of the Royal Society of London* 68: 253–261.

Darwin, L. 1929. "Memories of Down House." *The Nineteenth Century* 106: 118–123.

de Chadarevian, S. 1996. "Laboratory Science versus Country-House Experiments. The Controversy between Julius Sachs and Charles Darwin." *British Journal of the History of Science* 29: 17–41.

de Queiroz, A. 2005. "The Resurrection of Oceanic Dispersal in Historical Biogeography." *Trends in Ecology and Evolution* 20: 68–73.

de Queiroz, A. 2014. *The Monkey's Voyage: How Improbable Journeys Shaped the History of Life*. New York: Basic Books.

Desmond, A. 1989. *The Politics of Evolution: Morphology, Medicine, and Reform in Radical London*. Chicago: Chicago University Press.

Desmond, A., and J. Moore. 1991. *Darwin: The Life of a Tormented Evolutionist*. New York: Warner Books.

Desmond, A. and J. Moore. 2009. *Darwin's Sacred Cause: How a Hatred of Slavery Shaped Darwin's Views on Human Evolution*. Boston and New York: Houghton Mifflin Harcourt.

Dobbs, D. 2005. *Reef Madness: Charles Darwin, Alexander Agassiz, and the Meaning of Coral*. New York: Pantheon Books.

Edens-Meier, R., and P. Bernhardt, eds. 2014. *Darwin's Orchids: Then and Now*. Chicago: University of Chicago Press.

Ellison, A. M., and N. J. Gotelli. 2001. "Evolutionary Ecology of Carnivorous Plants." *Trends in Ecology and Evolution* 16: 623–629.

Evans, A. C. 1947. "A Method of Studying the Burrowing Activities of Earthworms." *Annals and Magazine of Natural History* 14: 643–650.

Evans, L. T. 1984. "Darwin's Use of the Analogy between Artificial and Natural Selection." *Journal of the History of Biology* 17 (1): 113–140.

Feller, C., G. G. Brown, E. Blanchart, P. Deleporte, and S. S. Chernyanskii. 2003. "Charles Darwin, Earthworms and the Natural Sciences: Various Lessons from Past to Future." *Agriculture, Ecosystems & Environment* 99: 29–49.

Fish, D. T. 1869. "A Chapter on Worms." *Gardeners' Chronicle and Agricultural Gazette* 20 (17 April): 417–418.

Fitzroy, R. 1839. *Proceedings of the Second Expedition, 1831–1836, Under the Command of Captain Robert Fitz-Roy, R.N.* Vol. 2 of Philip Parker King, Robert Fitzroy, and Charles Darwin, *Narrative of the Surveying Voyages of His Majesty's Ships* Adventure *and* Beagle: *Between the Years 1826 and 1836, Describing Their Examination of the Southern Shores of South America, and the* Beagle's *Circumnavigation of the Globe*. 3 vol. London: Henry Colburn.

Forbes, E. 1846. "On the Connexion between the Distribution of the Existing Fauna and Flora of the British Isles, and the Geological Changes which Have Affected Their Area, Especially During the Epoch of the Northern Drift." *Memoirs of the Geological Survey of Great Britain, and of the Museum of Economic Geology in London* 1: 336–432.

Frank, R. G. Jr. 1988. "The Telltale Heart: Physiological Instruments, Graphic

Methods, and Clinical Hopes." In *The Investigative Enterprise: Experimental Physiology in Nineteenth-Century Medicine*, edited by W. Coleman and F. L. Holmes, pp. 211–290. Berkeley: University of California Press.

Free, J. B. 1965. "The Ability of Bumblebees and Honeybees to Pollinate Red Clover." *Journal of Applied Ecology* 2: 289–294.

Freeman, R. B. 1968. "Charles Darwin on the Routes of Male Humble Bees." *Bulletin of the British Museum (Natural History), Historical Series* 3 (6): 179–188.

Freeman, R. B. 1978. *Charles Darwin: A Companion*. Folkestone, UK: Dawson and Archon.

Freeman, R. B., and P. J. Gautrey. 1969. "Darwin's *Questions about the Breeding of Animals*, with a Note on *Queries about Expression*." *Journal of the Society for the Bibliography of Natural History* 5 (3): 220–225.

Fries, E. M. 1850. "A Monograph of the *Hieracia*; Being an Abstract of Prof. Fries's '*Symbolae ad Historiam Hieraciorum*,' Translated and Abridged." *Botanical Gazette* 2: 85–92, 185–188, 203–219.

Gibson, T. C., and D. M. Waller. 2009. "Evolving Darwin's 'Most Wonderful' Plant: Ecological Steps to a Snap-Trap." *New Phytologist* 183: 575–587.

Gildenhuys, P. 2004. "Darwin, Herschel, and the Role of Analogy in Darwin's *Origin*." *Studies in History and Philosophy of Biological and Biomedical Sciences* 35: 593–611.

Gould, S. J. 1985. "Geoffroy and the Homeobox." *Natural History* 94 (11): 12–23.

Goulson, D. 2009. *Bumblebees: Behaviour, Ecology, and Conservation*. 2nd ed. Oxford: Oxford University Press.

Gray, A. 1858. "Note on the Coiling of Tendrils." *Proceedings of the American Academy of Arts and Sciences* 4 (May 1857–May 1860): 98–99.

Gray, A. 1861. *Natural Selection Not Inconsistent with Natural Theology. A Free Examination of Darwin's Treatise on the Origin of Species, and of Its American Reviewers*. Boston & London.

Hague, J. D. 1884. "A Reminiscence of Mr Darwin." *Harper's New Monthly Magazine* 69: 760–761.

Hales, T. C. 2000. "Cannonballs and Honeycombs." *Notices of the American Mathematical Society* 47 (4): 440–449.

Hanson, I., J. Djohari, J. Orr, P. Furphy, C. Hodgson, G. Cox, and G. Broadbridge. 2009. "New Observations on the Interactions between Evidence and the Upper Horizons of the Soil." In *Criminal and Environmental Soil Forensics*, edited by K. Ritz, L. Dawson, and D. Miller, 239–251. New York: Springer.

Hector, A., and R. Hooper. 2002. "Darwin and the First Ecological Experiment." *Science* 295: 639–640.

Hepburn, H. R., C. W. W. Pirk, and O. Duangphakdee. 2014. *Honey Bee Nests: Composition, Structure, Function*. Berlin: Springer.

Herbert, S. 2005. *Charles Darwin, Geologist*. Ithaca, NY: Cornell University Press.

Holmes, R. 2008. *The Age of Wonder: How the Romantic Generation Discovered the Beauty and Terror of Science*. New York: Pantheon.

Hooker, J. D. 1853. *Introductory Essay to the Flora of New Zealand*. London: Lovell Reeve.

Hutchinson, G. E. 1959. "Homage to Santa Rosalia, or Why Are There So Many Kinds of Animals?" *American Naturalist* 93: 145–159.

Johnson, D. L. 2002. "Darwin Would Be Proud: Bioturbation, Dynamic Denudation, and the Power of Theory in Science." *Geoarchaeology* 17 (1): 7–40.

Keith, A. 1942. "A Postscript to Darwin's *Formation of Vegetable Mould through the Action of Worms*." *Nature* 149: 716–720.

Keynes, R. L., ed. 1988. *Charles Darwin's* Beagle *Diary*. Cambridge: Cambridge University Press.

Keynes, R. L., ed. 2000. *Charles Darwin's Zoology Notes & Specimen Lists from H.M.S.* Beagle. Cambridge: Cambridge University Press.

Kingsley, C. 1855. *Glaucus; or the Wonders of the Shore*. Cambridge: Macmillan and Co.

Kirby, W. 1852. *On the Power, Wisdom, and Goodness of God as Manifested in the Creation of Animals, and in Their History, Habits, and Instincts*. New edition. Edited by Thomas Rymer Jones. 2 vols. London: Henry G. Bohn.

Knight, T. A. 1799. "An Account of Some Experiments on the Fecundation of Vegetables." *Philosophical Transactions of the Royal Society* 89: 195–204.

Knight, T. A. 1806. "On the Direction of the Radicle and Germen During the Vegetation of Seeds." *Philosophical Transactions of the Royal Society of London* 96: 99–108.

Kohn, D. 1985. "Darwin's Principle of Divergence as Internal Dialogue." In *The Darwinian Heritage* edited by D. Kohn, 245–257. Princeton, NJ: Princeton University Press.

Kohn, D., G. Murrell, J. Parker, and M. Whitehorn. 2005. "What Henslow Taught Darwin: How a Herbarium Helped to Lay the Foundations of Evolutionary Thinking." *Nature* 436: 643–645.

Kritsky, G. 1991. "Darwin's Madagascan Hawk Moth Prediction." *American Entomologist* 37: 206–209.

Kullenberg, B. 1961. "Studies in *Ophrys* Pollination." *Zoologiska Bidrag Uppsala* 34: 1–340.

Lankester, E. R. 1896. "[Recollections of] Charles Robert Darwin." *Library of the World's Best Literature Ancient and Modern*, vol. 2, edited by C. D. Warner, 4385–4393. New York: R. S. Peale & J. A. Hill.

Lankester, E. R. 1913. *Science From an Easy Chair; a Second Series*. New York: H. Holt and Co.

Lamarck, J.-P. 1809. *Philosophie zoologique, ou exposition des considérations relatives de l'histoire des animaux*. Paris.

Lamarck, J.-P. 1815–1822. *Histoire des animaux sans vertèbres*. Paris.

le Guyader, H. 2004. *Étienne Geoffroy Saint-Hilaire 1772–1844: A Visionary Naturalist*. Translated by M. Grene. Chicago: University of Chicago Press.

Li, J., M. A. Webster, M. C. Smith, and P. M. Gilmartin. 2011. "Floral Heteromorphy in *Primula vulgaris*: Progress towards Isolation and Characterization of the S Locus." *Annals of Botany* 108: 715–726.

Lindley, J. 1846. *The Vegetable Kingdom, or, The Structure, Classification, and Uses of Plants, Illustrated Upon the Natural System*. London: Bradbury and Evans.

Litchfield, H. E., ed. 1915. *Emma Darwin: A Century of Family Letters, 1792–1896*. 2 vols. London: John Murray.

Liu, K.-W., Z.-J. Liu, L. Huang, L.-Q. Li, L.-J. Chen, and G.-D. Tang. 2006. "Self-Fertilization Strategy in an Orchid." *Nature* 441: 945–946.

Livingstone, D. N. 1984. *Darwin's Forgotten Defenders: The Encounter between Evangelical Theology and Evolutionary Thought*. Vancouver: Regent College Publishing.

Love, A. C. 2002. "Darwin and Cirripedia Prior to 1846: Exploring the Origins of the Barnacle Research." *Journal of the History of Biology* 35: 251–289.

Loy, J. D., and K. M. Loy. 2010. *Emma Darwin: A Victorian Life*. Gainesville: University Press of Florida.

Lubbock, J. 1875. *On British Wild Flowers Considered in Relation to Insects*. London: Macmillan and Co.

Lyell, C. 1830–1833. *Principles of Geology*. 3 vols. London: John Murray.

McLauchlin, J. 2009. "Charles Darwin's Lawn Plot Experiment." *The London Naturalist* 88: 107–113.

Meysman, F. J. R., J. J. Middelburg, and C. H. R. Heip. 2006. "Bioturbation: A Fresh Look at Darwin's Last Idea." *Trends in Ecology and Evolution* 21 (12): 688–695.

Millett, J. R., I. Jones, and S. Waldron. 2003. "The Contribution of Insect Prey to the Total Nitrogen Content of Sundews (*Drosera* spp.) Determined *in situ* by Stable Isotope Analysis." *New Phytologist* 158: 527–534.

Milne-Edwards, H. 1851. *Introduction á la zoologie générale*. Paris: Victor Masson.

Mitra, O., M. A. Callaham Jr., M. L. Smith, and J. E. Yack. 2009. "Grunting for Worms: Seismic Vibrations Cause *Diplocardia* Earthworms to Emerge from the Soil." *Biology Letters* 5: 16–19.

Mori, A., D. A. Grasso, and F. Le Moli. 2000. "Raiding and Foraging Behavior of the Blood-Red Ant, *Formica sanguinea* Latr. (Hymenoptera, Formicidae)." *Journal of Insect Behavior* 13 (3): 421–438.

Müller. H. 1873. *Die Befruchtung der Blüten durch Insekten und die gegenseitigen*. Leipzig: Engelmann.

Müller, H. 1883. *The Fertilisation of Flowers*. Translated by D. W. Thompson. London: Macmillan.

Nelson, G. 1978. "From Candolle to Croizat: Comments on the History of Biogeography." *Journal of the History of Biology* 11 (2): 269–305.

Oldroyd, B. P., and S. C. Pratt. 2015. "Comb Architecture of the Eusocial Bees Arises from Simple Rules Used During Cell Building." *Advances in Insect Physiology* 49: 101–121.

Pasnau, R. O. 1990. "Darwin's Illness: A Biophychosocial Perspective." *Psychosomatics* 31: 121–128.

Pattison, A. 2009. *The Darwins of Shrewsbury*. Stroud, UK: The History Press.

Pauly, D. 2007. *Darwin's Fishes: An Encyclopedia of Ichthyology, Ecology, and Evolution*. Cambridge: Cambridge University Press.

Pearn, A. 2014. "The Teacher Taught? What Charles Darwin Owed to John Lubbock." *Notes and Records: The Royal Society Journal of the History of Science* 68: 7–19.

Poppinga, S., S. Hartmeyer, R. Seidel, and T. Masselter. 2012. "Catapulting Tentacles in a Sticky Carnivorous Plant." *PLoS One* 7 (9): e45735.

Reed, E. S. 1982. "Darwin's Earthworms: A Case Study in Evolutionary Psychology." *Behaviorism* 10: 165–185.

Reverat, G. 1960. *Period Piece: A Cambridge Childhood*. London: Faber & Faber.

Rice, B. A. 2006. *Growing Carnivorous Plants*. Portland, OR: Timber Press.

Rivadavia, F., K. Kondo, M. Kato, and M. Hasebe. 2003. "Phylogeny of the Sundews, *Drosera* (Droseraceae), Based on Chloroplast *RBCL* and Nuclear 18S Ribosomal DNA Sequences." *American Journal of Botany* 90: 123–130.

Roberts, M. 1996. "Darwin at Llanymynech: The Evolution of a Geologist." *British Journal for the History of Science* 29: 469–478.

Roberts, M. 2001. "Just before the *Beagle*: Charles Darwin's Geological Fieldwork in Wales, Summer 1831." *Endeavour* 25: 33–37.

Rudwick, M. J. S. 1974. "Darwin and Glen Roy: A 'Great Failure' in Scientific Method?" *Studies in the History and Philosophy of Science* 5: 97–185.

Ruricola. 1841. "Humble-Bees." *The Gardeners' Chronicle* 34 (24 July): 485.

Ruse, M. 1975. "Charles Darwin and Artificial Selection." *Journal of the History of Ideas* 36: 339–350.

Ruskin, J. 1906. *The Complete Works of John Ruskin*. Library edition. Edited by E. T. Cook and A. Wedderburn. London: George Allen / New York: Longmans, Green, and Co.

Schomburgk, R. 1837. "On the Identity of Three Supposed Genera of Orchidaceous Epiphytes, in a Letter to A. B. Lambert." *Transactions of the Linnean Society of London* 17: 551–552.

Schweber, S. S. 1980. "Darwin and the Political Economists: Divergence of Character." *Journal of the History of Biology* 13: 195–289.

Sebastian, P., H. Schaeffer, and S. S. Renner. 2010. "Darwin's Galapagos Gourd: Providing New Insights 175 Years After His Visit." *Journal of Biogeography* 37: 975–980.

Sebright, J. S. 1809. *The Art of Improving the Breeds of Domestic Animals, in a Letter Addressed to the Right Hon. Sir Joseph Banks, K.B.* London: John Harding.

Secord, J. A. 1981. "Nature's Fancy: Charles Darwin and the Breeding of Pigeons." *Isis* 72: 163–186.

Shepard, O., ed. 1961. *The Heart of Thoreau's Journals*. New York: Dover Publications.

Sinclair, G. 1826. "On Cultivating a Collection of Grasses in Pleasure-Grounds or Flower-Gardens [and on the Utility of Studying the Gramineae]." *The Gardener's Magazine and Register of Rural & Domestic Improvement* [*Loudon's Gardener's Magazine*] 1: 26–29, 112–116.

Slack, N. G. 2010. *G. Evelyn Hutchinson and the Invention of Modern Ecology*. New Haven, CT: Yale University Press.

Slaughter, T., ed. 1996. *Bartram: Travels and Other Writings*. New York: Library of America.

Sloan, P. R. 1985. "Darwin's Invertebrate Program, 1826–1836: Preconditions for Transformism." In *The Darwinian Heritage*, edited by D. Kohn, 71–120. Princeton, NJ: Princeton University Press and Nova Pacifica.

Sloan, P. R., ed. 1992. *The Hunterian Lectures in Comparative Anatomy, May and June 1837*. Chicago: University of Chicago Press and London: Natural History Museum Publications.

Smith, F. 1854. "Essay on the Genera and Species of British Formicidae." [Read

4 December 1854] *Transactions of the Entomological Society of London,* n.s. 3 (1854–1856): 95–135.

Smith, F. P., G. F. Nixon, A. V. Conway, J. C. King, S. D. Rosen, H. J. Roberts, A. V. Conway, A. J. Pelosi, and D. Adler. 1990. "Darwin's Illness." *Lancet* 336: 1139–1140.

Stauffer, R. C. 1960. "Ecology in the Long Manuscript Version of Darwin's 'Origin of Species' and Linnaeus' 'Oeconomy of Nature.'" *Proceedings of the American Philosophical Society* 104 (2): 235–241.

Stauffer, R. C., ed. 1975. *Charles Darwin's Natural Selection; Being the Second Part of his Big Species Book Written From 1856 to 1858.* Cambridge: Cambridge University Press.

Stott, R. 2003. *Darwin and the Barnacle.* London: Faber and Faber.

Tegetmeier, W. B. 1859. "On the Formation of the Cells of Bees." *Report of the 28th meeting of the British Association for the Advancement of Science, held at Leeds September 1858.* 28: 132–133.

Thierry, B. 2010. "Darwin as a Student of Behavior." *Comptes Rendus Biologies* 333: 188–196.

Thompson, J. V. 1830. "On the Cirripedes or Barnacles; Demonstrating their Deceptive Character; the Extraordinary Metamorphosis They Undergo, and the Class of Animals to which They Indisputably Belong." In *Zoological Researches, and Illustrations; or, Natural History of Nondescript or Imperfectly Known Animals, in a Series of Memoirs, Illustrated by Numerous Figures,* vol. 1, Pt 1, 1828–1830, edited by J. V. Thompson, 69–82, plates IX–X. Cork.

Thoreau, H. D. 1906. *The Writings of Henry David Thoreau (Walden edition).* Boston: Houghton, Mifflin and Co.

Thoreau, H. D. 1993. *Faith in a Seed: The Dispersion of Seeds and Other Late Natural History Writings.* Edited by B. P. Dean. Washington DC: Island Press.

Tóth, L. F. 1964. "What the Bees Know and What They Do Not Know." *Bulletin of the American Mathematical Society* 70: 468–481.

Uglow, J. 2002. *The Lunar Men: The Friends Who Made the Future.* London: Faber and Faber.

Wallace, A. R. 1863. "Remarks on the Rev. S. Haughton's Paper on the Bee's Cell, and on the Origin of Species." *Annals and Magazine of Natural History* 12 (3rd series): 303–309.

Wallace, A. R. 1867. "Creation by Law [Review of *The Reign of Law* by the Duke of Argyll, 1867]." *Quarterly Journal of Science* 4: 471–488.

Ward, A. 2016. "Evolution and Creation Debates." In *A Companion to the History of American Science*, edited by G. M Montgomery and M. A. Largent. Chichester: John Wiley & Sons.

Wasserthal, L. T. 1997. "The Pollinators of the Malagasy Star Orchids *Angraecum sesquipedale, A. sororium* and *A. compactum* and the Evolution of Extremely Long Spurs by Pollinator Shift." *Botanica Acta* 110: 343–359.

Waterhouse, G. R. 1835. "Bees." *Penny Cyclopaedia of the Society for the Diffusion of Useful Knowledge* 4: 149–156.

Wedgwood, B., and H. Wedgwood. 1980. *The Wedgwood Circle 1730–1897*. Don Mills, Ontario: Collier Macmillan Canada, Ltd.

Wedgwood, L. 1868. "Worms." *Gardeners' Chronicle and Agricultural Gazette* 19 (28 March): 324.

Westwood, J. O. 1835. "On the Supposed Existence of Metamorphosis in the Crustacea." *Philosophical Transactions of the Royal Society* 125: 312.

Whewell, W. 1840. *The Philosophy of the Inductive Sciences, Founded Upon Their History*. 2 vols. London: John W. Parker.

Wilkinson, J. 1820. *Remarks on the Improvement of Cattle, etc. In a Letter to Sir John Saunders Sebright Bart. M.D.* 3rd ed. Nottingham: H. Barnett.

Williams, N. 2005. "Darwin's Meadow Revisited." *Current Biology* 15: R480–481.

Wilson, L. G., 1970. *Sir Charles Lyell's Scientific Journals on the Species Question*. New Haven, CT: Yale University Press.

Winsor, M. P. 1969. "Barnacle Larvae in the Nineteenth Century: A Case Study in Taxonomic Theory." *Journal of the History of Medicine and Allied Sciences* 24 (3): 294–309.

van Wyhe, J., ed. 2009. *Charles Darwin's Shorter Publications, 1829–1883*. Cambridge: Cambridge University Press.

Yeates, G. W., and H. van der Meulen. 1995. "Burial of Soil-Surface Artifacts in the Presence of Lumbricid Earthworms." *Biology and Fertility of Soils* 19: 73–74.

INDEX

Note: Page numbers in *italics* refer to illustrations.